Oral Mucosa in Health and Disease

Lesley Ann Bergmeier
Editor

Oral Mucosa in Health and Disease

A Concise Handbook

Springer

Editor
Lesley Ann Bergmeier
Centre for Oral Immunobiology and Regenerative Medicine
Institute of Dentistry, Queen Mary School of Medicine and Dentistry
London, UK

ISBN 978-3-030-09615-1 ISBN 978-3-319-56065-6 (eBook)
https://doi.org/10.1007/978-3-319-56065-6

Softcover re-print of the Hardcover 1st edition 2018

Printed on acid-free paper

This Springer imprint is published by Springer Nature
The registered company is Springer International Publishing AG
The registered company address is: Gewerbestrasse 11, 6330 Cham, Switzerland

This book is dedicated to my late parents
Jean and Eric Bergmeier
Who made me curious, were the wind at my back
and whose love was the sun that shone warm upon my face.
And to my husband, Bob, for all his support without which
I would be lost

Contents

Introduction to the Oral Mucosa: Gatekeeper or Housekeeper?

The 1901 edition of Gray's Anatomy describes the mouth as part of the "Organs of Digestion" and goes on to list and describe the main structures and functions of the teeth and salivary glands with little or no discussion of the mucosa.

Box 1: Gray's Anatomy

The **alimentary canal** is a musculo-membranous tube about 30 feet long extending from the mouth to the anus, and lined throughout its entirety by mucous membrane…

The **mouth** (oral or buccal cavity) is placed at the commencement of the alimentary canal…

The *mucous membrane* lining the mouth is continuous with the integument at the free margin of the lips, and with the mucous lining of the pharynx behind… It is covered by stratified epithelium…

The **gums** are composed of a dense fibrous tissue, closely connected to the periosteum of the alveolar process, and surrounding the necks of the teeth.

They (the gums) are covered by a smooth and vascular mucous membrane, *which is remarkable for its limited sensibility…*

From: Gray's Anatomy: Facsimile of 15th edition (1901) Barnes and Noble 2010

One hundred and sixteen years later one could be forgiven for thinking that not much has changed when a simple PubMed search for papers and reviews on the "oral mucosa in health and disease" returns 49 papers that, while beginning with much promise of new work on the oral mucosa, often skim over what limited new knowledge is available and resort to descriptions of the gut! The oral mucosa is still a neglected topic of investigation in many areas of oral biology.

Yet, the final sentence quoted in Box 1 was highly prescient for its time as the major findings over the last century on the function of the oral mucosa have culminated in the knowledge that the oral mucosa represents a highly tolerogenic environment.

The tissues of the oral cavity have a variety of functions, from those associated with nutrition (masticatory, sensory, pain and temperature perception) to the barrier functions that defend against infection with pathogenic organisms (Chaps. 1, 4, and 10).

The oral mucosa is constantly exposed to antigenic stimulation in the form of foods, microbial antigens (commensals and pathogens) and inhaled materials, which have the potential to induce allergic reactions. However, in normal healthy individuals, acute inflammation is rarely seen and the mucosa is regarded as a tolerogenic environment where a functional homeostasis protects against pathological changes (Chaps. 5 and 8).

In recent years, it has become apparent that the relationship between the commensal microbial community (the MICROBIOME) in many mucosal tissues of the body has a significant influence on health and susceptibility to disease, especially chronic or autoimmune/autoinflammatory conditions. The oral cavity is no exception to this. Over 700 species have been identified in the oral microbiome, and the host response to some organisms (such as periodontal organisms) makes a significant contribution to health and/or disease progression (Chap. 10).

Genetic susceptibility to disease has been augmented by understanding the epigenetic effects of environmental challenges along with the activities of non-coding miRNAs and has gone some way in explaining the etiopathogenesis and behaviour of oral lesions (Chap. 9). New research on the processes of oral wound healing reveals complex pathways that are dependent on oral mucosal homeostasis (Chap. 6).

The purpose of this book is to review the current state of knowledge of the oral mucosa, to illustrate the bi-directional link between mouth and general health, and to signpost those changes in the mucosa that might be the first indications of developing or established systemic diseases.

Beyond the Hard Stuff

Beyond the teeth lies the highly dynamic and complex microenvironment of support structures and networks of cells and tissue that make up the **ORAL MUCOSA.** Local and systemic diseases frequently reflect changes in the homeostasis of the oral mucosa.

It is perhaps simplest to regard the soft tissue of the oral cavity as a barrier, cleansed by constant bathing with saliva containing antimicrobial agents, enzymes and antibodies that limit microbial growth in the mouth (Chap. 7). However, this belies the highly active and dynamic properties of these tissues in protecting the host from infection and the consequences of inflammation. This is most obviously observed in the natural history of periodontitis, from mildly inflammatory gingivitis to chronic inflammation and the development of the dysbiotic (dysregulated) microflora. An exaggerated host inflammatory response allows colonisation with "keystone" periodontal pathogens that alter the microenvironment leading to loss of attachment, bone destruction and tooth loss.

The luminal layers of the mucosa undergo constant renewal by desquamation, which acts to remove dead or damaged cells along with any adherent microorganisms. The presence of the mucous layer prevents penetration of most organisms, and the intricate association of adhesion molecules maintains the integrity of the tight junctions between epithelial cells, allowing passive diffusion of nutrients while excluding toxins by trapping them in the mucins where they are disposed of during desquamation (Chap. 3).

Although a small surface area compared with the gut mucosa, the oral mucosa is subdivided into different anatomical and functional regions. The cell populations within these regions can undergo diverse responses to both the commensal organisms and environmental challenges including ingested food and the disease risk factors associated with smoking, chewing betel quid and/or paan, alcohol and age-related degeneration. Despite these exposures, the oral mucosa of healthy individuals is a tolerogenic environment with tight control of the homeostasis existing between the commensals and the host responses (Chap. 5).

Inflammatory responses to infection and/or tissue damage act in concert with both the innate and adaptive immune systems to protect the oral mucosa. However, in several diseases with oral manifestations, it is clear that chronic inflammation, immune dysregulation and possible autoimmune mechanisms are responsible for the pathologies observed (Chaps. 8 and 10).

While the oral mucosa is frequently regarded as a part of the GI tract, with some similarities to skin, there are unique attributes to this tissue that warrant in-depth study. In the final chapter, we will attempt to summarise the types of experimental procedures currently being conducted to investigate the oral mucosa and speculate on the areas of investigation that might prove fruitful in elucidating the complex interactome of the oral cavity.

And Finally---

A recent review of the mucosal immune system entitled "The mucosal immune system: From dentistry to vaccine development" [1] contained an historical insight into the role that dental science and oral biology have played in furthering our understanding of the huge impact that the mucosal immune system has on health and well-being. The oral mucosa tends to be dismissed as just the entrance to the gastrointestinal tract, but this *Gatekeeper* tissue has a lot to teach us about the *Housekeeper* roles of both the oral mucosae and other mucosae in health and disease.

Reference

1. Kiyono H, Azegami T. The mucosal immune system: from dentistry to vaccine development. Proc Jpn Acad Ser B Phys Biol Sci. 2015;91(8):423–39.

1 Structure and Functions of the Oral Mucosa

Alan T. Cruchley and Lesley Ann Bergmeier

1.1 Introduction

Mucosal membranes are defined as moist linings of the body cavities including the gastrointestinal tract, the nasal passages, the vagina and other cavities that communicate with the exterior. The oral mucosa is a unique environment where the hard tissues of the teeth about the mucosal epithelium and a flourishing commensal microbiome contribute to homeostasis. The oral cavity is a dynamic environment that is subject to mechanical stresses (through eating and talking), but also the changes that are involved in consumption of hot or cold foods, rapid changes in local pH, sensory changes such as pain, and the unique sensations of taste and thirst. Reflexes such as swallowing, retching, gagging and salivating contribute to the complexity of the tissue environment.

The purpose of this chapter is to briefly review the structural features of the oral mucosa and place them in the context of the barrier and protective functions that maintain oral health and to act as a reference for the subsequent chapters on this mucosa in health and disease.

A.T. Cruchley
Centre for Teaching Innovation, Institute of Dentistry, Queen Mary School of Medicine and Dentistry, London, UK
e-mail: a.t.cruchley@qmul.ac.uk

L.A. Bergmeier (✉)
Centre for Oral Immunology and Regenerative Medicine, Institute of Dentistry, Queen Mary School of Medicine and Dentistry, London, UK
e-mail: l.a.bergmeier@qmul.ac.uk

1.2 Anatomy and Organisation

The oral mucosa is separated from the skin by the vermillion zone of the lips which is more deeply coloured than the rest of the oral mucosa. The colour is affected by several factors including the concentration and dilation state of blood vessels in the underlying connective tissues; the thickness of the epithelium; the degree of keratinisation and the amount of melanin pigment. The colour of the mucosa is of significant diagnostic importance. Inflamed mucosa, for example, will appear red while normal mucosa is pink. Oral pigmentation can also be increased as a result of systemic disease such as Addison's and Peutz-Jeghers diseases [1].

Other differences from skin include the moist nature of the mucosa and the absence of structures such as hair follicles, sweat glands and sebaceous glands. However, in some individuals Fordyce spots, a type of sebaceous gland, are found predominantly in the upper lip, the buccal and alveolar mucosa. There are also significant structures in the form of minor salivary glands in the oral mucosa.

In appearance, the mucosa is smoother than the skin except on the dorsal side of the tongue due to the papillae, the rugae of the hard palate and the stippling of the gingiva. In some

L.A. Bergmeier (ed.), *Oral Mucosa in Health and Disease*,
https://doi.org/10.1007/978-3-319-56065-6_1

individuals, there is a white line of keratinised tissue (linea alba) at the occlusal plane of the teeth which may be due to the abrasive effect of rough restorations or to cheek biting.

The firmness also differs from region to region, with the buccal mucosa and lips being loose and pliable while the hard palate and gingiva are firm.

The oral cavity can be divided into two regions—the outer oral vestibule found between the lips and cheeks on the outside and the maxillary and mandibular arches on the inside and the oral cavity proper situated within the dental arches (Fig. 1.1). The main structural features of the oral mucosa are the oral epithelium, lamina propria and submucosa. The oral epithelium is described as a stratified squamous epithelium and contains multiple types of cells with different morphologies arranged into discrete layers. The oral mucosa undergoes two distinct patterns of maturation resulting in the keratinised epithelium of the hard pallet and the gingivae and the non-keratinised epithelium of the sublingual and buccal mucosa.

The superior border of the latter is formed by the hard and soft palates and the inferior border by the tongue and the floor of mouth [2].

It is lined by a mucous membrane—the oral mucosa, which abuts on the skin at the mucocutaneous junction and the rest of the alimentary canal at the oropharynx. It forms a continuous lining, broken only at the junction between the gingiva and the tooth surface, and can be divided into masticatory mucosa, lining mucosa and specialised mucosa based on structure and function (Fig. 1.2). Masticatory mucosa is restricted to the hard palate and the gingiva, is tightly bound to the underlying tissues and possesses a tough keratinised surface to resist the loading and abrasive forces associated with mastication. Lining mucosa is generally non-keratinised and freely mobile and covers the cheeks, inner aspects of the lips, floor of the mouth and ventral surface of the tongue. The dorsum of the tongue forms the specialised mucosa because of the presence of numerous taste buds and sensory nerve endings [2].

1.3 Function of Oral Mucosa

The oral cavity is constantly exposed to a potentially damaging and rapidly changing environment, and the essential function of the oral

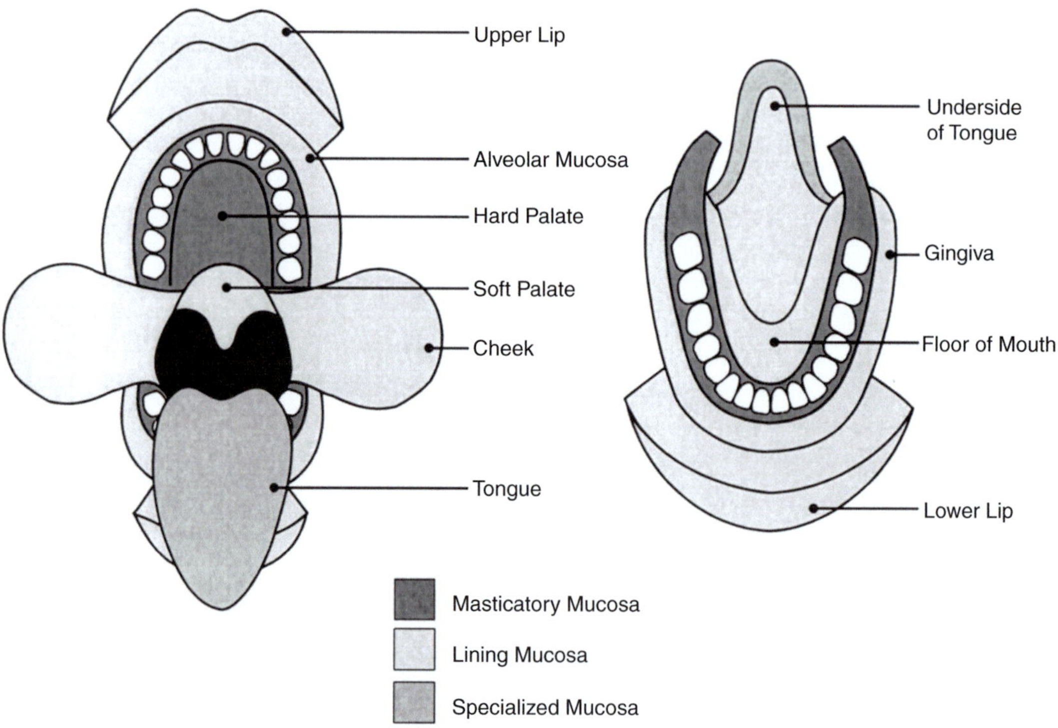

Fig. 1.1 The general anatomy of the oral cavity

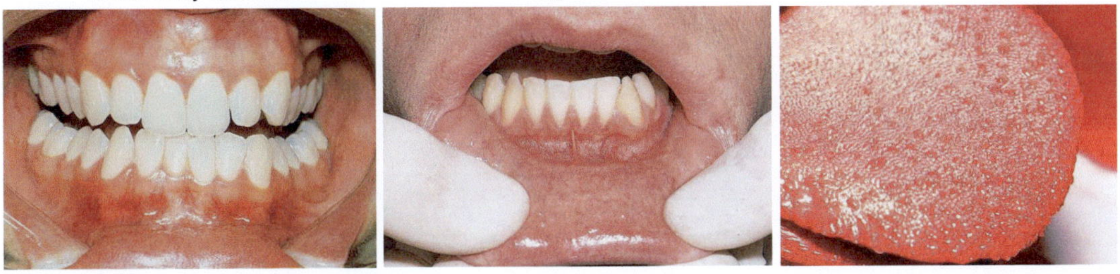

3 types of MUCOSA defined according to FUNCTION:

1. Masticatory Mucosa: 25% of *total mucosa.* Gingiva (free, attached and interdental) and hard palate. Primary mucosa that is in contact with food during mastication.
MASTICATORY MUCOSA IS USUALLY KERATINIZED.

2. Lining Mucosa: 60% of *total mucosa.* Covers the floor of mouth, ventral (underside) tongue, alveolar mucosa, cheeks, lips and soft palate.
Does not function in mastication and therefore has minimal attrition.
Non-keratinized; soft and pliable.

3. Specialised Mucosa: *15% of total mucosa.* Covers dorsal tongue and is composed of cornified epithelial papillae.

Fig. 1.2 The organisation of the oral mucosa; consisting of masticatory mucosa, 25%; lining mucosa 60%; specialised mucosa 15% [3]

mucosa is the defence and protection of the underlying tissues. This is achieved by:

1. Providing resistance to mechanical injury or insult
2. Resisting the entry of microorganisms
3. Providing a barrier to the permeability of noxious substances

This protective role is primarily mediated by the physical structure of the epithelium, the presence of immune-competent cells (Langerhans cells and lymphocytes) within the epithelium and lamina propria and the epithelial cells contributing to the protective capacity of the oral mucosa by sensing pathogens and by secretion of a variety of antimicrobial substances. In addition to protection, the oral mucosa also has important sensory functions including pain, touch, temperature unique to the oral cavity and taste. These are performed by a variety of specialised nerve endings, cells (Merkel cells) and cellular structures (taste buds) found in different areas of the oral mucosa. Human oral mucosa, unlike skin, is not thought to have an important thermal regulatory function, although in animals this role may be more significant, for example the tongue of a panting dog [2].

1.4 Structural Regions of the Mucosa

The two main tissue components of the oral mucosa are the epithelium and the lamina propria supported by a fibrous connective tissue (Fig. 1.3).

Microscopically the junction between the epithelium and the lamina propria (LP) appears to be distinct. However, the junction between the mucosa and the submucosal is more difficult to define. It is also much less organised when compared with the intestinal mucosa (Fig. 1.4).

In gut, there is a clear layer of smooth muscle and elastic fibres known as the *muscularis mucosae* (Fig. 1.4a) which is absent in the oral mucosa. In some areas of the oral cavity there is a layer of loose fatty or glandular connective tissues that contains both blood and nerve supplies and this area separates the mucosa from the bone or muscle that underlies these structures in the cheeks, the lips and parts of the hard palate. This is the *submucosa* and the composition of this area dictates the flexibility of the attachment to the underlying structures. However, in areas such as the gingiva and parts of the hard palate the *mucoperiosteum* reveals a direct attachment of the mucosa to the bone. The minor salivary

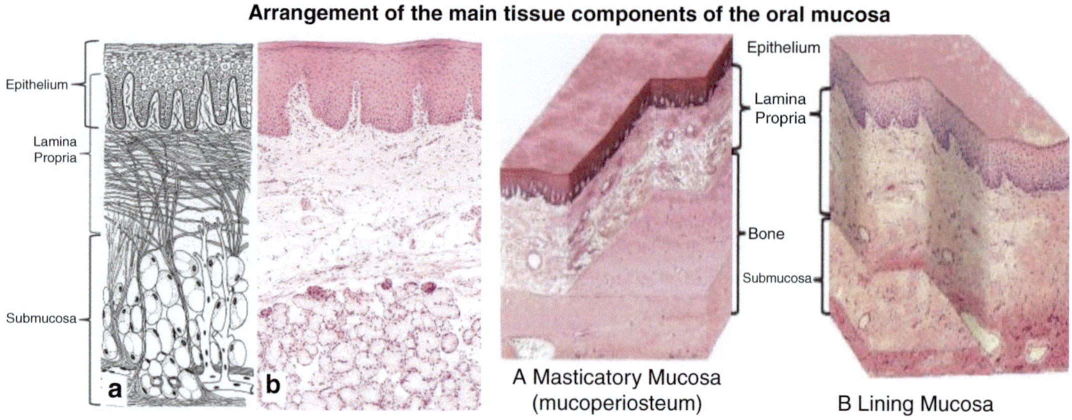

Fig. 1.3 The main tissue components of the oral mucosa are the epithelium and the lamina propria (**a**) cartoon of major tissue componenets, (**b**) Histology of the hard palate showing the components drawn in (**a**). Right hand panel (**A**) Masticatory Mucosa (mucoperiosteum). (**B**) Lining mucosa

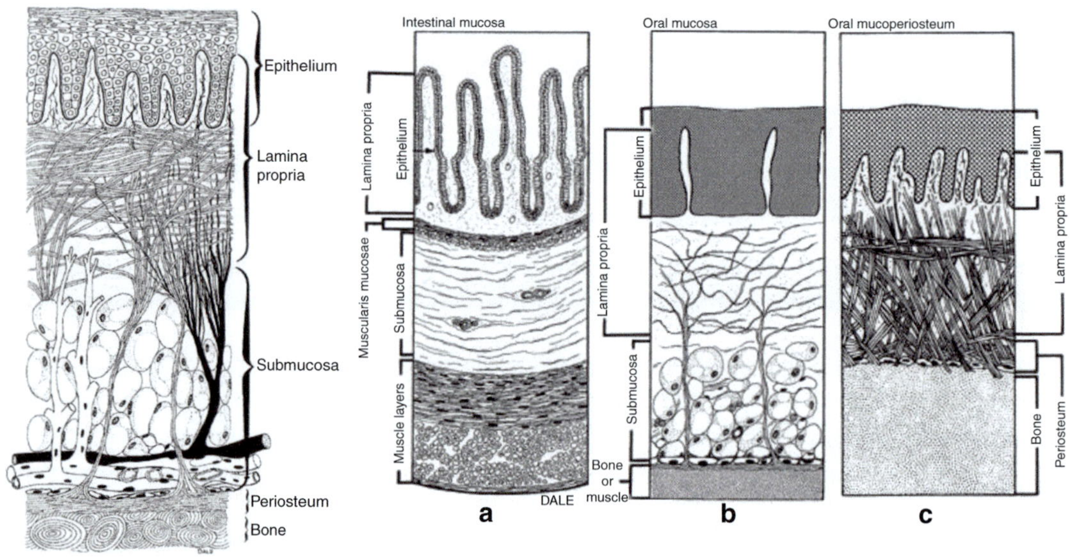

Fig. 1.4 Comparison of the arrangement of tissue layers in (**a**) intestinal mucosa, (**b**) oral mucosa and (**c**) oral mucoperiosteum

glands lie within the submucosa while the Fordyce spots, although few, lie within the lamina propria and are thought to produce sebum that lubricates the mucosa.

The immune cells lie within the lamina propria and there are scattered lymphoid loci. Waldeyer's ring consists of the lingual, palatine and pharyngeal tonsils. Along with the salivary glands, these are the major elements of the immune system within the oral cavity and are discussed in more detail in subsequent chapters (Chaps. 4 and 5).

1.5 Epithelium: Structure and Organisation

The oral epithelium is a stratified squamous epithelium and forms the main barrier between the oral environment and the deep tissues. The cells are tightly attached to one another and are layered from the basal lamina through spinous, granular and cornified layers (Fig. 1.5). The stratified squamous epithelium lining the oral mucosa shows some important regional differences in its

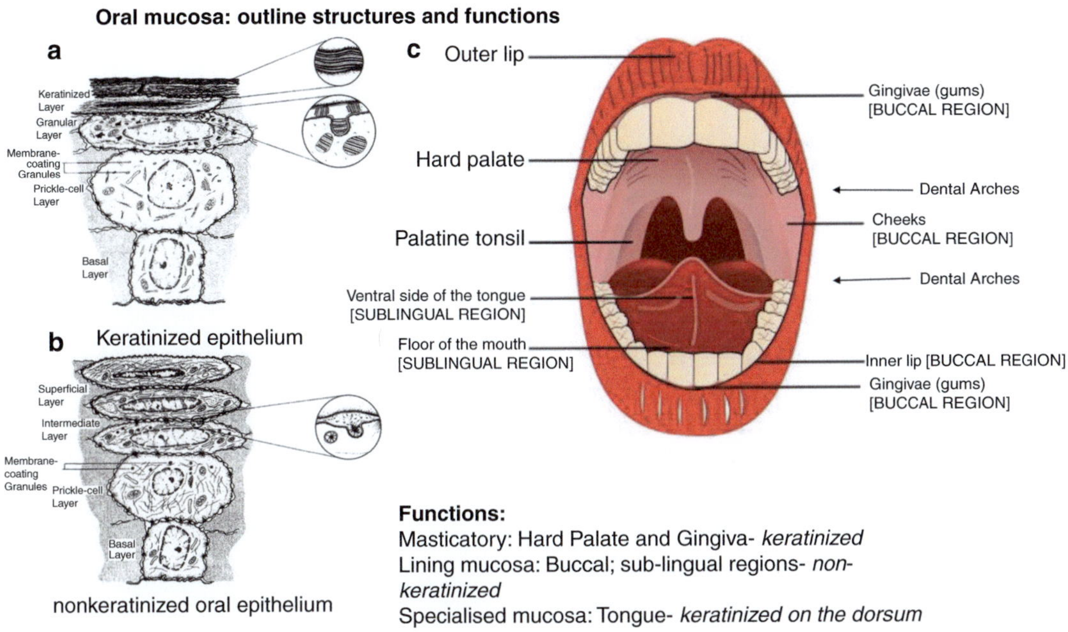

Fig. 1.5 Outline structure and function of keratinised (**a**) and non-keratinised mucosa (**b**). Regional distribution of functional tissues in the mouth (**c**). Reproduced with permission of the publishers

pattern of maturation reflecting the function of each of the major types of mucosa (Fig. 1.2).

The oral mucosa is a self-renewing tissue where cells in the deepest layers (the progenitor population) undergo mitotic cell division followed by terminal differentiation as the cells migrate to the surface (maturing population) and replace cells as they are shed from the surface. The transition from proliferation to differentiation is thought to be controlled by microRNAs [4].

The maturing cells generate a protective layer at the periphery known as the *cornified envelope.* This consists of keratins embedded into a protein matrix with a lipid envelope. In a series of maturation and synthesis steps the cytoplasmic face of the plasma membrane of the cells is replaced by this lipid envelope which eventually completely replaces the plasma membrane of the corneocytes and becomes coated with lipids consisting mainly of ceramides, cholesterol and free fatty acids which act as an essential water barrier. These corneocytes are tightly attached together by modified desmosomes (see Chap. 3) and undergo proteolytic degradation as the cells desquamate. It is now thought that there are two subsets of progenitor cells: a small population of cycling stem cells which maintain the proliferative potential of the tissue and a larger subset of amplifying cells which maintain the cells available for maturation. The progenitor cells lie in the basal layers of thin epithelium such as the floor of the mouth or in the lower layers of thick epithelium such as the buccal mucosa [5]. Oral keratinocyte progenitor cells have been characterised as relatively smaller than other cells in the tissue and by the expression of p75, a member of the tumour necrosis family (TNF) of proteins that mediate cell survival, apoptosis and intracellular signalling [6].

The turnover time for these cells differs from region to region and is dependent on keratinisation, with non-keratinised epithelium turning over faster than keratinised epithelium. These rates have been estimated at 25 days in the buccal mucosa and between 41 and 75 days in the gingiva. This compares with 52–75 days in skin and 4–14 days in the gut [7].

It should be noted that cancer chemotherapy drugs block mitotic division which can disrupt the process of epithelial turnover and result in damage to the oral epithelium with many patients developing oral ulcers.

Table 1.1 Characterisation of the layers of the masticatory and specialised mucosa (from Ten Cate's Oral Histology [67])

Regional variation in the oral mucosa: 1. Masticatory and specialised mucosa			
Masticatory mucosa	Covering epithelium	Lamina propria	Submucosa
Gingiva	Thick, orthokeratinised or parakeratinised. Stratified squamous epithelium often showing stippled surface	Long narrow papillae; dense collagenous connective tissue; not highly vascular but have long capillary loops with numerous anastomoses	No distinct layer, mucosa firmly attached by collagen fibres to cementum and periosteum of alveolar process (mucoperiosteum)
Hard palate	Thick, orthokeratinised (or parakeratinised in parts), stratified squamous epithelium with transvers palatine ridges (rugae)	Long papillae; thick dense collagenous tissue, especially beneath rugae, moderate vascular supply with short capillary loops	Dense collagenous connective tissues attaching mucosa to periosteum, fat and minor salivary glands packed into connective tissue in regions where mucosa overlies lateral palatine neurovascular bundles
Specialised mucosa			
Dorsal surface of the tongue	Thick keratinised and non-keratinised, stratified squamous epithelium: forms three types of lingual papillae—some bear taste buds	Long papillae: minor salivary glands in posterior portion; richly innervated especially near taste buds. Capillary plexus in papillary layer. Large vessels lying deeper	No distinct layer; mucosa is bound to connective tissue surrounding musculature of the tongue

The *keratinising epithelium* of the masticatory mucosa has a similar structure to that of skin, in which the epithelial cells (keratinocytes) undergo terminal differentiation to form corneocytes. These anucleate cells are densely packed with keratin fibres, which become tightly bound together to form the stratum corneum. This provides the main barrier to mechanical insult, and bacterial and chemical damage (Table 1.1).

Basal keratinocytes are cuboidal, basophilic cells which become larger and paler as they leave the basal layer to form the spinous cell layer (stratum spinosum), where epithelial stability is maintained by numerous desmosomal interconnections together with the characteristic arrangement of tonofilaments in bundles. These insert into the desmosomes and serve to distribute the stresses associated with friction. Towards the surface the cells become flattened and acquire keratohyalin granules associated with the tonofilaments (stratum granulosum). The granules contain filaggrin, which acts as the matrix in which the tonofilaments become embedded to form the corneocyte. This general pattern of keratinisation is known as *orthokeratinisation*.

In the masticatory mucosa, a variation of keratinisation known as *parakeratinisation* occurs where the cells retain shrunken or pyknotic nuclei. This is a normal process in the oral mucosa and is most commonly observed in the gingiva; however, in skin it is associated with psoriasis.

Incomplete keratinisation can also occur, where the outermost layer appears to have become rehydrated and resembles the deeper layers. Again, no pathology is associated with this morphology.

In *non-keratinising epithelia*, although the changes associated with terminal differentiation are less striking, a similar maturation pathway has been identified. These have been divided into stratum basale, stratum suprabasale, stratum filamentosum and stratum distendum more accurately reflecting the ultrastructure of the different layers [8] than the original classification of basal, prickle, intermediate and superficial layers, although the latter terminology has persisted. Overall, fewer tonofilaments are present and their distribution is random compared to that seen in a keratinising epithelium and, while there is some cell flattening, loss of organelles characteristic of orthokeratinisation does not occur. The random distribution of tonofilaments together with the presence of smaller and fewer desmosomes in buccal mucosa compared to palate reduces the resistance to mechanical damage, but allows the mucosa to fulfil its lining function (Table 1.2).

These morphological changes are also accompanied by changes in the biochemical and structural composition of keratinocytes including keratins, keratin-associated proteins such as filaggrin and involucrin, and cell surface carbohydrates [9, 10]. Keratins are a family of 20 different proteins, which are the product of two distinct gene families (type I and type II), and their expression shows a site-specific and differentiation-specific pattern [2, 9, 11]. All basal cells in stratified squamous epithelia express keratins 5 and 14 while keratinised sites (for example, epidermis and hard palate) express the differentiation-specific keratins 1 and 10 in the suprabasal layers. Non-keratinising epithelia such as buccal mucosa, however, express keratins 4 and 13 in suprabasal cells. Filaggrin is involved in the aggregation and packing of keratin filaments in the stratum corneum, while terminal differentiation is associated with the production of a thickened cornified cell membrane, formed from soluble precursors such as involucrin. Cell surface carbohydrates such as the ABO blood group antigens are expressed in a differentiation-specific pattern in epidermis and oral epithelia [12–15], with basal cells expressing shorter precursor structures, while expression in the superficial layers exhibits a stepwise increase in complexity of the carbohydrate molecules [16, 17]. The chemical structure of the keratins differs between the layers of the epithelium and various patterns of maturation can be identified by the keratins that are present. Other markers of differentiation

Table 1.2 Characterisation of the non-keratinised lining mucosa and the keratinised lip zones

Regional variation in the oral mucosa: 2. Lining mucosa			
	Covering epithelium	Lamina propria	Submucosa
Soft palate	Thin, non-keratinised stratified squamous epithelium taste buds present	Thick with numerous snort papillae elastic fibres forming elastic lamina. Highly vascular, well-defined capillary network	Diffuse tissue containing many minor salivary glands
Ventral surface of the tongue	Thin, non-keratinised stratified squamous epithelium	Thin numerous short papillae and some elastic fibres; a few minor salivary glands: capillary network in suprabasal layer; reticular layer relatively avascular	Thin and irregular; may contain fat and small vessels: where submucosa is absent, mucosa is bound to connective tissue surrounding the tongue musculature
Floor of the mouth	Vary thin, non-keratinised stratified squamous epithelium	Short papillae; some elastic fibres; extensive vasculature with short anastomosing capillary loops	Loose fibrous connective tissue containing fat and minor salivary glands
Alveolar mucosa	Thin, non-keratinised stratified squamous epithelium	Short papillae, connective tissue contains many elastic fibre capillary loops close to surface supplied by vessels running superficially to the periosteum	Loose connective tissue with thick elastic fibres attaching it to periosteum of alveolar process: minor salivary glands present
Labial and buccal mucosa	Vary thick, non-keratinised stratified squamous epithelium	Long slender papillae; dense fibrous connective tissue with collagen and soma elastic fibres: rich vasculature giving off anastomosing loops into papillae	Mucosa firmly attached to underlying muscle by collagen and elastin; dense collagenous connective tissue with fat, maw salivary glands (sometimes sebaceous glands)
Lips: vermillion zone	Thin, orthokeratinised stratified squamous epithelium	Many narrow papillae. capillary loops close to the surface of papillary layer	Mucosa firmly attached to muscle; some sebaceous glands in vermillion border
Lips: intermediate zone	Thin, parakeratinised stratified squamous epithelium	Long, irregular papillae elastic fibres and collagen fibres in connective tissue	Minor salivary glands and fat in intermediate zone

Table 1.3 The maturation of epithelium in the oral mucosa (adapted from Ten Cate's Oral Histology)

Maturation of keratinised and non-keratinised epithelium			
Keratinised		Non-keratinised	
Features	Cell layer	Features	Cell layer
Cuboidal or columnar cells; contain bundles of tonofibrils and other cell organelles. Site of most cell division	Basal	Cuboidal or columnar cells contain separate tonofilaments and other cell organelles. Site of most active cell division	Basal
Large ovoid cells containing conspicuous tonofibril bundles; membrane-coating granules in upper part of layer	Prickle/ spinosum	Large ovoid cells with dispersed tonofilaments. Membrane-coating granules present in upper part of layer; numerous filaments	Prickle/ spinosum
Flattened cells containing conspicuous keratohyalin granules-associated with tonofibrils; membrane-coating granules fuse with cell membrane in upper part; internal membrane thickening occurs	Granular	Slightly flattened cells containing many dispersed tonofilaments and glycogen	Intermediate
Extreme flattening and dehydration of cells; loss of all organelles; cells filled with fibrillar substances; when pyknotic nuclei are found parakeratinisation occurs	Keratinised	Slightly flattened cells with dispersed filaments and glycogen; fewer organelles. but nuclei still present	Superficial

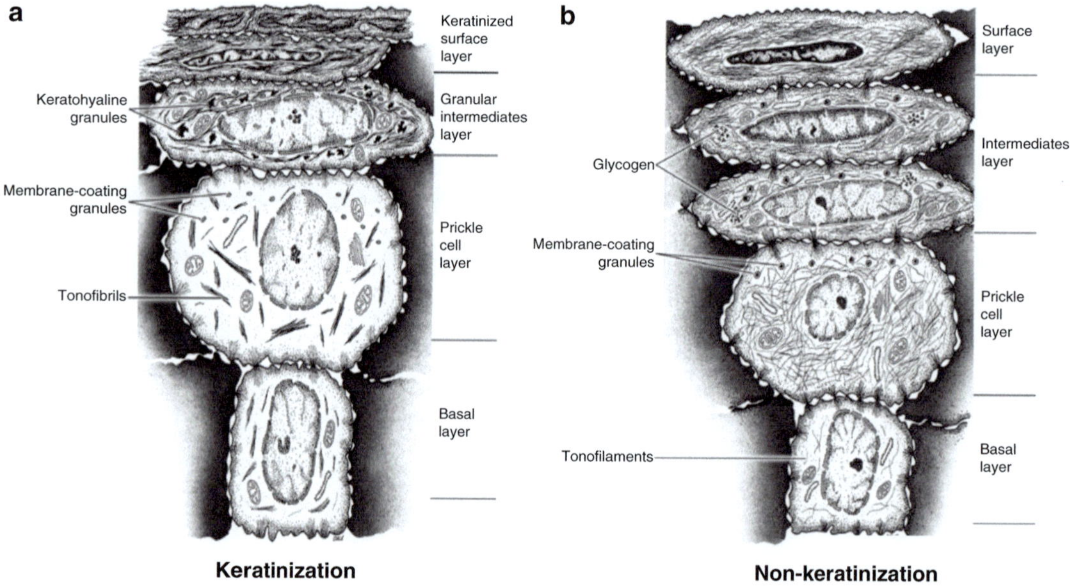

Fig. 1.6 Description of the cell layers and differential maturation in keratinised (**a**) compared with non-keratinised (**b**) mucosa (adapted from Squier and Brogden [2])

include Ki67 (a marker of cells which are actively cycling) [18] and E-cadherin (associated with desmosomes and epithelial cells [19]. Stages of maturation of the Keratinised and non-keratinised epithelium are described in Table 1.3.

In the upper part of the prickle cell layer (Fig. 1.6) an organelle known as the membrane-coating or lamellate granule is found. These small-membrane-bound structures contain glycolipid and are thought to originate from the Golgi apparatus. In keratinised epithelia, they are elongated and contain parallel lamellae, while in non-keratinised cells they appear as round structures with an amorphous centre. As the cells migrate to the surface the granules accumulate, become aligned to the superficial cell membrane and secrete lipids that contribute to the permeability barrier. The lipid content of the granule also differs between keratinised and non-keratinised epithelium which results in a differential permeability barrier between these two types of epithelium.

The physical structure of the epithelium, together with epithelial cell turnover and desquamation, serves to protect the tissue against mechanical insults and the ingress of microorganisms. This barrier property is dependent on the physical and chemical attachment between cells, especially on intracellular protein-carbohydrate complexes produced by the epithelial cell within the desmosomes and other junctional complexes. These interactions are more fully described in Chap. 3 (Wan).

1.6 Permeability

1.6.1 Evidence for Role of Membrane-Coated Granules and the Chemistry of the Barrier

The mechanisms of defence and protection referred to in the previous sections primarily serve to limit the colonisation, adherence and invasion of microorganisms. However, the oral environment contains numerous compounds that are potentially harmful if they gain access to the epithelial cells or the underlying connective tissue. Such compounds include products derived from oral organisms, e.g. toxins and antigens, and potential carcinogens introduced deliberately with food, alcohol or tobacco [7]. The permeability barrier in the oral mucosa is, however, not absolute and allows the passage of many substances, and this property has been utilised to understand the permeability barrier characteristics of oral mucosa [7] and more recently to exploit the differential permeability of different sites within the oral cavity for drug/vaccine delivery [20].

Much of our understanding of the epithelial permeability barrier has been gained from studies of human skin, following intradermal injection of horseradish peroxidase (HRP) [21]. The enzyme penetrated across the basal lamina and through the epidermis as far as the stratum granulosum. However, no HRP was detected in the stratum corneum.

Horseradish peroxidase and lanthanum, an electron-dense element with a smaller particle size than peroxidase, were subsequently used to demonstrate the location of the permeability barrier in keratinised and non-keratinised oral mucosa of a variety of animals, including rats, rabbits and monkeys [22, 23]. The tracers were applied topically or by sub-epithelial injection and microscopical examination revealed that the compounds did not penetrate areas corresponding to the outer quarter of either the keratinised or the non-keratinised epithelium, and indicated the presence of an intercellular barrier [23].

All the studies consistently demonstrated that the limit of penetration of the tracers coincided with the level where membrane-coating granules discharged their contents, suggesting that these organelles were important for the formation of the barrier.

Membrane-coating granules were first described in the epidermis by Selby [24], are present in all differentiating stratified squamous epithelia [21] and appear in the stratum spinosum of keratinised mucosa and the intermediate cell layers of non-keratinised oral mucosa [23, 25]. In keratinised epithelia, these intracellular organelles appear as ovoid, membrane-bounded organelles, 0.25 μm in length and containing a series of parallel internal lamellae, consisting of electron-dense and electron-lucent bands. The lamellae are bounded by an outer membrane, and which on extrusion from the cell rearrange to form sheets in the intercellular region. In non-keratinised epithelia the membrane-coating granules are usually spherical, membrane-enclosed vesicles. They are approximately 0.2 μm in diameter and contain an electron-dense amorphous core with radiating delicate strands (Fig. 1.7) [26].

The contents of epidermal membrane coating granules include acid hydrolases, which are asso-

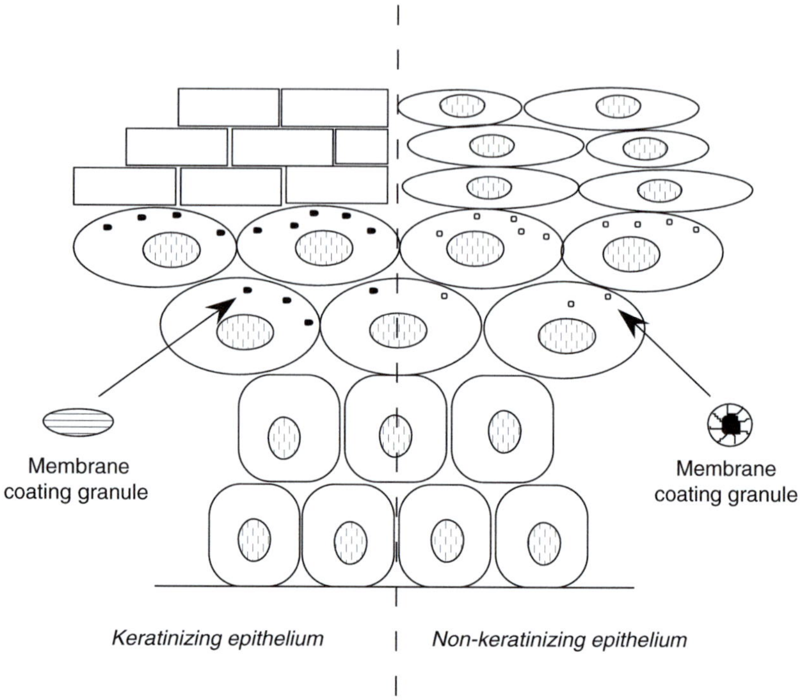

Fig. 1.7 Membrane-coating granules in keratinised and non-keratinised epithelia

ciated with the Golgi complex and significant amounts of sphingomyelin, phosphoglycerides, cholesterol, glucosylceramide, ceramides and some other neutral lipids. In particular, the granules contained an acyl glucosylceramide formed by a 30–34 carbon chain fatty acid attached to a sphingosine base linked to linoleic acid and glucose [2, 21].

The permeability barrier is dependent on the correct formation of the membrane-coating granules and differentiation of stratified squamous epithelium. When explants of keratinised oral epithelium (or skin) are held in a submerged liquid culture system differentiation is poor and the granules are not seen in ultrastructural studies [27]. However, when so-called raised or interfaced culture systems are used, the differentiation of the epithelium is restored and granules appear [28, 29].

In keratinised oral epithelia the barrier is represented by neutral lipid or ceramide [30] and the events that lead to the formation and extrusion of the contents of the membrane-coating granules appear to be similar to those seen in epidermis. The extruded lipids are organised into lamellae that form in the intercellular spaces of the stratum corneum (Fig. 1.8) and which were first identified by freeze fracture transmission electron microscopy [31].

In non-keratinised oral epithelia, the intercellular material also seems to be extruded by membrane-coating granules, although they differ morphologically from the membrane-coating granules found in keratinised epithelium. The intercellular substance is also different from that seen in keratinised epithelium, being amorphous and lacking the parallel lamellar structure evident between the corneocytes of the stratum corneum [21].

The importance of intercellular lipids is demonstrated by the impaired epidermal barrier that exists in fatty acid deficiency, which is associated with an abnormal organisation of intercellular lipids in the stratum corneum [32]. It has also been shown that treatment with lipid solvents increases the penetration of tracers through the intercellular regions of the stratum corneum [33].

The largest single class of lipids within the stratum corneum are the ceramides, which represent about 50% of the total lipid present. These can be separated into six different fractions (ceramides 1–6) each with a slightly different composition—ceramide 1 is derived from the acyl-glucosylceramide present in the membrane

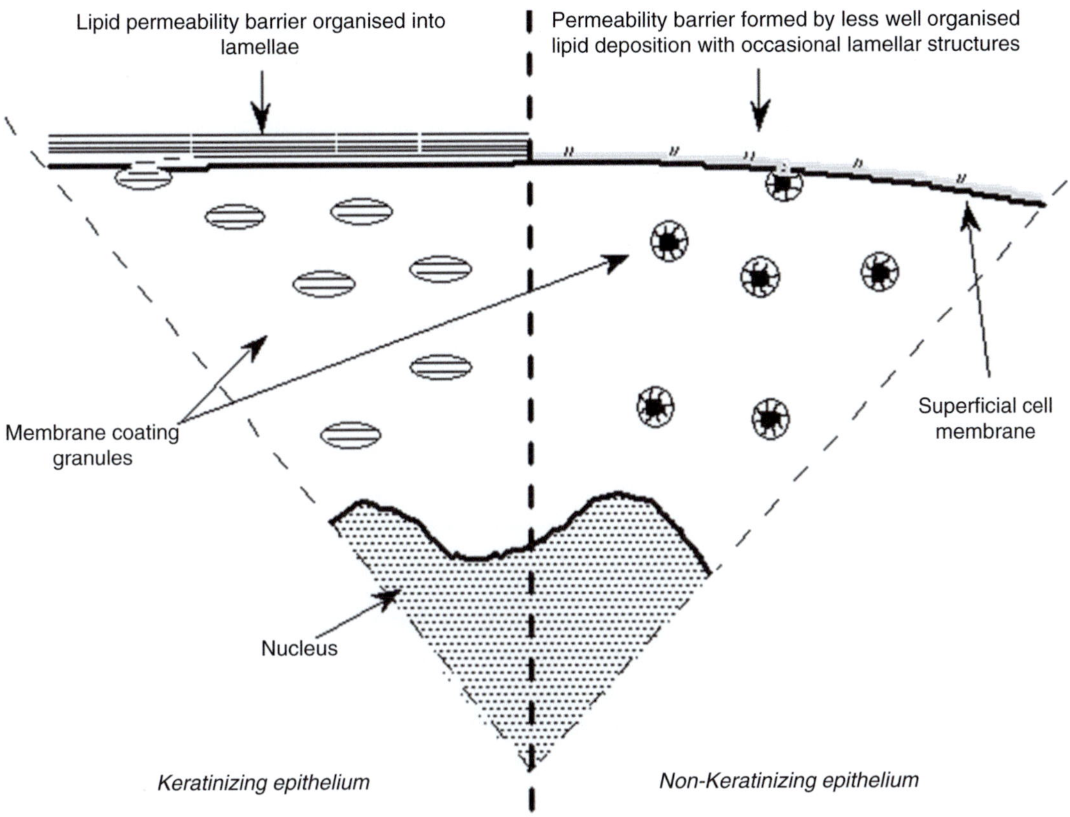

Fig. 1.8 The role of membrane-coating granules in the formation of the permeability barrier in keratinising and non-keratinising epithelia

granules and is thought to serve as a "molecular rivet" conferring stability to the intercellular lamellae, while ceramides 2–6 are thought to provide resistance to oxidative damage because they have a high degree of saturation [33].

Keratinised oral epithelium predominantly contained neutral liquids, such as acyl-ceramides and ceramides although the total quantity was 25–50% less than that found in epidermis. In contrast, epithelium from the non-keratinised oral regions, such as floor of mouth and buccal mucosa, contained no acylceramides or acyl-glucosylceramides and only small amounts of ceramide, but relatively high quantities of polar lipids, such as cholesterol sulphate and glycosyl-ceramides. These lipids were demonstrated, using histochemical staining techniques, in the intercellular spaces of the stratum corneum and the intercellular regions of non-keratinised oral epithelium [34].

Profiling the lipid content of different levels within porcine epidermis and comparing this with non-keratinised oral epithelium have revealed interesting differences that have profound effects on the barrier properties induced by these substances. Extracts of consecutive, horizontal, frozen sections showed that all phospholipids decreased in concentration towards the surface and were absent in the stratum corneum, while neutral lipids and ceramides increased. Glucosylceramides and acyl-glucosylceramides reached a peak concentration in the stratum granulosum and then decreased in the surface layers. Cholesterol sulphate reached a maximum concentration in the deeper stratum corneum and then abruptly decreased in the surface layer [35]. These changes in concentration are consistent with the formation of a neutral lipid, subsurface barrier to water by the hydrolysis of glucosylceramides into glucose and ceramides in keratinised tissue [36, 37]. In contrast, ceramides were absent from non-keratinised oral epithelium suggesting that there is no mechanism in this tissue for the conversion of glucosylceramides into ceramides and that whereas neutral lipids and ceramides contribute to the barrier function in keratinised tissue glycolipids may fulfil this function in non-keratinised regions [30, 35, 38].

In contrast to the intestinal lining, the oral epithelium does not have an absorptive function. However, the relative thickness in different parts of the mucosa has led to the exploitation of the sublingual mucosa as a rapid drug delivery surface, particularly with drugs such as nitroglycerin administered for angina pectoris. More recently the sublingual mucosa has been explored as a potential route for desensitisation in allergy and in animal models of autoimmune disease therapy [20, 39]. The oral mucosa has long been recognised as a potential immune-privileged site [40], while the distribution of antigen-presenting cells in the different mucosa has been interrogated as potential vaccination sites [41–43]. It has also been recently suggested that mechanical damage in the oral mucosa might influence the cytokine environment and lead to dysbiosis [44]. The inflammatory responses in the oral mucosa are further explored in subsequent chapters (Chaps. 4, 5, and 10). However, in health the permeability barrier is robust, as inflammatory responses are not generally seen and microbial colonisation is highly restricted.

1.7 Mucosal Pellicle

The surface of the oral mucosa is cleansed by constant bathing with saliva containing antimicrobial agents, enzymes and antibodies that limit microbial growth in the mouth [45]. Many of the components of saliva actively adhere to the mucosal surface and contribute to the barrier function. While a considerable amount is known about the protective function of the *acquired enamel pellicle* of the tooth surface, less is known about the *mucosal pellicle* of the desquamating oral epithelium (Hannig, [46]). The apical surface of the epithelium is covered by a "glycocalyx" consisting of a variety of glycoproteins and mucins which have been shown to differ slightly depending on the region of the mucosa and reflect the relative keratinisation of the epithelium. Many of the glycoproteins form protein-protein interactions that have been referred to as "heterotypic interactions" and there is some evidence that these complexes can act synergistically so that, for example, lysozyme bound to secretory IgA (SIgA) that is immobilised on the epithelial surface is a more efficient antimicrobial agent than when it is in solution in the saliva. Recent evidence suggests that the most abundant glycoproteins in the mucosal pellicle are the mucins such as MUC5B and MUC7 as well as SIgA [46–48]. The role of saliva is further explored by Saloom and Carpenter in Chap. 7.

1.8 Non-keratinocytes in the Oral Epithelium

About 10% of the cells in the oral epithelium consist of non-epithelial cells. Frequently described as "clear cells" they appear to have a halo around the nucleus in both ultrastructural and immunochemical studies. They include a variety of cells with different phenotypes and function: melanocytes, Merkel cells, Langerhans cells and lymphocytes (Table 1.4). A property that some have in common is the lack of desmosomes which makes them motile and is key to their functions.

Melanocytes dictate the colour of the epithelium (along with other factors) and the pigments most frequently contributing colour are melanin and haemoglobin. Melanin, produced by melanocytes in the basal layers of the epithelium, originates from the neural crest ectoderm and possesses long dendritic processes. The pigment is produced in the cytoplasm in small melanosomes and can be transferred into adjacent keratinocytes by the dendrites of the melanocytes [49]. The number of melanocytes in any given region of the mucosa does not differ in light or darkly pigmented individuals. However, the level of pigmentation is dependent on the relative rate of melanin production.

Merkel cells are situated in the basal layer of the oral epithelium and epidermis and are the

Table 1.4 Characteristics of non-keratinocytes in the oral epithelium (adapted from Ten Cate's Oral Histology)

Characteristics of non-keratinocytes of the oral epithelium				
Cell type	Position in the epithelial layers	Specific markers (staining)	Ultrastructure	Function
Melanocyte	Basal	Dopa oxidase-tyrosinase Silver stains	Dendritic: no desmosomes or tonofilaments Premelanosomes and melanosomes present	Synthesis of melanin pigment granules (melanosomes) and transfer to surrounding keratinocytes
Langerhans cells (LC)	Suprabasal (predominantly)	CD1a cell surface marker	Dendritic: no desmosomes or tonofilaments. Characteristic LC granule	Antigen trapping and processing. Very important antigen-presenting cell (APC)—able to push processes through the epithelial layer and sample the lumen
MerkeUel	Basal	Probably periodic acid-Schiff positive	Non-dendritic. sparse desmosomes/tonofilament; characteristic, electron-dense vesicles. Associated nerve axon	Tactile sensory cell
Lymphocyte	Variable: inflammation dependent	Cell surface antigen markers variable in all T cells: CD3^{+}; all B cells CD20^{+}	Large circular nucleus; scant cytoplasm with few organelles: No desmosomes/tonofilaments	Associated with inflammatory response and innate and adaptive immunity

only cells in the group of "clear cells" with occasional desmosomes and keratin tonofilament. These cells are not dendritic in character but have small membrane-bound vesicles in the cytoplasm often situated near a nerve fibre that is associated with the cell. Transmitter substances can be secreted from these vesicles across a synapse-like junction and generate nerve impulse. These cells appear to respond to touch and the granules have been shown by some groups to contain cytokeratin 20 and located in the more superficial layers of the palate [50]. More recently immunomodulatory functions have been suggested for Merkel cells [51].

1.9 Langerhans Cells (LC) and Dendritic Cells (DC)

The dendritic cells first described in the skin by Paul Langerhans in 1868 have now been well characterised in the oral mucosa [41, 43, 52]. These cells are the sentinels of the immune system and are the classical antigen-presenting cell [41, 53]. LCs are typically situated in the suprabasal layer and classically are HLA-DR positive and express CD1a. They are characterised by the presence of the rodlike Birbeck granule and the highest number of these cells are found in the non-keratinised mucosa of the sublingual region, the soft palate, the lip and the vestibule. Lower numbers are generally found in the hard palate and gingiva [41]. However, in the buccal mucosa numbers are increased in smokers [54]. These cells can sample the lumen of the mucosa by pushing their dendrites through the layers of epithelial cells. Their motility allows them to migrate to regional lymph nodes where they have the capacity to present antigens to naïve T cells and degenerate an immune response. The biochemical environment is important in driving immune responses, and the key messengers are the cytokines and chemokines. Keratinocytes have the capacity to secrete proinflammatory cytokines, including IL-8 (CCL8) and IL-1 which can lead to the recruitment of lymphocytes and polymorphonuclear leucocytes into the oral epithelium [52, 55]. The phenotype (and therefore function) of different subsets of dendritic cells has the subject of considerable investigation in recent years. Three distinct subsets were identified by Chalermsarp and Azuma [56] based on the expression of CD11c and CD207 surface markers. These three subsets represented a resident population, a newly recruited population and a slowly migrating population. All three were shown to have a mature phenotype with the potential for antigen presentation in the draining lymph nodes and expressed high levels of co-stimulatory molecules such as CD80/86 and MHCII. Using two different markers, namely CD236 and CD103, Aramaki et al. have been able to distinguish between resident LCs and resident DCs [57]. These are important distinctions as the potential for using the oral mucosa as a site of "intra-oral" vaccination is a topic of considerable current research. The priming of CD8+ T cells has been shown to be more robust when antigens are applied to the buccal mucosa compared with the sublingual lining mucosa [58]. This is also supported by extensive investigations by the Allam group (Chap. 5) who have demonstrated that tolerogenic T cells, Th1/Th17 cytokines and TLR2/4-expressing DCs predominate in buccal and sublingual regions. The expression of TLRs might be used as adjuvant targets in vaccination strategies [39, 59].

1.10 Lamina Propria (LP)

The connective tissue that supports the oral epithelium is the *lamina propria* and consists of cells, blood vessels, neural elements and support fibres held in amorphous ground structure. This element of the tissue can be divided into two parts: the superficial papillary layer and the netlike reticular layer. In the papillary layer, the collagen fibres are thin and arranged in a loose fashion while in the reticular layer they are bundled together and lie parallel to the plane of the surface. There is regional variation in the LP especially in terms of the proportion and composition of the cells in healthy compared with inflamed tissues (Table 1.5).

The other key inflammatory cells found in the oral mucosa are described in Table 1.5.

Fibroblasts: Responsible for elaboration and turnover of fibre and ground substance, these cells play a key role in maintaining tissue integrity including wound healing [60–62].

Macrophages: Functions of these cells include the ingestion of damaged cell/tissues or foreign material. In the oral mucosa two special

Table 1.5 Cells of the lamina propria (from Ten Cate's Oral Histology)

The cells in the lamina propria (LP) of the oral mucosa			
Cell type	Morphology	Function	Distribution
Fibroblast	Stellate or elongated with abundant rough endoplasmic reticulum (ER)	Secretion of fibres and ground substances	Throughout the lamina propria (LP)
Histiocyte	Spindle-shaped or stelate; often dark-staining nucleus; many lysosomal vesicles	Resident precursor of macrophages (Mφ)	Throughout the LP
Macrophage	Round, pale-staining nucleus; contains lysosomes and phagosome vesicles	Phagocytosis, including antigen processing	Areas of chronic inflammation
Mast cell	Round with basophilic granules. Stains metachromatically	Secretion of inflammatory mediators and vasoactive agents (histamine, heparin. serotonin)	Throughout the LP, often subepithelial
Polymorphonuclear leukocyte (neutrophil)	Round. Characteristic lobed nucleus contains lysosomes and specific granules	Phagocytosis and cell killing	Areas of acute inflammation within LP; may be present in epithelium
Lymphocyte	Round. Dark-staining nucleus. Scant cytoplasm, some mitochondria	Humoral and eel mediated immunity (innate and adaptive)	Areas of acute and chronic inflammation
Plasma cell	Cartwheel nucleus. Intensely basophilic cytoplasm with abundant rough ER	Synthesis of immunoglobulins	Areas of chronic inflammation. Often perivascular
Endothelial cell	Associated with basal lamma. Contains numerous pinocytic vesicles	Lining blood and lymphatic channels	Lining vascular channels throughout LP

types of macrophages have been identified: the melanophage and the siderophage. The former is common in pigmented mucosa. Both cell types have functions associated with colour changes in the mucosa. These cells are capable of antigen processing and presentation and are therefore key to immune responses in the LP. They also produce cytokines and chemokines—the chemical messengers of the immune system—that stimulate fibroblast proliferation and collagen production and are an important part of the wound repair mechanisms (Chap. 6) [60, 63].

Mast cells: These cells are classically associated with allergic responses and contain granules that stain with basic dyes such as methylene blue. The main contents of these granules are histamine, heparin and cytokines such as TNF-α. It is thought that these cells are important in the transition from an acute to a chronic inflammatory response [64].

Inflammatory cells: Lymphocytes and plasma cells are found scattered throughout the LP in small numbers—in healthy tissue. In the main lymphoid tissues of the nasopharyngeal organs such as the tonsils there are large structurally organised accumulations of lymphoid cells (discussed in Chaps. 4 and 5). Inflammatory cells mostly appear in the LP as a result of an infection or injury and the nature and composition of any infiltration are dependent on the nature of the injury, although there are resident antigen-presenting cells (discussed in the section on Langerhans cells) which are vitally important in surveying the tissue for potential pathogenic changes, either as a result of infection or injury. In acute inflammatory conditions, the main infiltration is from polymorphonuclear leucocytes, while macrophages, monocytes and lymphocytes are associated with chronic inflammation. The recruitment of these cells is part of the innate immune response in the oral cavity [65] (Fig. 1.9).

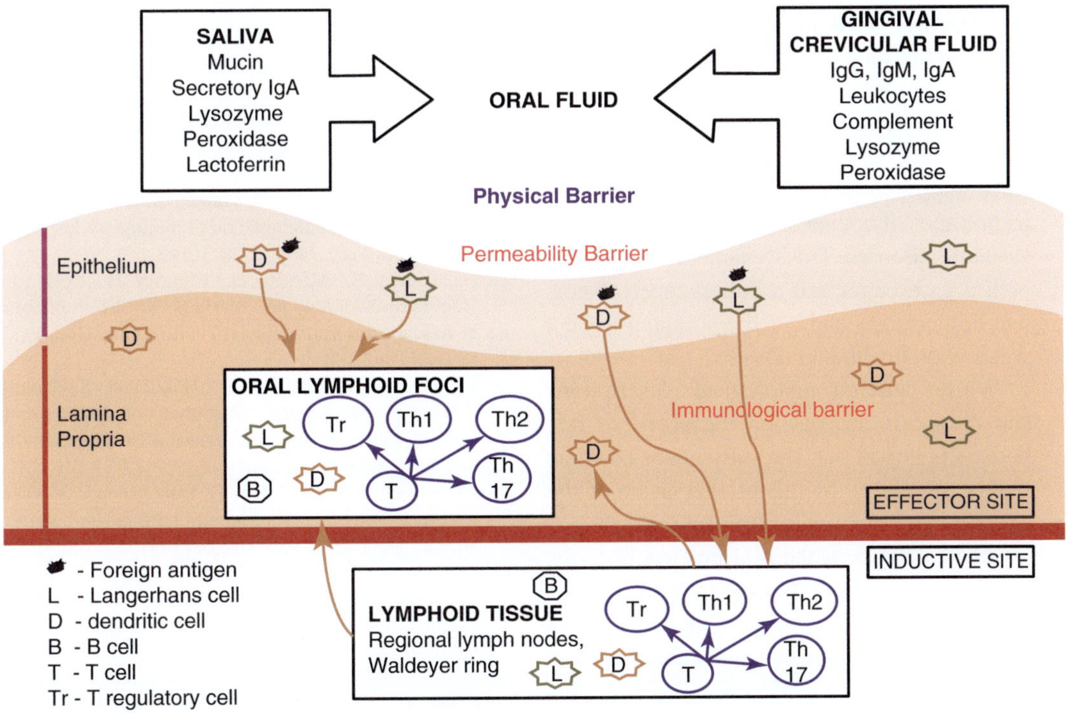

Fig. 1.9 The composite barrier: factors associated with oral immunity. The saliva contains secretory immunoglobulin A (sIgA), mucins and enzymes that protect the oral mucosa from bacterial colonisation. The gingival crevicular fluid that reaches the oral cavity contains leucocytes, IgG, IgM, IgA and a range of other agents that contribute to oral immunity. Langerhans cells and other myeloid dendritic cells, after capturing foreign antigens, migrate to immune-inductive sites (regional lymph nodes, Waldeyer's ring) where they prime immune-effector cells. In turn, these immune-effector cells migrate to the lamina propria (mucosal lymphoid foci) where they mediate either active immune responses or immune tolerance (adapted from Feller et al. [66] with permission of the publisher [66])

1.11 Protective Secretions and Barrier Function

Oral mucosal integrity is highly dependent on normal salivary gland function and many investigators have reported that salivary gland dysfunction predisposes the oral mucous membrane to disease. Whole saliva is made up of the secretions of three pairs of major salivary glands, the parotid, submandibular and sublingual, together with the secretions of the minor, or accessory, glands which are distributed in the mucosa of the cheeks, lips, hard and soft palates and tongue. In addition to these secretions, the total oral fluid includes the gingival crevicular fluid, which flows from the gingival cuff tissue into the oral cavity (Chap. 7: Saloom and Carpenter).

Conclusion

The purpose of this chapter was to *revise* the main structural and functional attributes of the oral mucosa as a protective and defensive environment and to enable readers to understand the context of the subsequent chapters.

The oral mucosa is fully exposed to the environment and as such is a gateway into body for infections but also expressions of pathologies associated with both local and systemic diseases. This dynamic epithelium is both a gatekeeper and a housekeeper of normal homeostasis and has the potential to be a window on health and disease.

While detailed histological descriptions and diagnosis are beyond the scope of this book it is hoped that the subsequent chapters will be supported by this brief revision of the structure and function of the oral mucosa.

Investigations of the oral mucosa in terms of cancer detection, disease biomarker discover and immune functions in health and disease are discussed in subsequent chapters.

References

1. Eisen D. Disorders of pigmentation in the oral cavity. Clin Dermatol. 2000;18(5):579–87.
2. Squier C, Brogden K. Human oral mucosa: develepement, structure and function. Hoboken: Wiley-Blackwell; 2011.
3. Collins LM, Dawes C. The surface area of the adult human mouth and thickness of the salivary film covering the teeth and oral mucosa. J Dent Res. 1987;66(8):1300–2.
4. Yi R, Poy MN, Stoffel M, Fuchs E. A skin microRNA promotes differentiation by repressing 'stemness. Nature. 2008;452(7184):225–9.
5. Nakamura T, Endo K, Kinoshita S. Identification of human oral keratinocyte stem/progenitor cells by neurotrophin receptor p75 and the role of neurotrophin/p75 signaling. Stem Cells. 2007;25(3):628–38.
6. Izumi K, Tobita T, Feinberg SE. Isolation of human oral keratinocyte progenitor/stem cells. J Dent Res. 2007;86(4):341–6.
7. Squier CA, Kremer MJ. Biology of oral mucosa and esophagus. J Natl Cancer Inst Monogr. 2001;(29):7–15.
8. Schroeder H. Differentiation of human oral startified epithelia. Basel: Krager; 1981.
9. Dale BA, Salonen J, Jones AH. New approaches and concepts in the study of differentiation of oral epithelia. Crit Rev Oral Biol Med. 1990;1(3):167–90.
10. Presland RB, Dale BA. Epithelial structural proteins of the skin and oral cavity: function in health and disease. Crit Rev Oral Biol Med. 2000;11(4):383–408.
11. Lindberg K, Rheinwald JG. Three distinct keratinocyte subtypes identified in human oral epithelium by their patterns of keratin expression in culture and in xenografts. Differentiation. 1990;45(3):230–41.
12. Dabelsteen E, Buschard K, Hakomori S, Young WW. Pattern of distribution of blood group antigens on human epidermal cells during maturation. J Invest Dermatol. 1984;82(1):13–7.
13. Dabelsteen E, Clausen H, Mandel U. Aberrant glycosylation in oral malignant and premalignant lesions. J Oral Pathol Med. 1991;20(8):361–8.
14. Dabelsteen E, Mandel U, Clausen H. Cell surface carbohydrates are markers of differentiation in human oral epithelium. Crit Rev Oral Biol Med. 1991;2(4):493–507.
15. Vedtofte P, Dabelsteen E, Hakomori S, Young WW. Changes in the expression of blood-group carbohydrates during oral mucosal development in human fetuses. Differentiation. 1984;27(3):221–8.
16. Clausen H, Vedtofte P, Moe D, Dabelsteen E. Keratin pattern in human and buccal and hard palate mucosa. Scand J Dent Res. 1983;91(5):411–3.
17. Clausen H, Vedtofte P, Moe D, Dabelsteen E, Sun T-T, Dale B. Differentiation-dependent expression of keratins in human oral epithelia. J Investig Dermatol. 1986;86(3):249–54.
18. Gibbs S, Ponec M. Intrinsic regulation of differentiation markers in human epidermis, hard palate and buccal mucosa. Arch Oral Biol. 2000;45(2):149–58.
19. Gasparoni A, Squier CA, Fonzi L. Cytokeratin changes in cell culture systems of epithelial cells isolated from oral mucosa: a short review. Ital J Anat Embryol. 2005;110(2):75–82.
20. Kraan H, Vrieling H, Czerkinsky C, Jiskoot W, Kersten G, Amorij JP. Buccal and sublingual vaccine delivery. J Control Release. 2014;190:580–92.

21. Wertz PW, Squier CA. Cellular and molecular basis of barrier function in oral epithelium. Crit Rev Ther Drug Carrier Syst. 1991;8(3):237–69.
22. Squier CA, Hopps RM. A study of the permeability barrier in epidermis and oral epithelium using horseradish peroxidase as a tracer in vitro. Br J Dermatol. 1976;95(2):123–9.
23. Squier CA, Rooney L. The permeability of keratinized and nonkeratinized oral epithelium to lanthanum in vivo. J Ultrastruct Res. 1976;54(2):286–95.
24. Selby CC. An electron microscope study of thin sections of human skin. II. Superficial cell layers of footpad epidermis. J Invest Dermatol. 1957;29(2):131–49.
25. Squier CA. The permeability of keratinized and nonkeratinized oral epithelium to horseradish peroxidase. J Ultrastruct Res. 1973;43(1):160–77.
26. Squier CA. Membrane coating granules in nonkeratinizing oral epithelium. J Ultrastruct Res. 1977;60(2):212–20.
27. Squier CA, Fejerskov O, Jepsen A. The permeability of a keratinizing squamous epithelium in culture. J Anat. 1978;126(Pt 1):103–9.
28. Lillie JH, MacCallum DK, Jepsen A. Growth of stratified squamous epithelium on reconstituted extracellular matrices: long-term culture. J Invest Dermatol. 1988;90(2):100–9.
29. Madison KC, Swartzendruber DC, Wertz PW, Downing DT. Lamellar granule extrusion and stratum corneum intercellular lamellae in murine keratinocyte cultures. J Invest Dermatol. 1988;90(2):110–6.
30. Squier CA, Cox PS, Wertz PW, Downing DT. The lipid composition of porcine epidermis and oral epithelium. Arch Oral Biol. 1986;31(11):741–7.
31. Breathnach AS. Application of the freeze-fracture replication technique to investigative dermatology. Br J Dermatol. 1973;88(6):563–74.
32. Elias PM, Brown BE. The mammalian cutaneous permeability barrier: defective barrier function is essential fatty acid deficiency correlates with abnormal intercellular lipid deposition. Lab Investig. 1978;39(6):574–83.
33. Squier CA, Cox P, Wertz PW. Lipid content and water permeability of skin and oral mucosa. J Invest Dermatol. 1991;96(1):123–6.
34. Squier CA, Cox P, Hall BK. Enhanced penetration of nitrosonornicotine across oral mucosa in the presence of ethanol. J Oral Pathol. 1986;15(5):276–9.
35. Cox P, Squier CA. Variations in lipids in different layers of porcine epidermis. J Invest Dermatol. 1986;87(6):741–4.
36. Wertz PW, Downing DT. Glycolipids in mammalian epidermis: structure and function in the water barrier. Science. 1982;217(4566):1261–2.
37. Wertz PW, Kremer M, Squier CA. Comparison of lipids from epidermal and palatal stratum corneum. J Invest Dermatol. 1992;98(3):375–8.
38. Dawson DV, Drake DR, Hill JR, Brogden KA, Fischer CL, Wertz PW. Organization, barrier function and antimicrobial lipids of the oral mucosa. Int J Cosmet Sci. 2013;35(3):220–3.
39. Allam J-P, Novak N. Immunological mechanisms of sublingual immunotherapy. Curr Opin Allergy Clin Immunol. 2014;14(6):564–9.
40. Novak N, Haberstok J, Bieber T, Allam JP. The immune privilege of the oral mucosa. Trends Mol Med. 2008;14(5):191–8.
41. Cutler CW, Jotwani R. Dendritic cells at the oral mucosal interface. J Dent Res. 2006;85(8):678–89.
42. Dutzan N, Konkel JE, Greenwell-Wild T, Moutsopoulos NM. Characterization of the human immune cell network at the gingival barrier. Mucosal Immunol. 2016;9(5):1163–72.
43. Hovav A-H. Dendritic cells of the oral mucosa. Mucosal Immunol. 2014;7(1):27–37.
44. Dutzan N, Abusleme L, Bridgeman H, Greenwell-Wild T, Zangerle-Murray T, Fife ME, Bouladoux N, Linley H, Brenchley L, Wemyss K, Calderon G, Hong BY, Break TJ, Bowdish DM, Lionakis MS, Jones SA, Trinchieri G, Diaz PI, Belkaid Y, Konkel JE, Moutsopoulos NM. On-going mechanical damage from mastication drives homeostatic Th17 cell responses at the oral barrier. Immunity. 2017;46(1):133–47.
45. Amerongen AV, Veerman EC. Saliva—the defender of the oral cavity. Oral Dis. 2002;8(1):12–22.
46. Hannig C, Hannig M, Kensche A, Carpenter G. The mucosal pellicle—an underestimated factor in oral physiology. Arch Oral Biol. 2017;80:144–52.
47. Gibbins HL, Proctor GB, Yakubov GE, Wilson S, Carpenter GH. SIgA binding to mucosal surfaces is mediated by mucin-mucin interactions. PLoS One. 2015;10(3):e0119677.
48. Gibbins HL, Yakubov GE, Proctor GB, Wilson S, Carpenter GH. What interactions drive the salivary mucosal pellicle formation? Colloids Surf B Biointerfaces. 2014;120:184–92.
49. Lin JY, Fisher DE. Melanocyte biology and skin pigmentation. Nature. 2007;445(7130):843–50.
50. Tachibana T, Kamegai T, Takahashi N, Nawa T. Evidence for polymorphism of Merkel cells in the adult human oral mucosa. Arch Histol Cytol. 1998;61(2):115–24.
51. Lucarz A, Brand G. Current considerations about Merkel cells. Eur J Cell Biol. 2007;86(5):243–51.
52. Jotwani R, Cutler CW. Multiple dendritic cell (DC) subpopulations in human gingiva and association of mature DCs with CD4+ T-cells in situ. J Dent Res. 2003;82(9):736–41.
53. Steinman RM. Dendritic cells: understanding immunogenicity. Eur J Immunol. 2007;37(Suppl 1):S53–60.
54. Boyle JO, Gümüs ZH, Kacker A, Choksi VL, Bocker JM, Zhou XK, Yantiss RK, Hughes DB, Du B, Judson BL, Subbaramaiah K, Dannenberg AJ. Effects of cigarette smoke on the human oral mucosal transcriptome. Cancer Prev Res (Phila). 2010;3(3):266–78.
55. Suchett-Kaye G, Morrier JJ, Barsotti O. Interactions between non-immune host cells and the immune system during periodontal disease: role of the gingival keratinocyte. Crit Rev Oral Biol Med. 1998;9(3):292–305.

56. Chalermsarp N, Azuma M. Identification of three distinct subsets of migrating dendritic cells from oral mucosa within the regional lymph nodes. Immunology. 2009;127(4):558–66.
57. Aramaki O, Chalermsarp N, Otsuki M, Tagami J, Azuma M. Differential expression of co-signal molecules and migratory properties in four distinct subsets of migratory dendritic cells from the oral mucosa. Biochem Biophys Res Commun. 2011;413(3):407–13.
58. Nudel I, Elnekave M, Furmanov K, Arizon M, Clausen BE, Wilensky A, Hovav AH. Dendritic cells in distinct oral mucosal tissues engage different mechanisms to prime CD8+ T cells. J Immunol. 2011;186(2):891–900.
59. Allam J-P, Duan Y, Winter J, Stojanovski G, Fronhoffs F, Wenghoefer M, Bieber T, Peng W-M, Novak N. Tolerogenic T cells, Th1/Th17 cytokines and TLR2/TLR4 expressing dendritic cells predominate the microenvironment within distinct oral mucosal sites. Allergy. 2011;66(4):532–9.
60. Atamas SP. Complex cytokine regulation of tissue fibrosis. Life Sci. 2002;72(6):631–43.
61. Kataoka M, Kido J, Shinohara Y, Nagata T. Drug-induced gingival overgrowth—a review. Biol Pharm Bull. 2005;28(10):1817–21.
62. Rinn JL, Wang JK, Liu H, Montgomery K, van de Rijn M, Chang HY. A systems biology approach to anatomic diversity of skin. J Invest Dermatol. 2008;128(4):776–82.
63. Atamas SP, Luzina IG, Dai H, Wilt SG, White B. Synergy between CD40 ligation and IL-4 on fibroblast proliferation involves IL-4 receptor signaling. J Immunol. 2002;168(3):1139–45.
64. Walsh LJ. Mast cells and oral inflammation. Crit Rev Oral Biol Med. 2003;14(3):188–98.
65. Dongari-Bagtzoglou A, Fidel PL Jr. The host cytokine responses and protective immunity in oropharyngeal candidiasis. J Dent Res. 2005; 84(11):966–77.
66. Feller L, Altini M, Khammissa RA, Chandran R, Bouckaert M, Lemmer J. Oral mucosal immunity. Oral Surg Oral Med Oral Pathol Oral Radiol. 2013;116(5):576–83.
67. Nanci A. St Louis Missouri: Elsevier Mosby; 2013.

2 Cell-Cell Interactions in the Oral Mucosa: Tight Junctions and Gap Junctions

Hong Wan, Hanan Gadmor, and Louise Brown

2.1 General Introduction to Cell-Cell Interactions

Cell adhesions are crucial in many aspects of cell and tissue biology and coordinate in various processes such as morphogenesis, tumour metastasis and tissue repair. Normally, they are required in tissue development and in establishment of apical-basal axis of cell polarity by generating physical and molecular asymmetry at both surface and intracellular structures, which leads to the formation of apical membrane in the non-contacting (free) cell surface, and the basolateral membrane along the contacting cell surface. A typical feature of epithelia is that the cells are tightly attached to each other and to the extracellular substrate, with little extracellular space, by numerous cell-cell and cell-matrix adhesion complexes, through a collection of glycol proteins and cytoplasmic plaque proteins, and these specific interactions enable epithelial cells to maintain polarity. The epithelial component of oral mucous membrane in the outermost layer is composed of stratified squamous epithelium that serves as a major portal for microbial invasion and provides robust protection against mechanical stress. The integrity of such a mucous barrier is maintained by physical interaction of highly specialised junctional complexes between adjacent cells as well as between cells and basement membrane that are important for safeguarding of systemic and oral health. Like epidermis, the major cell type in oral mucosa is keratinocytes and these cells are arranged as four classified sub-layers, namely, the basal layer, spinous layer, granular layer and stratum corneum. The intercellular adhesions are stable in mature epithelia and become modulated during regenerative processes such as wound healing or pathogenic processes such as cancer progression and invasion. In these latter situations, the junctions become highly dynamic that enables their rearrangements in order to facilitate cell migration, morphogenesis and tissue development. It has been widely accepted that the formation, maturation and homeostasis of epithelia require dynamic coordination between assembly and disassembly of the intercellular junctions, and dysfunction of any aspects is often associated with dedifferentiation and malignance and other diseases such as pemphigus. Emerging evidence suggests that the precisely controlled mechanisms of cell adhesion can in addition serve as crucial regulators for other downstream processes, such as proliferation, differentiation, migration and wound healing.

According to the functions, the cell adhesions in epithelia are classified into three groups of junction complexes, i.e. tight junction (also

H. Wan (✉) • H. Gadmor • L. Brown
Centre for Immunobiology and Regenerative Medicine, Institute of Dentistry, Barts and The London School of Medicine and Dentistry, London, UK
e-mail: h.wan@qmul.ac.uk

L.A. Bergmeier (ed.), *Oral Mucosa in Health and Disease*,
https://doi.org/10.1007/978-3-319-56065-6_2

known as occluding junction), gap junction (also known as channel-forming or communicating junction) and anchoring junction, the latter of which includes adherens junction (AJ), desmosome (DM), focal adhesion and hemi-desmosome (Fig. 2.1). The main feature of anchoring junctions is that this group is composed of members of the cadherin and integrin superfamilies of transmembrane proteins that are linked to cytoskeleton through a collection of cytoplasmic plaque proteins, and they are especially abundant in tissues, such as oral mucosa, skin and heart, that experience extensive mechanical stress. Overall, cell adhesions between cells, as well as between cell and the extracellular matrix which will not be discussed here, play a crucial role in various cellular processes, including cell recognition, positioning, differentiation, embryonic development and wound healing in adult tissues. This chapter focuses on the cell-cell interactions,

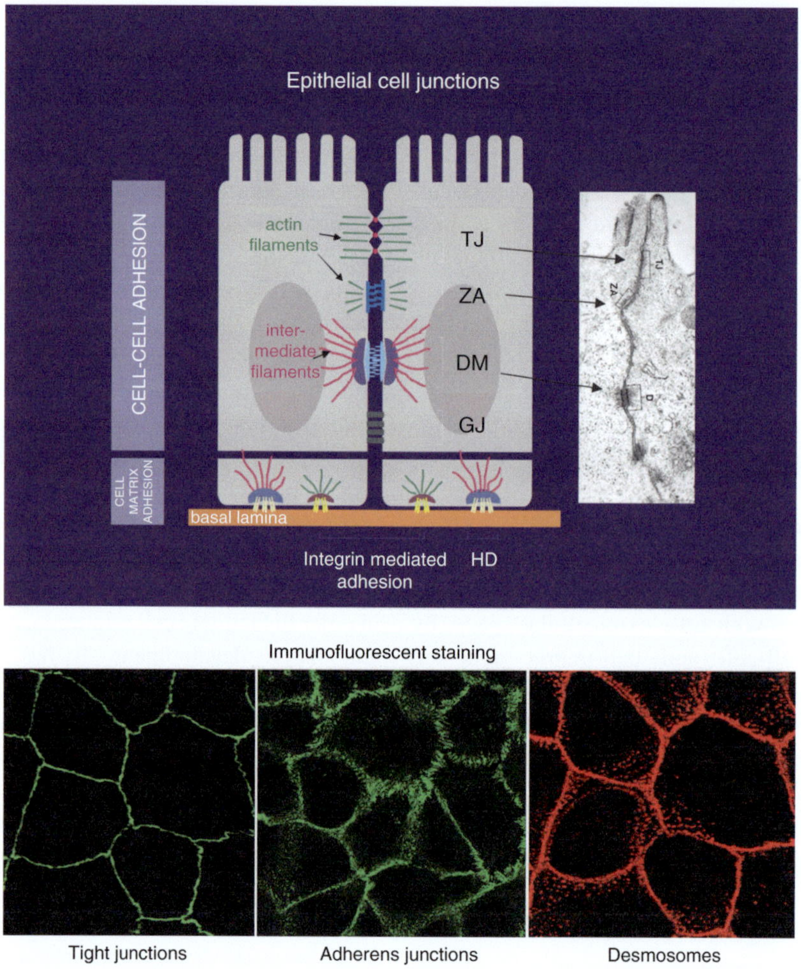

Fig. 2.1 Top panel shows a schematic diagram that illustrates the typical arrangement of major cell-cell and cell-matrix junctions in intestinal epithelial cells. An electron micrograph with corresponding cell-cell junctions is shown on the right. While the tight junction (TJ) and adherens junction (AJ) associate with the actin microfilaments desmosome (DM) links to keratin intermediate filaments. Gap junction (GJ) does not link to cytoskeleton. Below is the immunofluorescent staining of three distinct intercellular junctions in MDCK epithelial cells, i.e. ZO-1 staining for TJ (green), E-cadherin staining for AJ (green) and desmoplakin staining for DM (red). TJs appear as a continuous linear pattern whereas AJs and DMs show more broad peripheral distributions with either diffuse or more punctate staining pattern

in particular the tight junction, with a brief introduction about the gap junction. Adherens junction and desmosome are further discussed in Chap. 4.

2.2 Tight Junction

2.2.1 Tight Junction Function and Structure

Tight junctions (TJs), or *zonula occludens*, form the closest circumferential contact of the plasma membrane between adjacent cells and are found in the apical region of polarised epithelial cells. TJs are found in tissues that are involved in polarised secretions and absorption functions and maintaining barriers between blood and interstitial fluids. TJ is the unique feature of epithelial and endothelial cells and does not exist in other cell type. The epithelium of the small intestine provides a good example of TJ structure, location and function (Fig. 2.1) and this epithelium contains a single layer of tall, columnar shaped cells, specialised for uptake of nutrients from the internal cavity, or lumen, of gut. The principal function of TJ has two features, one of which is that TJ creates a permeability barrier between the neighbouring cells and regulates the flux of ions and nonelectrolytes through the paracellular space, the function called *the gate function* [1]. The degree of such a transepithelial resistance varies between different epithelial tissue organs in the body. Besides, TJ also provides a boundary near the apex of lateral plasma membrane that separates the lipid membrane bilayer into distinct apical and basolateral domains, and this function, known as *the fence function*, prevents the mixture of lipids and proteins between the two membrane domains and enables cells to exert functions in a polarised fashion including protein membrane trafficking, differentiation, morphogenesis and transport. Specific ion channels and pumps are usually localised in the surface domain that is exposed either to the luminal space or serosal space. It has been shown experimentally in Madin-Darby canine kidney (MDCK) cells that these two functions of TJ can be uncoupled simply by depletion of ATP using a combination of glycolytic and mitochondrial inhibitors, and the decline of the energy status, i.e. the ATP/ADP ratio, led to marked decrease in transepithelial resistance without affecting TJ protein expression and distribution as well as the geometrical complexity of TJ strands [2]. It was reasoned that this separation could be due to small but critical rearrangement between TJ strands located in opposing cells that induces larger changes in transepithelial resistance (Mandel Nature 1993). Initially it was thought that TJ is made by a fusion between outer leaflets of lipid bilayer on the adjacent cells, but later TJ was found to contain discrete ion selective pores through the extracellular portion of the junctions [3]. The structure of TJs has been greatly influenced by their appearance in transmission electron microscopy or in freeze-fracture electron microscopy. In transmission electron micrograph, TJs appear as a sealing point where the outer leaflets of two adjacent plasma membranes are tightly opposed. When visualised by freeze-fracture electron microscopy, TJs look as a network of sealing ridges or strands on P-face and complementary furrows on E-face in glutaraldehyde-fixed specimens [1]. The strands observed in freeze-fracture microscopy seem to correspond to the fusion sites shown by transmission microscopy.

2.2.2 Tight Junction Proteins

TJ consists of the transmembrane proteins and other adhesion molecules located at the cytoplasmic site, including ZO-1, ZO-2, ZO-3, AF6, Par3, cingulin, 7H6, MUPP1 and Rab13, that link TJ protein complexes to the actin cytoskeleton [4–6]. The first identified TJ protein is ZO-1 that was discovered in 1986 by raising monoclonal antibodies against a mouse liver TJ fraction [4, 7]. Since then, many proteins have been found to be associated with TJs using the same approach [4, 8]. Some of these proteins are restricted to TJs while others may have wider distribution. For example, ZO-1 is found to be located not only in TJs but also in some cadherin junctions as well as in nucleus, whereas ZO-2 shows largely the TJ restriction. The transmembrane proteins in TJ include at least three distinct proteins, named occludin, claudin, junctional adhesion molecules (JAM)/coxsackievirus and adenovirus receptor (CAR), and most of them were identified in 1990s. Claudins form a complex with occludin and/or JAM family members. While occludin and claudin share a common membrane topology containing four transmembrane domains and two extracellular loops, JAM and CAR belong to type I transmembrane glycoprotein and possess immunoglobulin (Ig)-like domains in the extracellular region [5]. The molecular weight of occludin is around 60–82 kDa. In terms of amino acid sequence, occludin does not show homology to known proteins whereas claudins are homologous to a family of four transmembrane domain proteins, such as RVP-1, TMVCF, CPE-R and BEC1 [8]. Studies suggest that the phosphorylation of occludin is required for its assembly in TJs. The phosphorylated forms of occludin exhibit relatively higher molecular weights and are associated with cytoskeleton which are insoluble in non-ionic detergent such as NP-40 [9, 10]. The C-terminal domain of occludin interacts with ZO-1 and ZO-3 that is thought to be important in mediating its basolateral targeting. However, the knockout experiments in mouse embryonic cells show that occludin is not necessarily required for TJ formation. Furthermore, it was demonstrated that occludin-deficient epithelial cells also exhibit normal TJ protein localisation, morphology, polarity and barrier function [8, 11]. This raises concern that occludin may not be the key protein of TJ.

In contrast, claudins have been identified to be crucial factors for epithelial barrier and transport based on numerous in vitro and in vivo studies and are believed to be the core protein in paired TJ strands [12, 13]. The claudin family consists of 24 proteins with molecular weights ranging from 20 to 27 kDa and these family proteins exhibit distinct tissue- and development-specific distributions [5]. They all contain four transmembrane domains, two extracellular loops and a short carboxyl intracellular tail. The extracellular loops of claudins on adjacent cells mediate direct cell-cell interaction, while those expressed in the same cell interact through their intracellular N-terminal domains. The C-terminus of claudins contains highly conserved PDZ-binding motifs that link them to the TJ PDZ-containing proteins including ZO-1, ZO-2 and ZO-3. The knockout and knockdown animal studies suggest that claudin family members exhibit redundant and compensatory functions since one knockout gene may not show the defected phenotype as expected [14]. It has been thought that claudins are likely arranged in clusters and collectively act as a functional entity suitable for the interaction of proteins within TJ strands [13]. Thus a function compensation of different members within the claudin family can occur when a single gene in the clusters is missing, as indicated by knockout animal work [5]. In principle, this family can be divided into three functional groups, namely, claudins with sealing function (claudin-1, -3, -4, -5, -8, -11, -14 and -19), claudins providing paracellular permeability (claudin-2 and -10) and claudins with ambiguous function (claudin-7, -12, -15 and -16), the latter of which is based on the observation that both permeability-enhancing and permeability-restricting effects have been reported for these proteins in the literature [5]. The interaction between claudins within clusters likely involves both *cis*- and *trans*-interactions via a variety of combinations. Whereas *cis* would be interaction of claudins within one cell membrane of a single cell (side by side), *trans* would describe the interaction with a putative multimer of the neighbouring cell (head to head). Using L-fibroblasts (which do not originally contain any TJ proteins) with stable expression of different members of claudin family, it is found that claudin-1 forms *trans*-interaction with claudin-3 and so as for claudin-2 and claudin-3, but no interaction was detected between claudin-1 and -2 [12], indicating heterogeneous clusters and specific binding of claudins within the TJ strands. The configuration of different claudins could contribute to various transepithelial resistances and paracellular permeability observed in different tissues and organs.

Both JAM and CAR belong to a growing superfamily of immunoglobulin-like surface molecules, many of which have been localised to sites of cell-cell contact and appear to function in cell adhesion or intercellular recognition. The immunoglobulin-like superfamily (IgSF) proteins all contain various numbers of Ig-like domains in their N-terminus. JAM and CAR share similar structural homology with two Ig-like domains, followed by a single-pass transmembrane domain and a small cytoplasmic tail [5]. There are five JAMs identified so far, namely JAM-A, JAM-B, JAM-C, JAM-4 and JAM-L, and their molecular weights range between 36 and 60 kDa. In addition to their expression in epithelial and endothelial cells, JAMs are also found on the surface of blood cells including leucocytes, platelets and erythrocytes, and have been recognised to be involved in a variety of cellular processes, including TJ assembly, leucocyte transmigration, platelet activation, angiogenesis and virus binding [15, 16]. Among five JAMs, JAM-A, -B, and -C are more closely related to each other. In contrast, JAM-L resembles other IgSF proteins such as CAR. The cytoplasmic domain of JAM-A, -B, and -C is short with only about 40 amino acids long, whereas those of JAM-4 and -L are considerably larger with 105 and 98 residues, respectively. Except for JAM-L, all other JAMs contain C-terminal PDZ-binding motifs that appear to facilitate interactions with other TJ-associated scaffold proteins such as ZO-1 and AF6. Besides, the cytoplasmic tails of JAMs also contain consensus phosphorylation sites that may serve as substrates for PKC, PKA and casein kinase II, and the phosphorylation at specific sites may play a role for targeting JAMs to cell-cell junctions [16]. Despite the compelling evidence implicating JAMs in intercellular junction formation, little is known about the mechanisms by which this might occur.

CAR, a ~46 kDa protein of TJ, is also known to mediate viral attachment and infection. CAR was first identified as a cellular protein involved in attachment and infection by group B coxsackieviruses (CVB) and later found to be an adenovirus (Ad) receptor [17]. As mentioned above, CAR belongs to IgSF family and the cytoplasmic domain contains a phosphorylation site and a hydrophobic motif that interacts with PDZ-domain proteins such as ZO-1 and MUPP1. Virus entry into polarised epithelium required disruption of TJs. In polarised epithelial cells, CAR is colocalised with ZO-1, and its expression in TJs is to enhance their barrier function and limit virus infection as well as reduce passage of macromolecules and ions across epithelial surfaces [18]. It is also seen to colocalise with some non-TJ proteins such as β-catenin and α-actinin-4. CAR is expressed in a wide range of tissues, including the liver, intestines, lung, heart, brain and pancreas, with high levels in the testis and prostate [18]. The biological roles of CAR remain poorly understood, but emerging evidence suggests that it may function during embryonic development and in regulating cell proliferation. High levels of CAR expression were found in the embryonic brain and heart but with significant dropping after birth [17]. In support, specific CAR deletion in the heart early in embryonic life leads to severe cardiac abnormalities and death in utero, suggesting that CAR is essential for normal cardiac development [19]. Other evidence suggests that CAR may be essential for early development of the central nervous system and other tissues [17]. As a junctional component of the intercalated disc, recent studies also have implicated CAR in cardiac remodelling and electrical conductance between atria and ventricle [20].

2.2.3 Tight Junction Formation and Maintenance

The assembly of TJ is driven by initial interaction of occludin and claudins via their extracellular domains, and this leads to recruitment of other adhesion proteins, including signal molecules, at their cytoplasmic site. The key organisational molecules in TJ are ZO (*zonula occludens*) proteins (ZO1–3) which are large scaffolding proteins and provide a structural support beneath the plasma membrane for TJ assembly. ZO proteins consist of a string of the protein-binding domains, typically including three PDZ domains (each with approximately 80 amino acids), an SH3 domain (around 60-amino-acid motif) and a GK domain [1, 21]. These domains enable the proteins to bind to each other and also with other partners that together in turn link to the actin cytoskeleton and regulate signal propagation to the cell interior. Tyrosine phosphorylation of ZO-1 via the EGF receptor-mediated Src signalling is required for TJ assembly and this was particularly demonstrated in A431 cells treated with EGF [21, 22]. In the absence of EGF, ZO-1 was found to be located diffusively along lateral cell borders as well as in micro-spikes and ruffles. After EGF treatment, ZO-1 was relocated towards apical position where it colocalised with the actin ring [21]. In this model, the EGF-induced transient tyrosine phosphorylation of ZO-1 causes its recruitment with actin into a focused position corresponding to an apical TJ. On the other hand, tyrosine phosphorylation of ZO-1 and ZO-2 is also found to be correlated with a temporal drop in transepithelial resistance of TJs, suggesting that the phosphorylation of these scaffolding proteins is associated with dynamic assembly process or modulation of TJs.

In addition to Src signalling, peri-junctional actin-myosin contraction, protein kinase C (PKC) and isoforms of heterotrimeric G proteins have also been shown to be involved in physiological regulation of TJs [4, 22–24]. Briefly, changes in the intracellular calcium induced by various stimuli result in an increase of paracellular permeability and very often this increased paracellular permeability can be blocked by the use of PKC inhibitors. In addition, depletion of extracellular calcium from confluent epithelial culture results in disassembly of both TJs and AJs. It was thought that the action of PKC is mediated through AJ formation rather than directly on the TJ proteins. The assembly of AJs is known to trigger initial and first cell-cell contact event that is prerequisite for other junction formation (discussed below). It has long been known that integrity and stability of TJs are dependent on the extracellular calcium and this could be due to indirect effects of calcium on AJs as well as the actin cytoskeleton rather than direct effects on TJs. MDCK cells have been extensively utilised for the study of TJ formation and treatment of MDCK cells with cytochalasins, drugs that disrupt actin filaments, caused gradual increase in paracellular permeability, and prevented the normal rapid recovery of transepithelial resistance in cells that were subjected to calcium depletion (TJ disruption) and then repletion (induction of cell-cell interaction and junction formation) [25]. There are two actin populations observed in MDCK, the apical ring and basal actin filament bundles that are associated with focal adhesion. It has been demonstrated in a study based on cytochalasin D treatment of MDCK cells that apical actin ring is more important in the regulation of TJ permeability [26]. The epithelial cell adhesion molecule uvomorulin or L-CAM, which undergoes a calcium-dependent change in molecular conformation, was thought to be responsible, at least in part, for the calcium sensitivity of TJ formation [25]. Internal calcium stores are also required for the maintenance of TJs [27, 28] and in contrast PKC does not seem to be in this case as PKC inhibitors showed minimal effects on already established TJs [4]. Furthermore, a heterotrimeric G protein α-subunit is found to localise at TJs and to functionally regulate transepithelial resistance in MDCK cells [29]. Other signalling molecules, such as small GTPases, Rac1 and RhoA as well as Rab13, are also demonstrated to be involved in both *gate* and *fence functions* of TJs and are shown to be colocalised with ZO-1 and TJs in a variety of epithelial and endothelial cells, respectively [30, 31]. Taken together, these findings suggest that TJ is a very dynamic structure and its assembly, functions and maintenance are involved in many molecular processes including protein phosphorylation, intracellular and extracellular calcium, actin-myosin contraction and several associated signalling pathways.

2.2.4 Tight Junction in Epidermis and Oral Mucosa

Most studies on TJs are based on simple epithelia and their derived cell lines, such as intestine, renal and airways. Relatively less work has been done in stratified squamous epithelial tissues, in particular oral mucosa. Some pioneer studies are performed in the skin and cultured keratinocytes [32, 33]. In 2002, Furuse et al. published the claudin-1-deficient mouse model reporting that animals died within 1 day of birth with wrinkled skin and severe defect of epidermal barrier function, albeit the layered organisation of keratinocytes appeared to be normal [33]. This study provides the first evidence of crucial role of TJs in the mammalian epidermal barrier and demonstrates directly continuous TJs in this tissue. Thereafter, several TJ proteins are found localised at the apical site of the lateral plasma membrane in the stratum granulosum of epidermis and they are occludin, claudin-1, -4 and 7, JAM-A and TJ plaque proteins, ZO-1 and -2, cingulin as well as cell polarity proteins, Par3 and aPKC [34]. However, it was found that the structure of TJs is, especially in adult skin, less complex than in other epithelia; however, their exact localisation within the stratum granulosum of epidermis remains not fully characterised. In addition to granular cells, claudin-1 and -4 are also found to be diffusely distributed along the plasma membranes of keratinocytes in deeper layers, but these diffusely distributed claudins are thought not to be constituted to TJ strands per se, nor to be directly involved in the barrier function of the epidermis. Similar finding was also shown in simple epithelial cells without proper TJ formation in some cell types. In the cultured keratinocytes, TJs and their constitutive protein expression are shown to be responsive to calcium that induces cell stratification and differentiation [32], analogous to that of cadherin-mediated AJ formation. For detail of phenotypic differences of knockout and knockdown in mouse and man and for direct comparison of TJs and their protein expression between skin and intestine, please refer this review [34].

Relatively limited studies are performed on the functional significance of TJs in oral mucosa, albeit some TJ proteins, such as claudins, occludin, JAM-A and ZO-1, are reported to be expressed in salivary glands [11]. In the rat large salivary glands, claudin-3 was detected in the acinar cells and intercalated ducts whereas claudin-4 was principally expressed by the striated and interlobular ducts [35]. Occludin was ubiquitously detected in the duct system. In the mouse submandibular gland, claudins including caludin-3 to -8, -10 and -11 are found exhibiting differential expression pattern in the developing epithelium [36]. While claudin-3, -5, and -7 were restricted to the luminal cells of the ducts, claudin-4 was found in the ducts at all the developmental stages. The expression of claudin-6 and -8 was detected in the ducts at E14 and E16 but after birth, only claudin-8 was detectable. Claudin-10 and -11 were found in the terminal tubules at and after E16, and in addition claudin-16 was also detected in human major salivary glands (parotid, submandibular and sublingual glands), as well as in their excretory ducts where it shows colocalisation with ZO-1 and occludin [37].

2.2.5 Tight Junction Proteins in Oral Cancer

Studies on TJs and the barrier function in oral mucosa begin emerging [38, 39]. Since TJs play a crucial role in cell-cell adhesion, cell differentiation and polarity, it is not surprising that loss of TJ protein expression occurs in various human cancers including oral squamous cell carcinomas (OSCC). Defect of TJs due to the loss of TJ molecules in cancer cells can stimulate dedifferentiation process and drive cell detachment from primary tumour, leading to distant dissimilation that is a hallmark of cancer. Thus many studies have established that loss of TJ proteins is associated with carcinogenesis, recurrence and poor patient survival [32, 40–45]. However, this paradigm has recently been challenged by many other reports that the overexpression of CAR, JAM-A, JAM-C and several claudins has been shown to promote tumorigenesis in specific cancers [5]. Thus it has been proposed that aberrant TJ protein expression, rather than exclusively TJ protein loss, may promote tumorigenesis [5]. Although the mechanisms of their roles in cancer remain unclear, it was reasoned that TJ proteins likely have some non-adhesion functions such as regulating intracellular signalling that controls proliferation and migration and these additional functions contribute to the tumorigenesis and progression. Evidently, optimal expression of TJ proteins is the key in maintaining normal physiological function and any imbalance could have pathological consequences.

It has been shown that overexpression of claudin-1 is associated with angiolymphatic and perineural invasion, consistent with aggressive tumour behaviour and with advanced-stage disease in OSCC [46]. On the other hand, another study on SCC of oral cavity has indicated that loss of claudin-1 expression is correlated with clinical stage and poor differentiation status of oral cancer with the highest levels in well-differentiated OSCCs and almost negative staining in poorly differentiated tumours [47]. In vitro study in various OSCC cell lines with different invasion activities suggests that the action of claudin-1 in promoting cancer cell invasion is through a mechanism of activating MT1-MMP and MMP-2 which causes enhanced cleavage of extracellular matrix protein laminins, and knockdown of claudin-1 suppresses the invasion of OSCC cells and decreases the activation of MMP-2 [48].

Some JAMs have also been implicated in a variety of pathologic processes involving cellular adhesion. In addition, JAM-A has been shown to be a receptor for reovirus. Deregulation of JAM-A is found to be associated with various cancers. While a strong correlation between JAM-A protein upregulation and poor prognosis was observed in breast cancer patients, paradoxically, downregulation of JAM-A has also been shown in breast tumour progression [49–51]. Another JAM reported to be related to cancer is JAM-C and its primary role in aiding cancer progression is involved in the promotion of cancer cell migration and angiogenesis rather than directly influencing tumour cell proliferation or survival [5]. CAR acts as a receptor for coxsackie- and adenoviruses and its upregulation is found in various cancers with the levels of expression positively correlating with tumour grade and metastasis. One study based on an OSCC cell line suggests that a critical role of CAR in cancer progression is probably through a mechanism of the negative regulation of apoptotic pathway and promoting cancer cell growth and survival [52]. It was shown in the study that such a growth regulation is via the specific interaction of CAR with Rho-associated protein kinase (ROCK) and this interaction causes inhibition of ROCK activity that in turn facilitates cell-cell adhesion and stability that is required for cell growth and survival. Knockdown of CAR results in growth suppression and anoikis of SCC cells due to cell dissociation caused by abnormal distribution of E-cadherin [52].

2.3 Gap Junction

Intercellular communication is important in controlling homoeostasis in organisms and in permitting responses to external stimuli, and gap junction functions for such a purpose and serves to facilitate direct intercellular communication [53, 54]. Gap junctions have a pore size of about 1.4 nm and are tightly packed, which allows transfer of inorganic ions (including Ca^{2+} and Mg^{2+}) and other small molecules of <1 kDa (such as cAMP, cGMP and ATP), but not macromolecules such as proteins or nucleic acids, between neighbouring cells. Conventional electron microscopy and X-ray crystallographic studies show that gap junctions appear as a patch of a hexagonal array where the membranes of two neighbouring cells are separated by a uniform narrow gap of about 2–4 nm [1]. Gap junctions do not open all the time; instead, they flip between open and close states, depending upon the external stimuli or controlled by multiple factors such as calcium concentration, pH, trans-junctional membrane potential and protein phosphorylation [55]. Again, each gap junction plaque is a dynamic structure that can readily assemble, disassemble or be remodelled.

Gap junctions contain hydrophilic membrane channels that bridge gaps between adjacent cells so as to create direct channels from the cytoplasm of one to that of the other, and they do not link directly to any cytoskeleton. The gap junction channels are formed by structure known as connexons in the plasma membrane of adjacent cells, and each connexon is composed of six connexin subunits [1]. The connexon can be homomeric (identical subunits) or heteromeric with different connexin subunits (Cxs), and the bridged connexon channels between adjacent cells can be either homotypic or heterotypic too [1]. There are

21 connexin proteins identified in man, and they are named according to their mass; for example Cx43 is approximately 43 kDa. Connexins are found in majority of tissue types and among them Cx43 is the most ubiquitously expressed [54]. Multiple connexin proteins are found expressed in a single tissue type and such diversity of the expression is likely to confer different properties to gap junctions. Connexin proteins are transmembrane proteins, each of which contains four transmembrane domains, two extracellular loops and one intracellular loop, and N- and C-terminus both exposed to cytoplasm [1]. Phosphorylation of C-terminus of connexins is important in gap junction assembly, trafficking, channel gating and turnover, and two kinases Src and PKC are well known to be involved in their phosphorylation and subsequent interaction with other junctional proteins [55]. The non-bridged connexon, arranged by six connexin subunits or hexagonal, forms the hemichannel that is believed to involve in paracrine signalling by enabling the transfer of molecules between the cell and the extracellular environment [54]. In addition to communication, gap junctions also provide a form of cell-cell adhesion that supplements the cadherin- and claudin-mediated adhesions we discussed earlier.

A number of studies have indicated a close association between gap junctions and cadherin-based AJs. Inhibition of AJs can impair gap junction formation and disrupt cell-cell coupling or communication, suggesting that gap junctions are dependent upon the proper formation of AJs. Furthermore, Cx43 has been shown to colocalise and/or interact with ZO-1, β-catenin and p120 as well as classical cadherins, and depletion of N-cadherin results in altered subcellular distribution of p120 [55]. All these findings indicate a crosstalk between gap junctions and AJs or even TJs. There are reports suggesting that Cx43 may also associate with cytoskeletal proteins in some specific cell types and direct interaction of Cx43 with microtubules has been demonstrated in a couple of studies in the literature [55]. Furthermore, Cx43 is shown to interact with caveolin-1, a scaffolding protein in a specialised lipid raft known as caveolae that serves as the platform for a number of signal molecules.

Gap junctions play an important role in regulating growth and development. They are present in most animal tissues and organs, including epithelia and heart as well as connective tissues. In skin, gap junctions are found in the basal, spinous and granular layers of human epidermis, but not in the stratum corneum [54]. Similar feature could be seen in oral mucosa although no report has directly demonstrated it. Different isoforms of connexins can be found at distinct locations in the epidermis, and this may be due to different functional roles of isoforms at different stages of keratinocyte differentiation. In mice connexin proteins, such as Cx26, Cx30, Cx31.1, Cx32 and Cx43, are found in the gingival epithelial cells and buccal mucosa, and during wound healing these connexins are rapidly downregulated at the wound edge [56, 57]. An in vitro study shows that both Cx26 and Cx43 are expressed in normal oral epithelial cells but only the membrane expression of Cx43 is found in OSCC cell lines and no Cx26 expression was detected at all in cancer cell lines [58]. High membrane expression of Cx43 has also been implicated as an independent prognostic marker in OSCC and was shown to be correlated with poor cancer patient survival, but no such correlation was found for Cx26 and Cx45 expression [59]. Mutations in the Cx26, Cx30, Cx31, Cx32 and Cx43 genes have been shown to be associated with a wide range of human diseases including syndromic and skin conditions as well as hear loss, and these findings underscore the important roles of gap junctions in multiple tissues and organs [54, 55]. The mechanisms by which Cx mutants cause diseases in vivo are largely unknown and are likely to differ depending upon the different genes as well as the nature of the mutation. It is suggested that the following steps are possibly involved in disease pathology: (1) accumulation of a Cx mutant in the cytoplasm, (2) a Cx mutant exerting a dominant-negative effect, (3) loss of GJ function and (4) aberrant hemichannels [54].

Conclusion

Epithelial cells are attached to each other by numerous intercellular junctions, including TJs, AJs, DMs and gap junctions. These junctions differ in their structures, functions, molecular compositions and tissue presentations. In simple epithelial cells, TJs are located in the apical aspect of lateral membrane, and serve as a tight barrier for the molecules within the plasma membrane of lipid bilayer and concomitantly control the paracellular flux of ions and nonelectrolytes. In stratified epithelia, TJs are restricted to the granular layer and provide protection against all sorts of external stimuli, bacteria invasion and excessive water loss. Gap junctions, on the other hand, couple the neighbouring cells to each other and allow direct communication through the connexon channels between cells. Both AJs and DMs belong to a functional group of anchoring junction which couple the cytoskeletal networks of adjacent cells and serve as scaffolding for the maintenance of epithelial architecture and structural integrity (see Chap. 4). All these junctions are not static structures and in fact they are able to undergo modulation during tissue repair, development or metastatic transition. Because of their important role in tissue integrity and homeostasis, deregulation of intercellular junctions is frequently found in human pathological conditions, including autoimmune, infectious and hereditary diseases as well as cancers. Downregulation of these junctions and their constitutive proteins often is associated with tumorigenesis. However, emerging evidence also suggests the upregulation of some junctional proteins in cancers, and the pro-cancerous roles of these proteins are likely associated with their non-junctional functions involving their cell signalling activity.

References

1. Alberts B, Johnson A, Lewis J, Morgan D, Raff M, Roberts K, Walter P. Molecular biology of the cell. In: Alberts B, editor. 6th ed; 14 A.D. p. 1036–1090.
2. Mandel LJ, Bacallao R, Zampighi G. Uncoupling of the molecular 'fence' and paracellular 'gate' functions in epithelial tight junctions. Nature. 1993;361:552–5.
3. Gumbiner BM. Breaking through the tight junction barrier. J Cell Biol. 1993;123:1631–3.
4. Denker BM, Nigam SK. Molecular structure and assembly of the tight junction. Am J Phys. 1998;274:F1–9.
5. Leech AO, Cruz RG, Hill AD, Hopkins AM. Paradigms lost-an emerging role for over-expression of tight junction adhesion proteins in cancer pathogenesis. Ann Transl Med. 2015;3:184.
6. Niessen CM. Tight junctions/adherens junctions: basic structure and function. J Invest Dermatol. 2007;127:2525–32.
7. Stevenson BR, Siliciano JD, Mooseker MS, Goodenough DA. Identification of ZO-1: a high molecular weight polypeptide associated with the tight junction (zonula occludens) in a variety of epithelia. J Cell Biol. 1986;103:755–66.
8. Citi S, Cordenonsi M. Tight junction proteins. Biochim Biophys Acta. 1998;1448:1–11.
9. Sakakibara A, Furuse M, Saitou M, ndo-Akatsuka Y, Tsukita S. Possible involvement of phosphorylation of occludin in tight junction formation. J Cell Biol. 1997;137:1393–401.
10. Wong V. Phosphorylation of occludin correlates with occludin localization and function at the tight junction. *Am.* J Physiol. 1997;273:C1859–67.
11. Baker OJ. Tight junctions in salivary epithelium. J Biomed Biotechnol. 2010;2010:278948.
12. Furuse M, Sasaki H, Tsukita S. Manner of interaction of heterogeneous claudin species within and between tight junction strands. J Cell Biol. 1999;147:891–903.
13. Markov AG, Aschenbach JR, Amasheh S. Claudin clusters as determinants of epithelial barrier function. IUBMB Life. 2015;67:29–35.
14. Kage H, Flodby P, Gao D, Kim YH, Marconett CN, DeMaio L, Kim KJ, Crandall ED, Borok Z. Claudin 4 knockout mice: normal physiological phenotype with increased susceptibility to lung injury. Am J Physiol Lung Cell Mol Physiol. 2014;307:L524–36.
15. Luissint AC, Nusrat A, Parkos CA. JAM-related proteins in mucosal homeostasis and inflammation. Semin Immunopathol. 2014;36:211–26.
16. Mandell KJ, Parkos CA. The JAM family of proteins. Adv Drug Deliv Rev. 2005;57:857–67.
17. Coyne CB, Bergelson JM. CAR: a virus receptor within the tight junction. Adv Drug Deliv Rev. 2005;57:869–82.
18. Cohen CJ, Shieh JT, Pickles RJ, Okegawa T, Hsieh JT, Bergelson JM. The coxsackievirus and adenovirus receptor is a transmembrane component of the tight junction. Proc Natl Acad Sci U S A. 2001;98:15191–6.
19. Chen JW, Zhou B, Yu QC, Shin SJ, Jiao K, Schneider MD, Baldwin HS, Bergelson JM. Cardiomyocyte-specific deletion of the coxsackievirus and adenovirus receptor results in hyperplasia of the embryonic left ventricle and abnormalities of sinuatrial valves. Circ Res. 2006;98:923–30.
20. Fischer R, Poller W, Schultheiss HP, Gotthardt M. CAR-diology--a virus receptor in the healthy and diseased heart. J Mol Med. 2009;87:879–84.
21. Anderson JM, Fanning AS, Lapierre L, Van Itallie CM. Zonula occludens (ZO)-1 and ZO-2:

membrane-associated guanylate kinase homologues (MAGuKs) of the tight junction. Biochem Soc Trans. 1995;23:470–5.
22. Van Itallie CM, Balda MS, Anderson JM. Epidermal growth factor induces tyrosine phosphorylation and reorganization of the tight junction protein ZO-1 in A431 cells. J Cell Sci. 1995;108(Pt 4):1735–42.
23. Dodane V, Kachar B. Identification of isoforms of G proteins and PKC that colocalize with tight junctions. J Membr Biol. 1996;149:199–209.
24. Stuart RO, Nigam SK. Regulated assembly of tight junctions by protein kinase C. Proc Natl Acad Sci U S A. 1995;92:6072–6.
25. Gumbiner B. Structure, biochemistry, and assembly of epithelial tight junctions. *Am.* J Physiol. 1987;253:C749–58.
26. Stevenson BR, Begg DA. Concentration-dependent effects of cytochalasin D on tight junctions and actin filaments in MDCK epithelial cells. J Cell Sci. 1994;107(Pt 3):367–75.
27. Stuart RO, Sun A, Bush KT, Nigam SK. Dependence of epithelial intercellular junction biogenesis on thapsigargin-sensitive intracellular calcium stores. J Biol Chem. 1996;271:13636–41.
28. Stuart RO, Sun A, Panichas M, Hebert SC, Brenner BM, Nigam SK. Critical role for intracellular calcium in tight junction biogenesis. J Cell Physiol. 1994;159:423–33.
29. Denker BM, Saha C, Khawaja S, Nigam SK. Involvement of a heterotrimeric G protein alpha subunit in tight junction biogenesis. J Biol Chem. 1996;271:25750–3.
30. Jou TS, Schneeberger EE, Nelson WJ. Structural and functional regulation of tight junctions by RhoA and Rac1 small GTPases. J Cell Biol. 1998;142:101–15.
31. Zahraoui A, Joberty G, Arpin M, Fontaine JJ, Hellio R, Tavitian A, Louvard D. A small rab GTPase is distributed in cytoplasmic vesicles in non polarized cells but colocalizes with the tight junction marker ZO-1 in polarized epithelial cells. J Cell Biol. 1994;124:101–15.
32. Brandner JM, Kief S, Grund C, Rendl M, Houdek P, Kuhn C, Tschachler E, Franke WW, Moll I. Organization and formation of the tight junction system in human epidermis and cultured keratinocytes. Eur J Cell Biol. 2002;81:253–63.
33. Furuse M, Hata M, Furuse K, Yoshida Y, Haratake A, Sugitani Y, Noda T, Kubo A, Tsukita S. Claudin-based tight junctions are crucial for the mammalian epidermal barrier: a lesson from claudin-1-deficient mice. J Cell Biol. 2002;156:1099–111.
34. Brandner JM, Schulzke JD. Hereditary barrier-related diseases involving the tight junction: lessons from skin and intestine. Cell Tissue Res. 2015;360:723–48.
35. Peppi M, Ghabriel MN. Tissue-specific expression of the tight junction proteins claudins and occludin in the rat salivary glands. J Anat. 2004;205:257–66.
36. Hashizume A, Ueno T, Furuse M, Tsukita S, Nakanishi Y, Hieda Y. Expression patterns of claudin family of tight junction membrane proteins in developing mouse submandibular gland. Dev Dyn. 2004;231:425–31.
37. Kriegs JO, Homann V, Kinne-Saffran E, Kinne RK. Identification and subcellular localization of paracellin-1 (claudin-16) in human salivary glands. Histochem Cell Biol. 2007;128:45–53.
38. Matsui T, Amagai M. Dissecting the formation, structure and barrier function of the stratum corneum. Int Immunol. 2015;27:269–80.
39. Yoshida K, Yokouchi M, Nagao K, Ishii K, Amagai M, Kubo A. Functional tight junction barrier localizes in the second layer of the stratum granulosum of human epidermis. J Dermatol Sci. 2013;71:89–99.
40. Higashi Y, Suzuki S, Sakaguchi T, Nakamura T, Baba S, Reinecker HC, Nakamura S, Konno H. Loss of claudin-1 expression correlates with malignancy of hepatocellular carcinoma. J Surg Res. 2007;139:68–76.
41. Kominsky SL, Argani P, Korz D, Evron E, Raman V, Garrett E, Rein A, Sauter G, Kallioniemi OP, Sukumar S. Loss of the tight junction protein claudin-7 correlates with histological grade in both ductal carcinoma in situ and invasive ductal carcinoma of the breast. Oncogene. 2003;22:2021–33.
42. Li D, Mrsny RJ. Oncogenic Raf-1 disrupts epithelial tight junctions via downregulation of occludin. J Cell Biol. 2000;148:791–800.
43. Martin TA, Mansel RE, Jiang WG. Loss of occludin leads to the progression of human breast cancer. Int J Mol Med. 2010;26:723–34.
44. Resnick MB, Konkin T, Routhier J, Sabo E, Pricolo VE. Claudin-1 is a strong prognostic indicator in stage II colonic cancer: a tissue microarray study. Mod Pathol. 2005;18:511–8.
45. Tokes AM, Kulka J, Paku S, Szik A, Paska C, Novak PK, Szilak L, Kiss A, Bogi K, Schaff Z. Claudin-1, −3 and −4 proteins and mRNA expression in benign and malignant breast lesions: a research study. Breast Cancer Res. 2005;7:R296–305.
46. Dos Reis PP, Bharadwaj RR, Machado J, Macmillan C, Pintilie M, Sukhai MA, Perez-Ordonez B, Gullane P, Irish J, Kamel-Reid S. Claudin 1 overexpression increases invasion and is associated with aggressive histological features in oral squamous cell carcinoma. Cancer. 2008;113:3169–80.
47. Ouban A, Ahmed A. Analysis of the distribution and expression of claudin-1 tight junction protein in the oral cavity. Appl Immunohistochem Mol Morphol. 2015;23:444–8.
48. Oku N, Sasabe E, Ueta E, Yamamoto T, Osaki T. Tight junction protein claudin-1 enhances the invasive activity of oral squamous cell carcinoma cells by promoting cleavage of laminin-5 gamma2 chain via matrix metalloproteinase (MMP)-2 and membrane-type MMP-1. Cancer Res. 2006;66:5251–7.
49. Brennan K, McSherry EA, Hudson L, Kay EW, Hill AD, Young LS, Hopkins AM. Junctional adhesion molecule-a is co-expressed with HER2 in breast tumors and acts as a novel regulator of HER2 protein degradation and signaling. Oncogene. 2013;32:2799–804.

50. McSherry EA, McGee SF, Jirstrom K, Doyle EM, Brennan DJ, Landberg G, Dervan PA, Hopkins AM, Gallagher WM. JAM-A expression positively correlates with poor prognosis in breast cancer patients. Int J Cancer. 2009;125:1343–51.
51. Naik MU, Naik TU, Suckow AT, Duncan MK, Naik UP. Attenuation of junctional adhesion molecule-a is a contributing factor for breast cancer cell invasion. Cancer Res. 2008;68:2194–203.
52. Saito K, Sakaguchi M, Iioka H, Matsui M, Nakanishi H, Huh NH, Kondo E. Coxsackie and adenovirus receptor is a critical regulator for the survival and growth of oral squamous carcinoma cells. Oncogene. 2014;33:1274–86.
53. Aberle H, Butz S, Stappert J, Weissig H, Kemler R, Hoschuetzky H. Assembly of the cadherin-catenin complex in vitro with recombinant proteins. J Cell Sci. 1994;107(Pt 12):3655–63.
54. Scott CA, Kelsell DP. Key functions for gap junctions in skin and hearing. Biochem J. 2011;438:245–54.
55. Wei CJ, Xu X, Lo CW. Connexins and cell signaling in development and disease. Annu Rev Cell Dev Biol. 2004;20:811–38.
56. Davis NG, Phillips A, Becker DL. Connexin dynamics in the privileged wound healing of the buccal mucosa. Wound Repair Regen. 2013;21:571–8.
57. Hatakeyama S, Mikami T, Habano W, Takeda Y. Expression of connexins and the effect of retinoic acid in oral keratinocytes. J Oral Sci. 2011;53:327–32.
58. Frank DK, Szymkowiak B, Hughes CA. Connexin expression and gap junctional intercellular communication in human squamous cell carcinoma of the head and neck. Otolaryngol Head Neck Surg. 2006;135:736–43.
59. Brockmeyer P, Jung K, Perske C, Schliephake H, Hemmerlein B. Membrane connexin 43 acts as an independent prognostic marker in oral squamous cell carcinoma. Int J Oncol. 2014;45:273–81.

3 Anchoring Junctions in the Oral Mucosa: Adherens Junctions and Desmosomes

Hong Wan, Hanan Gadmor, and Louise Brown

3.1 Introduction

Cell adhesions are crucial in many aspects of cell and tissue biology and coordinate in various processes such as morphogenesis, tumour metastasis and tissue repair. Normally, they are required in tissue development and in establishment of apical-basal axis of cell polarity by generating physical and molecular asymmetry at both surface and intracellular structures, which leads to the formation of apical membrane in the non-contacting (free) cell surface, and the basolateral membrane along the contacting cell surface. A typical feature of epithelia is that the cells are tightly attached to each other and to the extracellular substrate, with little extracellular space, by numerous cell-cell and cell-matrix adhesion complexes, through a collection of glycol proteins and cytoplasmic plaque proteins, and these specific interactions enable epithelial cells to maintain polarity. The epithelial component of oral mucous membrane in the outermost layer is composed of stratified squamous epithelium that serves as a major portal for microbial invasion and provides robust protection against mechanical stress. The integrity of such a mucous barrier is maintained by physical interaction of highly specialised junctional complexes between adjacent cells as well as between cells and basement membrane that are important for safeguarding of systemic and oral health. Like epidermis, the major cell type in oral mucosa is keratinocytes and these cells are arranged as four classified sub-layers, namely, the basal layer, spinous layer, granular layer and stratum corneum. The intercellular adhesions are stable in mature epithelia and become modulated during regenerative processes such as wound healing or pathogenic processes such as cancer progression and invasion. In these latter situations, the junctions become highly dynamic that enables their rearrangements in order to facilitate cell migration, morphogenesis and tissue development. It has been widely accepted that the formation, maturation and homeostasis of epithelia require dynamic coordination between assembly and disassembly of the intercellular junctions, and dysfunction of any aspects is often associated with dedifferentiation and malignance and other diseases such as pemphigus. Emerging evidence suggests that the precisely controlled mechanisms of cell adhesion can in addition serve as crucial regulators for other downstream processes, such as proliferation, differentiation, migration and wound healing.

According to the functions, the cell adhesions in epithelia are classified into three groups of junction complexes, i.e. tight junction (also known as occluding junction), gap junction (also

H. Wan (✉) • H. Gadmor • L. Brown
Centre for Immunobiology and Regenerative Medicine, Institute of Dentistry, Bart and The London School of Medicine and Dentistry, London, UK
e-mail: h.wan@qmul.ac.uk

L.A. Bergmeier (ed.), *Oral Mucosa in Health and Disease*,
https://doi.org/10.1007/978-3-319-56065-6_3

known as channel-forming or communicating junction) and anchoring junction, the latter of which includes adherens junction (AJ), desmosome (DM), focal adhesion and hemi-desmosome (Fig. 2.1). The main feature of anchoring junctions is that this group is composed of members of the cadherin and integrin superfamilies of transmembrane proteins that are linked to cytoskeleton through a collection of cytoplasmic plaque proteins, and they are especially abundant in tissues, such as oral mucosa, skin and heart, that experience extensive mechanical stress. Overall, cell adhesions between cells, as well as between cell and the extracellular matrix which will not be discussed here, play a crucial role in various cellular processes, including cell recognition, positioning, differentiation, embryonic development and wound healing in adult tissues. This chapter focuses on the cell-cell interactions, in particular the adherens junction and desmosome.

3.2 Anchoring Junction: Adherens Junction

Adherens junctions (AJs), along with DMs, belong to the functional group of anchoring junctions that confer mechanical strength and integrity essential for the maintenance of tissue architecture. In general, these junctions exhibit high frequency of occurrence in epithelial and endothelial cells that are exposed to environmental mechanical stress. They are the evolutionarily conserved structure at the plasma membrane and mediate cell-cell adhesions in multicellular organisms. In contrast to DMs which link to intermediate filament cytoskeleton, AJs are physically coupled with the actin microfilaments, the thinnest filaments of the cytoskeleton, via cytoplasmic plaque proteins known as catenins and other signal molecules, and coordinate assembly and organisation of cortical actin throughout the stratified squamous epithelia. AJs are found in a uniform distribution along the plasma membrane and appear as a circumferential continuous *zonula adherens* subjacent to TJs in polarised simple epithelial cells (Fig. 2.1). A similar junctional structure is also observed in cardiomyocytes and is called *fascia adherens* which appears in ribbon-like pattern and does not completely encircle the cells. The maintenance and stabilisation of AJs require attachment of the core cadherin-catenin complex to the actin cytoskeleton.

3.2.1 Adherens Junction Function and Structure

The ability of cells to adhere and communicate to each other is recognised as a prerequisite for the formation and maintenance of a multicellular organism. Through cell-cell interactions, cells can decide whether to change shape, undergo polarisation, continue growing or switch on terminal differentiation, and AJs play a crucial role in the initial engagement of cell-cell interactions. Inhibition of both E- and P-cadherins in vivo not only impairs AJs, but also TJs and DMs as well as the cortical actin cytoskeleton [1, 2]. It is believed that AJs could bring the membranes of opposite cells into close proximity, thereby allowing desmosomal molecules and other signalling molecules to engage and cluster that facilitate other junction formation.

The primary function of AJs is to resist the external forces that pull cells apart [3]. At the same time, they must also be dynamic and adaptable, and be able to be modulated or repaired according to the forces acting on them. Besides imparting structural integrity, AJs are also involved in various functions such as the maintenance of tissue polarity, regulation of cell shape and transmission of signalling events. In the past years, cadherin-catenin complexes have been recognised as important regulators in cell-cell adhesion, embryonic development and adult tissue homeostasis, linking cell-cell adhesion to cell interior to multiple signalling networks. Thus, there is no doubt that the formation, maturation and homeostasis of tissues require dynamic coordination between assembly and disassembly of AJs and other junctional complexes. It has been well established that cadherin-mediated adhesion is crucial in cell sorting during embryonic development, and this property has also been directly demonstrated in cultured L-fibroblasts with exogenous expression of cadherins. If L-cells expressing different cadherins are mixed together, they sort out and aggregate separately, indicating that different cadherins preferentially bind to their own type [3]. AJs have also emerged to act as dynamic tension sensors and response to forces acting on them by modifying local actin and myosin behaviour in order to balance the forces on both sides of the junction, in which process α-catenin is recognised to play an important role [3]. In some cell types, actin contractility reduces cell-cell adhesion strength, particularly when cells are subjected to large forces which can cause the edges of cell-cell adhesions to peel them apart [3].

Because of the adhesive nature, it is widely accepted that AJs function as tumour suppressors by restricting cell growth through adhesion-mediated contact inhibition and decreased β-catenin signalling-dependent LEF-1 activity [4, 5]. In support, disruption of cell-cell adhesion has been shown to significantly contribute to uncontrolled cell proliferation and tumour cell dissemination. The tumour suppressor function of AJs is exemplified by the function of E-cadherin which, when disrupted, is associated with tumour development [6], and in addition the loss of E-cadherin is regarded as a hallmark of cancer. When expressed in cancer cell lines, E-cadherin inhibits growth and reduces invasiveness [7–9]. Likewise, the expression of desmosomal components in L-fibroblasts generates adhesion and inhibits the strongly invasive nature of these cells. Furthermore, the invasive phenotype of these fibroblasts was restored when desmosomal adhesion was inhibited [10].

The current accepted model is that AJs serve as a bridge connecting the actin cytoskeleton of neighbouring cells through direct interaction. Catenins and their interactions with the cytoskeleton are required for the clustering of cadherins into AJs. Deletion of the catenin-binding sites in cadherins, or α-catenin depletion, or abnormalities in the actin cytoskeleton all can abolish AJ assembly [11]. The components and schematic structure of AJ are shown in Fig. 3.1. Essentially, it contains classical cadherins, β- or γ-catenin and α-catenin that link to the actin cytoskeleton. Interaction between the actomyosin cytoskeleton and the AJs is prominently regulated by the mechanical forces and Rho family of small GTPases. Within each cell, a contractile bundle of actin filaments and myosin II lies adjacent to the adhesion belt, oriented parallel to the plasma membrane and tethered to it by the cadherins and

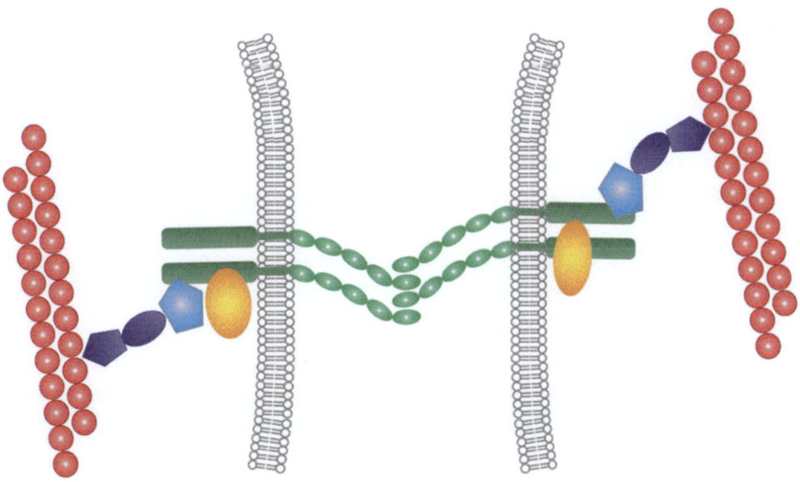

Fig. 3.1 Schematic diagram of the classical cadherin-catenin complex. Classical cadherins (green), which mediate calcium-dependent intercellular adhesion, are composed of an extracellular domain, a transmembrane domain and a cytoplasmic domain. The cytoplasmic domain of cadherin binds to β-catenin (blue) that in turn interacts with α-catenin (purple) and then the actin cytoskeleton. p120 (yellow) binds to the juxtamembrane domain of classical cadherins

their associated cytoplasmic plaque or adaptor proteins. The early study of AJ's ultrastructure in thin sections examined by transmission electron microscope was based on intact adult lens of various species including human and showed that AJs appear to have the same structural characteristics as the *zonula adherens*, except that they were macular contacts, not belts, and had a characteristic distribution in the 'intersections' where three hexagonal fibre cells met [12]. It was also shown that AJs and associated actin were distributed randomly along the entire cell membranes of both wide and narrow sides of cortical fibre cells. In vitro study of cultured mouse keratinocytes subjected to calcium switch (by raising Ca^{2+} concentration in the culture medium from 0.1 to 1 mM) showed AJ formation at 2 h after calcium addition with the appearance of electron-dense undercoat of the plasma membrane that links to actin microfilaments [13].

3.2.2 Adherens Junction Proteins

3.2.2.1 Classical Cadherins

The transmembrane core of AJs is composed of classical cadherin such as E-cadherin (~120 kDa protein), whose ectodomain binds Ca^{2+} to mediate *trans*-oligomers between cadherins on the surface of opposing cell. Classical cadherins are the single-pass transmembrane proteins, named for the tissue in which they were thought to be mainly expressed (e.g. E-cadherin in epithelium, P-cadherin in the placenta, N-cadherin in neuronal tissue). There are five motifs in the extracellular domain separated by flexible hinge regions. Ca^{2+} ions bind in the neighbourhood of each hinge, preventing it from flexing. As a result, the extracellular region forms a rigid, curved structure [3] (Fig. 3.1). To generate cell-cell adhesion, the cadherin domain at the N-terminal tip of one cadherin molecule binds the N-terminal domain from a cadherin molecule on opposing cells. Initial interaction between cadherin extracellular domains is weak, but strong cell-cell adhesion develops during lateral clustering of cadherins (Figs. 19–11 in [3]). Clustering of cadherins depends on their anchorage with the actin cytoskeleton via catenins. The adhesive function of cadherins is selective, which means that these molecules have homophilic binding preference [3]. In contrast to homophilic binding, recent studies also suggest a heterophilic cross-binding between different cadherin isoforms, such as N- and E-cadherins [14]. The clustered cadherins are stabilised by a number of cytoplasmic proteins, primarily comprised of armadillo family proteins such as β- and α-catenins as well as p120. p120 binds to the juxtamembrane domain of classical cadherin and β-catenin binds to the catenin adhesion site downstream of juxtamembrane domain (Fig. 3.2) that in turn binds to the actin-binding protein α-catenin [15]. At the cellular level, E-cadherin adhesion facilitates assembly of other specialised intercellular junctions (DMs, gap and tight junctions) necessary to form functional monolayers of epithelial cells [16, 17].

3.2.2.2 Armadillo Proteins

Armadillo (ARM)-repeat proteins form a large family with diverse and fundamental activities in many eukaryocytes. Since the mid-1980s, a growing number of related proteins have been identified based on sequence homologies. It has become increasingly clear that these family proteins combine structural roles as cell-contact and cytoskeleton-associated proteins and signalling functions by generating and transducing signals affecting the gene expression. Hence, these family proteins have both junction and nucleus distributions. A common feature of this family is that they share a central domain that is composed of a series of imperfect 45 amino acid repeats. In epithelia, the armadillo proteins include β-catenin, α-catenin, plakoglobin (γ-catenin), p120-catenin, plakophilins 1-3 (PKP1-3), p0071 (also known as PKP4) and δ-catenin. Together, they play an important role in tethering AJs to actin cytoskeleton and in regulating clustering of AJ components.

β-Catenin (~95 kDa) is an important member of the armadillo family, and binds to the distal part of the E-cadherin cytoplasmic tail (Fig. 3.2) and connects it to the actin cytoskeleton via α-catenin. In addition to its role in connecting E-cadherin with the actin-binding proteins, β-catenin has a

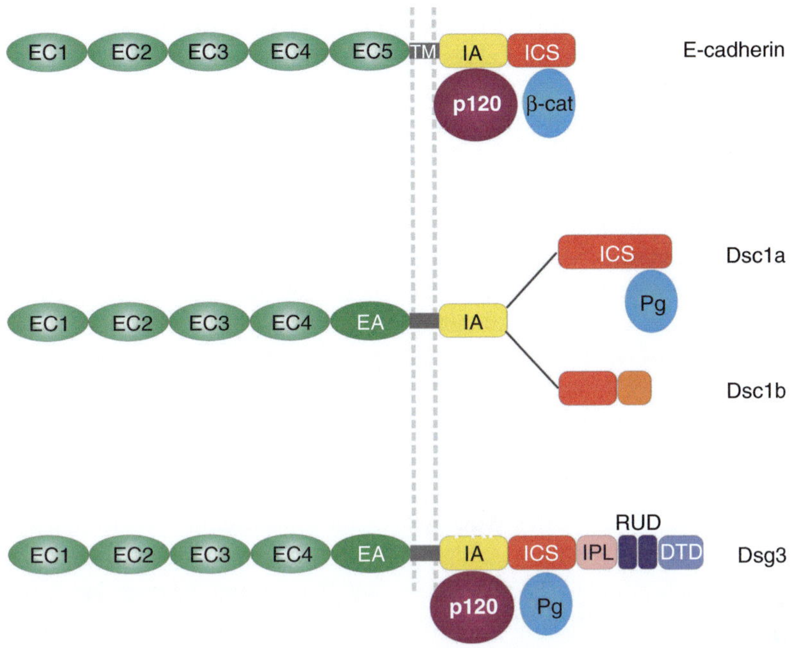

Fig. 3.2 Structure of the classical cadherin E-cadherin and the desmosomal cadherins (the two splice forms of Dsc1, and Dsg3). *EC1-EC5* five extracellular cadherin-like repeats, *EA* extracellular anchoring domain, *TM* transmembrane domain, *IA* intracellular anchoring domain, *ICS* intracellular cadherin-specific domain, *IPL* proline-rich linker domain, *RUD* repeating unit domain, *DTD* desmoglein-specific terminal domain. Both Dsc1 'a' isoform and Dsg3 contain an ICS which binds to plakoglobin (Pg). The Dsc1 'b' isoform also contains a shorter cytoplasmic region with unique sequences (orange). Comparing with Dsc1, the cytoplasmic tail of Dsg3 contains additional domains of IPL, RUD and DTD downstream of ICS

second role as a transcriptional co-activator in the Wnt signalling pathway. Abnormal Wnt signalling caused by mutation of β-catenin is frequently detected in cancer [18]. β-Catenin can translocate to the nucleus and participates in signalling that is mediated by Wnt growth factor receptor [19]. Therefore, the stability of β-catenin is a central control point to the Wnt signalling pathway. Interestingly, β-catenin stability is achieved only by its binding to the cadherin tail. In contrast, p120 is not subject to the same cadherin dependency and is stable in the cytosol when unbound by the cadherin tail [20, 21].

Plakoglobin (~82 kDa), also known as γ-catenin, is a major regulatory protein in both AJs and DMs. However, it exhibits higher affinity for desmosomal cadherins [22, 23] (Fig. 3.2) and therefore its signalling function is primarily to direct DM organisation. Like other junctional proteins, plakoglobin is subjected to modulation by tyrosine kinases and its downstream effect can affect the cadherin-mediated adhesion in both DMs and AJs [24]. Plakoglobin is formed by 12 ARM repeats that share 65% amino acid identity with β-catenin [25]. The central armadillo domain interacts with desmoplakin, which in turn tethers the intermediate filaments to desmosomal plaque. Plakoglobin also exhibits the AJ location where it binds to classical cadherins and links the cadherins to α-catenin and actin, similar to β-catenin function. However, its higher affinity for desmoplakin may explain why plakoglobin, but not β-catenin, locates to DMs [26]. It has been described that the association of plakoglobin with E-cadherin is necessary for DM assembly [27]. In an epithelial cell line that does not express classical cadherins DMs were unable to form, even though it retains the requisite desmosomal components. Introduction of E-cadherin and/or P-cadherin into this cell line did not restore the ability to organise DMs; however, overexpression of plakoglobin, along with

E-cadherin, did permit DM organisation [27]. These results suggest that plakoglobin plays an essential role in the crosstalk between E-cadherin-mediated AJs and DM formation. However, it was later found that this is not the case in mouse keratinocytes where desmosomal cadherins were still capable of clustering on the cell surface and forming DM-like structure in epidermis that has plakoglobin knockout [23]. It was postulated that this was due to the other armadillo family members, such as plakophilin 1 that can substitute for the loss of plakoglobin [28]. Nevertheless, this compensation was unsuccessful in the muscle cells of heart which does not express plakophilin 1 [29]. Another study based on immortalised HaCaT keratinocytes indicates that plakoglobin is responsible for stabilising cell-cell adhesion via inhibition of the p38 mitogen-activated protein kinase (p38 MAPK) activity [30], as well as Src signalling since keratinocytes with plakoglobin knockout exhibited increased activity of Src [31].

p120-Catenin also belongs to armadillo family and is characterised to bind the juxtamembrane domain in the cytoplasmic tail of cadherin (Fig. 3.2) but does not bind to α-catenin [32–34]. At present, four alternative splicing isoforms have been identified. p120 is known to participate in a wide range of biological processes such as influencing cell motility, cadherin clustering [34] and modulating the activities of Rho family GTPases in cell adhesion and reorganisation of actin cytoskeleton [35]. The major function of p120 is involved in stabilisation of cadherins at the junctions. Strong evidence suggests that p120 plays a key role in stabilising the E-cadherin-catenin complex. It has been shown that p120 protects E-cadherin at the cell surface from endocytosis and in turn strengthens the adhesiveness of AJs [36]. This finding is in line with a study by Reynolds and colleagues that showed that down-regulation of p120 is associated with the concomitant loss of E-cadherin in some cases of metastatic cancer [32]. Moreover, in the absence of p120, cadherins undergo internalisation from the cell surface and degradation in the lysosomes [37–40]. Binding of p120 to VE-cadherin inhibits its endocytosis, underscoring an important role of p120 for adhesive junction stability [41]. Recent studies have uncovered that p120 functions as a cap for preventing association of the classical cadherins with clathrin adaptor proteins, thus, as a consequence, preventing the clathrin-mediated internalisation of cadherins [42].

α-Catenin (~102 kDa) is an actin-binding protein, which shares an overall similarity with vinculin, another actin-binding protein, and differs considerably in sequence and structural organisation from the other catenins [43]. It harbours four distinct domains and the N-terminal domain interacts with β-catenin, while the C-terminus binds to α-actinin, vinculin and ZO-1, which in turn links the E-cadherin-catenin complex to the actin cytoskeleton [44]. Recent study found that α-catenin cannot simultaneously bind β-catenin and F-actin, and suggests that the oligomeric state of α-catenin dictates which partner it binds [45]. New roles for α-catenin begin to emerge in the regulation of actin assembly and dynamics through the Arp2/3 complex [43]. Besides serving as an essential component of AJs, α-catenin can also integrate adhesion with other essential cellular events. It has been demonstrated that ablation of α-catenin enhances the Ras-MAPK signalling pathway resulting in hyper-proliferation and defected cell polarisation of keratinocytes [46]. Given the nature of many associates of α-catenin, it seems likely that this protein plays an essential role in the assembly and organisation of AJs and actin cytoskeleton. Overall, catenins play a crucial role in linking cadherins to the actin cytoskeleton and control the adhesive function of AJs.

3.2.3 Adherens Junction Assembly and Regulation

AJs require calcium for their adhesion and maturation since cadherin-mediated adhesion is calcium dependent. During the early phases of calcium-induced intercellular junction formation, cells project membrane protrusions such as lamellipodia and filopodia at the leading edge of cells, and these membrane protrusions help to initiate cell-cell contacts and allow the formation of transient weak adhesion zipper. It is also suggested that the nectin-mediated calcium-independent adhesion is a prerequisite for AJ assembly and E-cadherin adhesion activity [47]. The adhesion zipper attaches the extending membrane to the extracellular matrix that is followed by clustering of AJ proteins E-cadherin and β/α catenin along the developing cell contacts and recruits them into the punctate structures known as adhesion puncta [48]. The highly regulated assembly of the AJ complex begins when β-catenin binds to the carboxy-terminus of E-cadherin [49]. When the cadherin–β-catenin complex reaches the plasma membrane, α-catenin is recruited from the cytosol and binds to the complex through β-catenin [50, 51]. α-Catenin can bind directly to actin filaments, or indirectly via the linker protein vinculin, which in turn binds VASP that interacts with actin [52]. Together, these cytoplasmic proteins (α, β and p120) that associate with cadherins provide anchorage to the actin cytoskeleton to form stable cell-cell contacts.

Regulation of cadherin-mediated adhesion and the associated AJs is thought to underlie the dynamics of the adhesive interactions between cells. Because AJs are required for strong cell-cell adhesion in tissues, the associated catenins have often been investigated as potential cytoplasmic targets for regulation. Changes in the composition of the complex, phosphorylation of components in the complex and alterations in the interaction of the complex with the actin cytoskeleton have all been suggested to play a role in regulation of adhesion [53]. Hence, there are different mechanisms described in the cadherin regulation. For example, tyrosine phosphorylation of the cadherin-catenin complex induced by Src family kinases has been implicated in the regulation of adhesion [54] and tyrosine phosphorylation of β-catenin is shown to correlate with the inhibition of cadherin-mediated adhesion [55–57]. The small GTPases, Rac, Rho and Cdc42, are also involved in cadherin-mediated adhesion [58]. The subfamily of small GTPases is well-known regulators in actin polymerisation and their membrane interactions [59] and thus it is not surprising that they are capable of influencing AJs or cadherin-mediated adhesion. However, the physiological or developmental roles of the small GTPases in the regulation of cadherin adhesion have not been fully elucidated, but overall the findings suggest that they may function in both the assembly and disassembly of AJs.

3.2.4 Adherens Junction in Cancer

The classical cadherin-mediated interactions are the key regulators of tissue architecture and dynamics, and thus they are essential for normal tissue morphogenesis and homeostasis. Disruption of the cadherin-mediated adhesions results in pathological abnormalities in different tissues, and deregulation of cadherins is often found in various human diseases, such as cancer, inflammation and some autoimmune diseases [60–63]. It is well known that down-regulation of E-cadherin and its associated AJs is a hallmark of epithelial to mesenchymal transition (EMT), a process in which cells lose their polarised epithelial phenotype and concomitantly acquire a migratory or mesenchymal cell characteristic [64]. Alteration of AJs and their constitutive molecules is a common event in cancer [65]. Many studies have confirmed that aberrant expression of E-cadherin is associated with invasiveness and metastasis potential in various tissue cancers including lung, prostate, gastric, breast and colon cancers [66]. Thus, it is believed that E-cadherin expression can serve as a significant prognostic marker for tumour progression and behaviour [67]. In line with those findings for E-cadherin, recent studies have identified mutations in the genes encoding α-catenin in head and neck SCC [68]. Further prognostic assessment to evaluate the impact of these α-catenin mutations revealed fourfold increase in mortality if tumours harbour these lesions [68]. These findings highlight an important role of AJs in the control of tumorigenesis.

3.3 Anchoring Junction: Desmosome

Desmosomes (DMs) serve localised adhesive function and connect the plasma membrane to intermediate filaments of adjacent cells, thus forming a structural network known as DM intermediate filament complex (DIFC), and providing strong mechanical strength to epithelial tissues [22]. The great adhesive strength of DMs makes it unique when compared with other intracellular junctions. Due to this property they are found in excess in tissues such as the skin and oral mucosa that must withstand extensive mechanical stress [22].

DM was first discovered by the Italian pathologist Giulio Bizzozero (1846–1901). During his examination of spinous layer of epidermis, he observed small dense nodules at the contact points between adjacent cells and this led him to the understanding that these structures are cell-cell adhesion contact points [69, 70]. Since this discovery, further experiments have been undertaken to provide more detailed information about the structure and function of the DMs.

3.3.1 Desmosome Function and Structure

The primary function of DMs is cell-cell adhesion and maintenance of tissue integrity. DMs can resist mechanical stress because they adopt to a strong adhesive state in which they are described to be hyper-adhesive and which distinguishes them from other intracellular junctions, such as TJs and AJs [22]. The key link between the DMs and the intermediate filaments of the cytoskeleton allows them to carry out their principal function. In fact, DM exists in two adhesive states, calcium dependent *(hypo-adhesive)* and calcium independent *(hyper-adhesive)*. During assembly and before its stabilisation or maturation, DM is calcium dependent. At this stage DM adhesion can be dissociated by extracellular calcium depletion. After formation, a state of hyper-adhesion is gradually achieved allowing the structure to carry out its principal function in maintaining tissue integrity [71, 72]. In this state DMs can no longer be dissociated by calcium depletion. In vitro, hyper-adhesion of DMs is achieved through the maintenance of confluent culture for a few days and occurs, as currently understood, without altering molecular composition of the DM, but rather through the molecular rearrangement of desmosomal cadherins through the entrapment of Ca^{2+} ions [71, 72]. The adhesive state of DMs in young cultures is reversible but ultimately replaced by calcium independence (hyper-adhesiveness). During wound healing, DMs shift to a calcium-dependent state in order to facilitate cell migration and re-epithelisation. The ability of DMs to modulate their adhesive state is essential as they need to be as dynamic as the tissues they support. In vitro studies show that activation of PKCα causes reversion of DMs to calcium dependence to facilitate epithelial remodelling in wound healing [71].

The role of intercellular junctions in cell differentiation is mainly seen in DMs, and the desmosomal cadherins in particular act as the surface receptors and regulate intracellular signalling that negatively or positively control differentiation [60]. DM composition varies with cell type and differentiation status. Such variations result in structural changes that are visible in DM appearance and size, and such tailoring is proposed to suit the specialised functions of the cells and tissues that possess them. This feature is well exemplified through the expression patterns of desmosomal components restricted to specific layers in the epidermis and oral mucosa [73]. Experimental studies have shown that the absence or mutations of desmosomal cadherins result in abnormal proliferation and differentiation [22, 70]. An example of this is the ectopic expression of Dsg2 in the mouse suprabasal epidermis that augments various signalling pathways downstream of EGFR, resulting in suppressed differentiation, impaired apoptosis and premalignant papillomas [74]. In contrast, Dsg1 is found to promote keratinocyte differentiation by attenuating MAPK/ERK signalling via cooperating with Erbin [75]. Furthermore, recent studies have revealed that Dsg3 acts as a key regulator for various signal pathways such as Src, small Rho GTPases, the ERM protein ezrin and transcription factor activator protein-1 (AP-1), all of which are involved in actin-based cell shape change and migration and invasion [76–81]. Moreover, Dsg3 is found to crosstalk with E-cadherin and regulate its junction assembly during the process of calcium-induced cell-cell contact [79, 80]. Desmosomal cadherins can also regulate proliferation and differentiation by influencing its cytoplasmic plaque proteins, such as plakoglobin which also has a role in the Wnt/β-catenin signalling pathway. In the absence of β-catenin, plakoglobin can play a unique role in Wnt signalling that differs from that triggered by β-catenin [22, 82].

DMs have a very characteristic and highly organised electron-dense structure with size less than 1 μm in diameter [22]. Based on electron micrograph, a DM can be divided into three morphologically characteristic zones: the extracellular core domain, outer dense plaque (ODP) and inner dense plaque (IDP) [83] (Fig. 3.3). The extracellular adhesion core is made of the electron-dense material known as *desmoglea* which is bisected by an electron-dense midline [83]. *Desmoglea* spans 20–35 nm in diameter and consists of the extracellular domains of desmosomal cadherins. In the cytoplasmic site of the plasma membrane, there are two symmetrical, highly electron-dense plaques from adjacent cells that surround a shared *desmoglea*, and each can be further divided into two outer dense plaque (ODP) and inner dense plaque (IDP), both approximately 15–20 nm in thickness [70, 83]. The ODP contains the intracellular domains of desmosomal cadherins and the adaptor proteins, plakophilins and plakoglobin as well as the head domain of desmoplakin. The IDP consists of C-terminus of desmoplakin and keratin intermediate filaments in epithelial cells.

3.3.2 Desmosomal Proteins

Desmosome proteins are essentially comprised of three families. As illustrated in Figs. 3.2 and 3.3, a characteristic DM is composed of transmembrane cadherins, desmogleins (Dsgs) and desmocollins (Dscs), which mediate direct adhesion between adjacent cells; Armadillo family proteins, plakoglobin and plakophilin, which associate with the cytoplasmic tail of cadherins; and plankin family proteins, including desmoplakin which anchors the stress-bearing intermediate filaments to the DMs. The cytoplasmic tails of the desmosomal cadherins directly interact with plakoglobin, which in turn associates with the N-terminal globular head of desmoplakin and plakophilins in the ODP. The C-terminal tail of

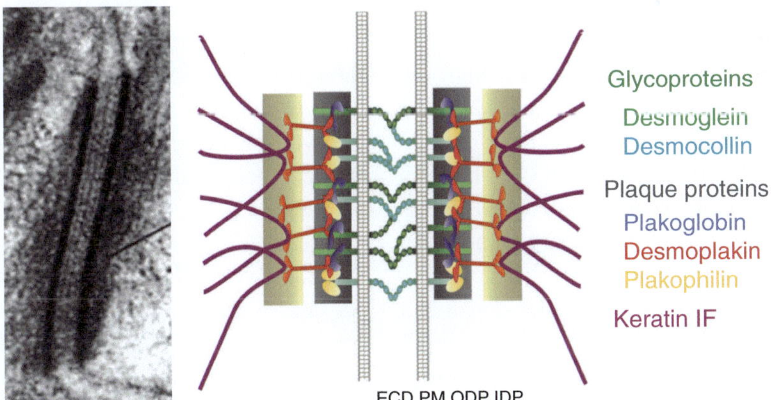

Fig. 3.3 Transmission electron micrograph (left) and molecular map of a desmosome (right). The electron micrograph: Fawcett DW, The Cell: An Atlas of Fine Structure, WB Saunders, Philadelphia, 1966, p. 371 (http://www.columbia.edu/itc/hs/medical/sbpm_histology_old/lab/micro_popup41.html). Desmosome is a multi-protein complex composed of desmosomal cadherins, armadillo family proteins (plakoglobin and plakophilins) and plakin family members (desmoplakin). They are arranged as two symmetrical electron-dense plaques, inner dense plaque (IDP) and outer dense plaque (ODP), which facilitate the association between the C-terminus of desmosomal cadherins and the intermediate filaments of cytoskeleton. The extracellular core domain (ECD) of the desmosome is bisected by an electron-dense midline as shown in the electron micrograph. PM, plasma membrane

desmoplakin extends to the IDP where it interacts with the intermediate filaments of the cytoskeleton. Collectively, these molecular components are arranged in a well-ordered array and form structure as described above by conventional electron microscopy [22, 70, 83].

3.3.2.1 Desmosomal Cadherins

The core adhesion proteins in DMs are the desmosomal cadherins, Dsgs and Dscs, both of which belong to cadherin superfamily of glycoproteins and have the molecular weights of about 100–160 kDa. There are seven desmosomal cadherins identified in human, three Dscs and four Dsgs [22]. These proteins help to form a bridge by heterophilic/homophilic binding of their extracellular domains between two neighbouring cells. Both Dscs and Dsgs are required for the formation and normal function of DMs. The genes of Dscs and Dsgs are all clustered together on opposite sides of a central region on chromosome 18 [22]. Like the classical E-cadherin, desmosomal cadherins are single-pass transmembrane proteins with their C-terminus located in the cytoplasmic plaque and N-terminal domains exposed on the cell surface (Fig. 3.3) where they connect the two halves of the DM together in intercellular space. All desmosomal cadherins are synthesised with N-terminal signal and pro-peptides that are cleaved during protein maturation [22]. The extracellular domains of Dsgs and Dscs contain five cadherin repeats, each with a Ca^{2+}-binding site. A cell adhesion recognition (CAR) site in each protein is located in the first motif that directly contributes to the adhesive function of desmosomal cadherins in a calcium-dependent manner. The cytoplasmic tails contain two well-characterised domains, intracellular anchorage (IA) domain and intracellular cadherin segment (ICS) with the former binding to p120 and latter to plakoglobin. With regard to Dscs, there are two splice isoforms ('a' and 'b') for each protein that are encoded by each desmocollin gene. Both isoforms possess an IA domain but only 'a' isoforms have an ICS domain. The cytoplasmic tails of Dsgs exhibit more variation than Dscs and consist of additional unique region, including intracellular proline-rich linker (IPL) domain, a repeat unit domain (RUD) and a glycine-rich desmoglein terminal domain (DTD) [22], and the functions of these additional domains remain not characterised. In general, the intracellular domains of desmosomal cadherins are believed to play roles in signal transduction via clustering with various but distinct signal molecules for each cadherin. A recent study indicated that the unique region of Dsg2 is involved in Dsg2 stabilisation at the cell surface through a mechanism of inhibiting its internalisation and facilitating Dsg tail-tail interactions [84]. In support, a Dsg2 mutant, identified in arrhythmogenic right ventricular cardiomyopathy (ARVC) patients, led to a loss of Dsg2 tail self-association and exhibited rapid endocytosis in cardiac muscle cells [84].

Calcium is an essential requirement to trigger junctional assembly of desmosomal cadherins, like the classical cadherins. Numerous studies have shown the importance of calcium at this process and how the cadherins are distributed during cell junction formation. In an early study, Watt et al. [85] used human keratinocytes to investigate calcium-induced DM formation and discovered that after 15 min of calcium addition the desmosomal proteins were re-localised at the cell periphery and continued to assemble there for at least 2 h. In addition, calcium ions are also used to monitor the DM status in vitro and in vivo and those that resist calcium depletion are characterised as the DMs with hyper-adhesion [71, 72].

3.3.2.2 Armadillo Family Proteins

As described in 3.2.2.2, the Armadillo family contains many proteins which share common feature with variable number of arm repeats, and the members of this family in DMs include plakoglobin and plakophilins. They are found in the ODP of the DM, where plakophilins are located closer to the plasma membrane than plakoglobin [83]. The main function of these proteins is to facilitate the cytoplasmic associations and clustering of the desmosomal cadherins and to recruit and stably connect with desmoplakin in DMs that in turn provide strong cell-cell adhesion and cytoskeletal attachment [22, 70]. Plakophilins are closely related to p120 while plakoglobin is more closely related to β-catenin [60, 70]. Although plakoglobin is localised in both DMs and AJs, it has a greater affinity for desmosomal cadherins than for E-cadherin [86]. Serving as a linker protein, plakoglobin binds directly to desmoplakin and plakophilins through its central armadillo domain, which anchors DMs to the intermediate filaments. Its interaction with tyrosine kinases has been shown to modulate cadherin-dependent adhesion [24]. Plakoglobin is also reported to be involved in mediating intracellular signalling events associated with pathogenesis of pemphigus vulgaris [87, 88] and its binding with Dsg3 is important for the incorporation of Dsg3 to DMs [89]. In addition, plakoglobin is found to be functionally related to β-catenin, both of which are involved in the regulation of β-catenin/TCF signalling pathway [90, 91] and may compensate for each other's adhesive function. However, β-catenin is restricted to AJs only. Therefore, it is believed that plakoglobin is vital for the cross communication between DMs and AJs [27]. The vital role of plakoglobin in DM assembly in vivo was first examined in the study of plakoglobin knockout in mice which died between 12 and 16 days of development as a result of defects in the heart function and stability [29]. The DMs in these mice were not detected and the remaining junctional structures were dramatically altered, leading to change in the distribution of desmoplakin and Dsgs.

The plakophilin (PKP) family consists of four members, PKP1–3 and p0071 (PKP4), and have molecular weights ranging between 80 and 130 kDa. They come in two isoforms 'a' and 'b' due to alternative splicing [92]. All plakophilins exhibit the dual localisations of DMs and the nucleus, and are regulated by 14-3-3 protein which retains cytoplasmic localisation. The N-terminal head of plakophilins mediates interactions with desmoplakin, plakoglobin and desmosomal cadherins as well as facilitates their transport to cell-cell contacts. PKP1 is essential for proper attachment of intermediate filaments to the DMs [93, 94] and has the ability to recruit desmoplakin to the cell borders [24]. PKP2 is found to be associating with RNA polymerase III [95] and is thought to be involved in β-catenin signalling [96]. Overall, it is known that plakophilins reinforce junctional stability through lateral interactions between junctional components and have a high functional conservation with p120. Other studies have raised the possibility that plakophilins may be involved in nuclear function [97].

3.3.2.3 Plakin Family Proteins

The plakin family consists of a group of large structural proteins with molecular weight ranging between 210 and 530 kDa. This family includes desmoplakin, plectin, envoplakin and periplakin and among them desmoplakin is the ubiquitous and most abundant component of the DM. Unlike plakoglobin, desmoplakin is localised predominantly in the IDP, further away from the membrane. It directly links the cytoskeletal networks to the plasma membrane and regulates binding of actin and intermediate filaments [98]. Desmoplakin consists of a globular head and tail domain which flank a central α-helical coiled-coil rod domain that mediates its dimerisation. The N-terminal of desmoplakin targets to the ODP, where it binds to plakoglobin/plakophilins and desmosomal cadherin complexes and clusters them into organised patches [99, 100]. The C-terminal tail of desmoplakin has distinct plakin repeat domains, which is thought to regulate the binding of desmoplakin to intermediate filaments. There are two isoforms of desmoplakin (Dp), DpI and DpII, generated by alternative splicing. These two isoforms are identical in amino acid sequence with the exception that DpII contains only one-third of the central alpha-helical rod domain present in DpI. Recently, another minor isoform of desmoplakin, named desmoplakin Ia (DSPIa), has been identified and it is also produced by alternative splicing of the desmoplakin gene [101]. Desmoplakin is the predominant plaque protein required for DM assembly, intermediate filament association and regulation of DM and AJ localisation. Previous study has shown the importance of desmoplakin in connecting intermediate filaments to the plasma membrane and demonstrated that when endogenous desmoplakin was depleted the cells were not able to be attached to one other by intermediate filament bundles [102].

3.3.3 Desmosomal Protein Expression in Oral Mucosa

The expression of desmosomal proteins varies depending upon the tissue and cell type as well as their differentiation status. With regard to desmosomal cadherins, for example, Dsg2 and Dsc2 are widely expressed in all DM-baring tissues and all seven desmosomal cadherins identified in man are expressed in the epidermis [73]. Expression of Dsg2 and Dsg3 is restricted to the lower compartment of the epidermis, whereas high expression levels of Dsg1 are found in the upper layers. Dsg4 is predominantly present in the granular layer and in the hair follicle [22, 70] (Fig. 3.4a). This complex expression pattern of desmosomal cadherins indicates that DMs are biochemically and functionally distinct. Different expression patterns are also observed for plakin family and armadillo proteins between different tissues (Fig. 3.4a). Overall, more uniform expression is seen for desmoplakin and plakoglobin than plakophilins.

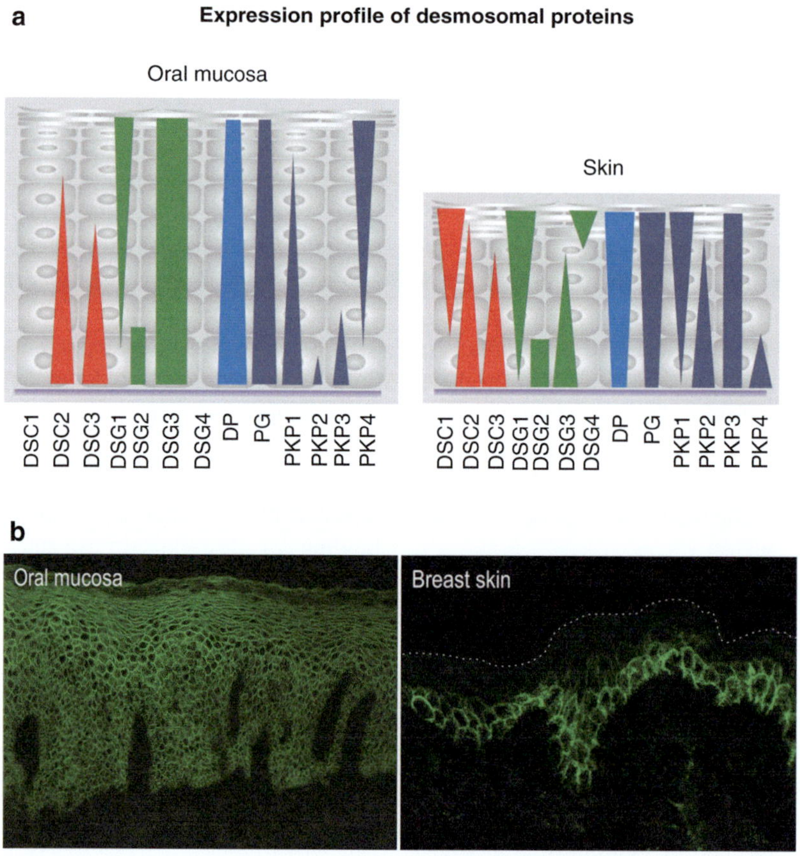

Fig. 3.4 Expression of the desmosomal proteins in oral mucous membrane and epidermis. (**a**) Direct comparison of the desmosomal protein expression in two different tissues. In oral mucosa, no expression of Dsc1, Dsg4 and PKP2 was found and Dsg2 levels were very low (reproduced from publication [73]). (**b**) Immunofluorescent staining of Dsg3 in two different epithelial tissues and the distinct expression patterns are observed. While uniform expression of Dsg3 is seen across the entire stratified epithelium in the oral mucous membrane, this protein is largely restricted to the basal and immediate suprabasal layers in the epidermis. Note that the images are displayed at different magnifications and the dot line in the right indicates the skin surface

In oral mucosa, a distinct expression pattern is seen for Dsg3, in particular, with the uniform appearance across the entire stratified squamous epithelium, in contrast to that in epidermis (Fig. 3.4b). Why Dsg3 displays such a distinct expression pattern remains not understood. It was thought that this could be due to enhanced cell turnover in oral mucosa or signal pathways regulated by Dsg3 that are required in this specific tissue. In addition, less Dsg1/Dsg2 and little or no expression of Dsc1, Dsg4 and PKP2 were found in oral mucosa as compared to epidermis (Fig. 3.4a). With regard to the plaque proteins, PKP3 levels are relatively low and PKP4 is only present in differentiated cells in oral epithelium.

3.3.4 Desmosome Assembly and Regulation

Initiated by cell-cell contact and AJ formation, DMs assemble to confer and stabilise strong intracellular adhesion [22, 27, 71]. The precise sequence of events of DM assembly has been hindered by the relative insolubility of the junctions. However, technological advances including live-cell imaging and fluorescent reporters have begun to overcome this obstacle. As currently understood, DM assembly occurs in several phases [103]. Firstly, the desmosomal cadherins assemble at the plasma membrane. Specifically, Dsc-enriched vesicles initiate assembly [104], and this is followed by Dsg-enriched vesicles that are transported to the plasma membrane where they are associated with desmoplakin. To complete DM assembly, targeting and tethering of these vesicles into the plasma membrane have been indicated to be facilitated through the Sec3 exocyst protein complex [105]. The next phase is the cytoplasmic plaque assembly, and this phase is already initiated during Dsg recruitment to the plasma membrane, where desmoplakin in concert with Dsg3 accumulates at the cell cortex. This is quickly followed by the recruitment of non-membrane-bound desmoplakin-containing particles that are associated with the intermediate filament cytoskeleton. Once at the cell cortex desmoplakin is translocated to cell-cell contacts where it binds to the C-terminus of desmosomal cadherins via plakoglobin in the ODP in a microtubule- and actin-dependent manner [106]. The translocation of desmoplakin-associated complexes is regulated and promoted by PKP2 which acts as a scaffold for a complex containing plakin and PKCα which modulates the interaction between desmoplakin and intermediates filaments. The contribution of plakoglobin and other PKPs to the DM assembly is still unclear although both have been shown to be required for normal DM plaque assembly [29, 94, 98, 107].

While being essential in providing mechanical strength, DMs themselves are dynamic structures with changes in their molecular composition and structure during processes such as epithelial differentiation and stratification, tissue remodelling and wound healing [108]. DMs are regulated at both transcriptional and post-transcriptional levels. Transcriptionally, understanding of how DMs are regulated is still limited. However, preliminary studies based on the characterisation of desmosomal cadherin promoters revealed that desmosomal cadherins are regulated by different promoters and the specific expression of each cadherin can be achieved without affecting the others [108]. On a post-transcriptional level, signals generated by cell-cell contacts and junction assembly provoke protein phosphorylation that in turn positively and negatively regulate DM formation. It has been shown that phosphorylation of serine residues of Dsc promotes plakoglobin association and Dsg binding to confer the DM formation in response to calcium [109]. In addition, PKCα has been shown to have a pivotal role in the regulation of DMs, with its activation promoting DM formation in a calcium- and AJ-independent manner [110, 111]. PKCα is also involved in desmoplakin trafficking for its junctional incorporation through a PKP2-dependent mechanism [112]. The activity of PKCα is directly associated with the adhesive state of DMs with its suppression rendering DMs in a status of hyper-adhesion, whereas PKCα activation promotes its calcium dependence [71]. Furthermore, both intracellular calcium and extracellular calcium are required for DM junction formation and maintenance [113, 114]. As mentioned above, DM assembly is dependent on AJ formation, with the relationship between these junctions attributed to plakoglobin [1, 115]. Recent studies however suggest that Dsg3 also plays a crucial role in crosstalk between these two junctions and in DM formation [79–81, 116]. In contrast, there is also evidence that suggests that AJs are not necessarily for DM assembly [1, 111]. Thus, DM assembly may depend on cell context and junction protein expression.

3.3.5 Desmosome-Related Diseases

The importance of desmosomal roles in cell-cell adhesion and tissue integrity is highlighted by many human diseases. However, it is often less clear whether the symptoms that occur in DM human diseases arise primarily from loss of adhesion or from modulation of signalling pathways involving DMs. Thus, the altered cell signals may themselves cause loss of desmosomal adhesion or indeed changes in tissue phenotype [22]. Accumulated evidence suggests that DMs are not simply to mechanically join cells together and in fact they play a crucial role in cell signalling and in regulating signal transductions that control cell proliferation, differentiation, morphogenesis and motility. Deregulation of these pathways is often found in pathological conditions such as cancer, autoimmune and inherited diseases that share some common features in DM alterations.

3.3.5.1 Genetic and Infectious Diseases

Disruption of the desmosomal cadherins and cytoplasmic plaque proteins caused by mutations in the desmosomal genes or the bacteria infection can have significant clinical consequences with defect manifested in the skin and heart. For instance, mutations in Dsg1 lead to skin disorder such as striate palmoplantar keratoderma [24], whereas mutations in Dsg2 and desmoplakin are associated with heart disorder known as arrhythmogenic right ventricular cardiomyopathy (ARVC) [117]. ARVC, that causes arrhythmias and sudden cardiac death, is characterised by fibro-fatty replacement of cardiac myocytes. Mutations in five desmosomal genes, Dsc2, Dsg2, desmoplakin, plakoglobin and PKP2, are shown to be the causative factors and account for 50–70% of ARVC cases. These include non-syndromic ARVC mutations affecting all domains of desmoplakin, which appear to be inherited in a dominant manner [118]. Both autosomal dominant and recessive mutations in Dsc2 can cause ARVC without a cutaneous or hair phenotype [119]. Mutations leading to the loss of Dsg4 are responsible for disruptions in hair follicle differentiation [120]. For more information about genetic DM diseases see [121–123].

There are also some infectious diseases that lead to disruption of desmosomal cadherins. Bullous impetigo and Staphylococcal scalded skin syndrome (SSSS) are caused by infectious bacteria toxins produced by some strains of pathogenic bacteria such as *Staphylococcus aureus* that target specifically the extracellular domain of Dsg1, which is also the target of autoantibodies in pemphigus foliaceus. Thus blisters formed in these diseases are identical to pemphigus foliaceus in tissue specificity and histology. The bacterial proteases attack and cleave the extracellular domain of Dsg1 leading to blister formation just below the stratum corneum [124]. There are three types of exfoliative toxins produced by *Staphylococcus aureus* (ETA, ETB and ETD), all of which specifically target Dsg1, with ETA being the most common. For review of DMs in acquired diseases see [122].

3.3.5.2 Pemphigus (Autoimmune Disease)

Pemphigus is a life-threatening, autoimmune disease characterised by the loss of keratinocyte adhesion within the spinous layer of stratified epithelia, in a process called 'acantholysis', which clinically manifests as blistering of the skin and oral mucosa. There are two major types of pemphigus, pemphigus foliaceus and pemphigus vulgaris. Pemphigus foliaceus is caused by autoantibody binding to Dsg1, and is characterised by blistering in the upper granular layers of the epidermis. Pemphigus vulgaris can be divided into two subtypes, (1) a mucosal dominant type which involves oral lesions with little or no skin involvement and is caused mainly by anti-Dsg3 autoantibodies with blistering located between the basal and immediate suprabasal layers of stratified epithelia, and (2) a mucocutaneous type, which is characterised by both oral and skin lesions, has the presence of autoantibodies against both Dsg3 and Dsg1. Paraneoplastic pemphigus (PNP) is another type of pemphigus where autoantibodies target the desmosomal and hemidesmosomal proteins [125]. Pemphigus vulgaris is the most common form of pemphigus accounting for up to 80% of pemphigus cases with pemphigus foliaceus and PNP being less severe and least common, respectively.

Several hypotheses have been proposed to explain the mechanisms by which pemphigus manifests. Since the discovery of circulating autoantibodies targeting Dsg3 and Dsg1 in the sera of pemphigus patients, it was proposed that these autoantibodies mechanically inhibit the adhesive function of Dsg1 and Dsg3 by the mechanism of steric hindrance. The distinct expression patterns of Dsg1 and Dsg3 between the skin and oral mucosa may account for clinical manifestations of blistering locations in pemphigus [126]. In epidermis, the expression of Dsg3 is restricted to the basal layer and is gradually replaced by Dsg1 in the upper compartment of the epidermis. In contrast, in oral mucosa Dsg3 is uniformly present across the entire stratified epithelium with limited and restricted Dsg1 in superficial layer [73]. Based on these distinct expression patterns of Dsg1 and Dsg3, a compensation theory was proposed, i.e. Dsg3 compensates, where possible, for the loss of Dsg1 and vice versa [126]. In the case of pemphigus foliaceus where Dsg1-mediated adhesion is compromised, the uniform expression of Dsg3 throughout oral mucosa compensates its loss of function. However, in the epidermis, Dsg3 expression is limited to the basal layers and is unable to compensate for the loss of Dsg1 in the stratum granulosum. As a result, blistering occurs in the skin at the superficial layers of the epidermis with no mucosal involvement. In pemphigus vulgaris where autoantibodies target Dsg3, Dsg1 is unable to compensate for the loss of Dsg3-mediated adhesion in lower compartment of oral mucosa resulting in mucosal dominant pemphigus. When autoantibodies to both Dsg1 and Dsg3 are in circulation neither is able to compensate, resulting in mucocutaneous pemphigus. Although a large body of evidence supports the 'desmoglein compensation' hypothesis, clinical studies have found that the autoantibody titres do not always correlate with the clinical presentation, suggesting that other factors, such as intracellular signalling, may play a role in the pathogenesis of pemphigus acantholysis.

Many studies based on anti-Dsg3 autoantibodies suggest the possible signalling mechanism involved in pemphigus acantholysis [122]. Treatment of keratinocyte culture with anti-Dsg3 autoantibodies has shown to trigger a series of intracellular events that cause disruption of cell-cell adhesions and promote rearrangement of cortical actin filaments [127–129]. These intracellular events include the phosphorylation of Dsg3 and its dissociation from DMs, increased intracellular calcium concentrations and activation of various signalling molecules such as Pg, PKC, p38 MAPK, heat-shock protein p27, Src and c-Myc [127–129]. Taken together these findings not only affirm the signalling capabilities of Dsg3 but also strongly suggest that the signalling processes are involved in pemphigus pathogenesis.

3.3.5.3 Desmosomes in Cancer

Traditional views consider DMs, and other intercellular junctions, as having a protective function in carcinogenesis through cell-adhesion-mediated contact inhibition. In accordance, the loss of cell-cell adhesion is an essential event in EMT. Indeed, loss of DMs and their constitutive proteins has been found in various tumours in the body. EMT is a biologic process that transforms polarised epithelial cells through multiple morphological and biochemical changes to become mesenchymal in phenotype with characteristics such as enhanced migration and invasion capability, resistance to apoptosis and increased production of extracellular matrix components. Accumulating evidence suggests that EMT is a prerequisite to pathological processes including cancer progression and metastasis. Loss of AJs and DMs occurs in epithelial cells to allow cellular dissociation in the conversion from benign to metastatic tumours [130]. Specific to DMs, the loss or reduction of junction components including Dsg1-3, Dsc1-3, plakoglobin and plakophilins has been observed in the development and/or progression of SCCs of the skin, head and neck, and prostate, correlating with increased metastasis and poor prognosis [131].

Paradoxically, several independent studies have shown an increase of some desmosomal genes and proteins, such as Dsg2, Dsg3 and PKP3, in cancers and this overexpression is associated with enhanced tumour progression [76, 132, 133]. For instance, Dsg3 is reported to be upregulated in SCC of the head and neck, oesophageal and lung as well as in cancer cell lines derived from head and neck SCC [76]. Furthermore, the levels of overexpression correlate with clinical stage of disease and regional lymph node metastasis. Dsg3 silencing suppressed cancer cell growth, migration and invasion in vitro and in vivo [134]. Analysis of clinical samples and cancer cell lines showed elevated Dsg3 expression in cancers of the colon, oesophagus, stomach, pancreas and skin [76]. Furthermore, Dsg3 has been identified as an accurate biomarker for the detection of metastatic spread of SCC and ancillary marker to separate SCC from other subtypes of lung cancer [76]. Together, these findings support a procancerous role for Dsg3, and it is believed that this role may not be associated with its adhesive function in DMs but rather with its additional signalling function beyond the junctions, i.e. Dsg3 acting as a key regulator in the control of actin-based cellular processes. For more detail on the role of cell adhesion in cancers see [65, 131].

3.4 General Conclusions on Cell-Cell Interactions

Epithelial cells are attached to each other by numerous intercellular junctions, including TJs, and gap junctions as described in Chap. 2 as well as AJs and DMs, described in this chapter. These junctions differ in their structures, functions, molecular compositions and tissue presentations. In simple epithelial cells, TJs are located in the apical aspect of lateral membrane, and serve as tight barrier for the molecules within the plasma membrane of lipid bilayer and concomitantly control the paracellular flux of ions and nonelectrolytes. In stratified epithelia, TJs are restricted to the granular layer and provide protection against all sorts of external stimuli, bacteria invasion and excessive water loss. Gap junctions, on the other hand, couple the neighbouring cells to each other and allow direct communication through the connexon channels between cells (see Chap. 2).

Both AJs and DMs belong to a functional group of anchoring junction which couple the cytoskeletal networks of adjacent cells and serve as scaffolding for the maintenance of epithelial architecture and structural integrity. All these junctions are dynamic not static structures and in fact they are able to undergo modulation during tissue repair, development or metastatic transition. Because of their important role in tissue integrity and homeostasis, deregulation of intercellular junctions is frequently found in human pathological conditions, including autoimmune, infectious and hereditary diseases as well as cancers. Downregulation of these junctions and their constitutive proteins often is associated with tumorigenesis. However, emerging evidence also suggests the upregulation of some junctional proteins in cancers, and the procancerous roles of these proteins are likely associated with their non-junctional functions involving their cell signalling activity.

References

1. Lewis JE, Jensen PJ, Johnson KR, Wheelock MJ. E-cadherin mediates adherens junction organization through protein kinase C. J Cell Sci. 1994;107(Pt 12):3615–21.
2. Tinkle CL, Pasolli HA, Stokes N, Fuchs E. New insights into cadherin function in epidermal sheet formation and maintenance of tissue integrity. Proc Natl Acad Sci U S A. 2008;105:15405–10.
3. Alberts B, Johnson A, Lewis J, Morgan D, Raff M, Roberts K, Walter P. Molecular biology of the cell. Alberts B. editor. 6th ed.; 14 A.D. p. 1036–1090.
4. Novak A, Dedhar S. Signaling through beta-catenin and Lef/Tcf. Cell Mol Life Sci. 1999;56:523–37.

5. Porfiri E, Rubinfeld B, Albert I, Hovanes K, Waterman M, Polakis P. Induction of a beta-catenin-LEF-1 complex by wnt-1 and transforming mutants of beta-catenin. Oncogene. 1997;15:2833–9.
6. Pecina-Slaus N. Tumor suppressor gene E-cadherin and its role in normal and malignant cells. Cancer Cell Int. 2003;3:17.
7. Frixen UH, Behrens J, Sachs M, Eberle G, Voss B, Warda A, Lochner D, Birchmeier W. E-cadherin-mediated cell-cell adhesion prevents invasiveness of human carcinoma cells. J Cell Biol. 1991;113:173–85.
8. Vleminckx K, Vakaet L Jr, Mareel M, Fiers W, van Roy F. Genetic manipulation of E-cadherin expression by epithelial tumor cells reveals an invasion suppressor role. Cell. 1991;66:107–19.
9. Watabe M, Nagafuchi A, Tsukita S, Takeichi M. Induction of polarized cell-cell association and retardation of growth by activation of the E-cadherin-catenin adhesion system in a dispersed carcinoma line. J Cell Biol. 1994;127:247–56.
10. Tselepis C, Chidgey M, North A, Garrod D. Desmosomal adhesion inhibits invasive behavior. Proc Natl Acad Sci U S A. 1998;95:8064–9.
11. Stappert J, Kemler R. A short core region of E-cadherin is essential for catenin binding and is highly phosphorylated. Cell Adhes Commun. 1994;2:319–27.
12. Lo WK. Adherens junctions in the ocular lens of various species: ultrastructural analysis with an improved fixation. Cell Tissue Res. 1988;254:31–40.
13. O'Keefe EJ, Briggaman RA, Herman B. Calcium-induced assembly of adherens junctions in keratinocytes. J Cell Biol. 1987;105:807–17.
14. Niessen CM, Gumbiner BM. Cadherin-mediated cell sorting not determined by binding or adhesion specificity. J Cell Biol. 2002;156:389–99.
15. Shapiro L, Weis WI. Structure and biochemistry of cadherins and catenins. Cold Spring Harb Perspect Biol. 2009;1:a003053.
16. Gumbiner B. Cadherins: a family of Ca^{2+}-dependent adhesion molecules. Trends Biochem Sci. 1988;13:75–6.
17. Wheelock MJ, Jensen PJ. Regulation of keratinocyte intercellular junction organization and epidermal morphogenesis by E-cadherin. J Cell Biol. 1992;117:415–25.
18. Austinat M, Dunsch R, Wittekind C, Tannapfel A, Gebhardt R, Gaunitz F. Correlation between beta-catenin mutations and expression of Wnt-signaling target genes in hepatocellular carcinoma. Mol Cancer. 2008;7:21.
19. Barth AI, Nathke IS, Nelson WJ. Cadherins, catenins and APC protein: interplay between cytoskeletal complexes and signaling pathways. Curr Opin Cell Biol. 1997;9:683–90.
20. Ireton RC, Davis MA, van Hengel J, Mariner DJ, Barnes K, Thoreson MA, Anastasiadis PZ, Matrisian L, Bundy LM, Sealy L, Gilbert B, van Roy F, Reynolds AB. A novel role for p120 catenin in E-cadherin function. J Cell Biol. 2002;159:465–76.
21. Thoreson MA, Anastasiadis PZ, Daniel JM, Ireton RC, Wheelock MJ, Johnson KR, Hummingbird DK, Reynolds AB. Selective uncoupling of p120(ctn) from E-cadherin disrupts strong adhesion. J Cell Biol. 2000;148:189–202.
22. Garrod D, Chidgey M. Desmosome structure, composition and function. Biochim Biophys Acta. 2008;1778:572–87.
23. Lie PP, Cheng CY, Mruk DD. The biology of the desmosome-like junction a versatile anchoring junction and signal transducer in the seminiferous epithelium. Int Rev Cell Mol Biol. 2011;286:223–69.
24. Getsios S, Huen AC, Green KJ. Working out the strength and flexibility of desmosomes. Nat Rev Mol Cell Biol. 2004;5:271–81.
25. Taneyhill LA. To adhere or not to adhere: the role of Cadherins in neural crest development. Cell Adhes Migr. 2008;2:223–30.
26. Choi HJ, Gross JC, Pokutta S, Weis WI. Interactions of plakoglobin and beta-catenin with desmosomal cadherins: basis of selective exclusion of alpha- and beta-catenin from desmosomes. J Biol Chem. 2009;284:31776–88.
27. Lewis JE, Wahl JK III, Sass KM, Jensen PJ, Johnson KR, Wheelock MJ. Cross-talk between adherens junctions and desmosomes depends on plakoglobin. J Cell Biol. 1997;136:919–34.
28. Caldelari R, de Bruin A, Baumann D, Suter MM, Bierkamp C, Balmer V, Müller E. A central role for the armadillo protein plakoglobin in the autoimmune disease pemphigus vulgaris. J Cell Biol. 2001;153:823–34.
29. Ruiz P, Brinkmann V, Ledermann B, Behrend M, Grund C, Thalhammer C, Vogel F, Birchmeier C, Gunthert U, Franke WW, Birchmeier W. Targeted mutation of plakoglobin in mice reveals essential functions of desmosomes in the embryonic heart. J Cell Biol. 1996;135:215–25.
30. Spindler V, Dehner C, Hubner S, Waschke J. Plakoglobin but not desmoplakin regulates keratinocyte cohesion via modulation of p38MAPK signaling. J Invest Dermatol. 2014;134:1655–64.
31. Todorovic V, Desai BV, Patterson MJ, Amargo EV, Dubash AD, Yin T, Jones JC, Green KJ. Plakoglobin regulates cell motility through Rho- and fibronectin-dependent Src signaling. J Cell Sci. 2010;123:3576–86.
32. Reynolds AB. p120-catenin: past and present. Biochim Biophys Acta. 2007;1773:2–7.
33. Thoreson MA, Reynolds AB. Altered expression of the catenin p120 in human cancer: implications for tumor progression. Differentiation. 2002;70:583–9.
34. Yap AS, Niessen CM, Gumbiner BM. The juxtamembrane region of the cadherin cytoplasmic tail supports lateral clustering, adhesive strengthening, and interaction with p120ctn. J Cell Biol. 1998;141:779–89.

35. Noren NK, Liu BP, Burridge K, Kreft B. p120 catenin regulates the actin cytoskeleton via Rho family GTPases. J Cell Biol. 2000;150:567–80.
36. Yap AS, Crampton MS, Hardin J. Making and breaking contacts: the cellular biology of cadherin regulation. Curr Opin Cell Biol. 2007;19:508–14.
37. Davis MA, Ireton RC, Reynolds AB. A core function for p120-catenin in cadherin turnover. J Cell Biol. 2003;163:525–34.
38. Miyashita Y, Ozawa M. Increased internalization of p120-uncoupled E-cadherin and a requirement for a dileucine motif in the cytoplasmic domain for endocytosis of the protein. J Biol Chem. 2007;282:11540–8.
39. Peifer M, Yap AS. Traffic control: p120-catenin acts as a gatekeeper to control the fate of classical cadherins in mammalian cells. J Cell Biol. 2003;163:437–40.
40. Xiao K, Allison DF, Buckley KM, Kottke MD, Vincent PA, Faundez V, Kowalczyk AP. Cellular levels of p120 catenin function as a set point for cadherin expression levels in microvascular endothelial cells. J Cell Biol. 2003;163:535–45.
41. Nanes BA, Chiasson-MacKenzie C, Lowery AM, Ishiyama N, Faundez V, Ikura M, Vincent PA, Kowalczyk AP. p120-catenin binding masks an endocytic signal conserved in classical cadherins. J Cell Biol. 2012;199:365–80.
42. Xiao K, Garner J, Buckley KM, Vincent PA, Chiasson CM, Dejana E, Faundez V, Kowalczyk AP. p120-catenin regulates clathrin-dependent endocytosis of VE-cadherin. Mol Biol Cell. 2005;16:5141–51.
43. Perez-Moreno M, Fuchs E. Catenins: keeping cells from getting their signals crossed. Dev Cell. 2006;11:601–12.
44. Kobielak A, Fuchs E. Alpha-catenin: at the junction of intercellular adhesion and actin dynamics. Nat Rev Mol Cell Biol. 2004;5:614–25.
45. Rangarajan ES, Izard T. Dimer asymmetry defines alpha-catenin interactions. Nat Struct Mol Biol. 2013;20:188–93.
46. Vasioukhin V, Bauer C, Degenstein L, Wise B, Fuchs E. Hyperproliferation and defects in epithelial polarity upon conditional ablation of alpha-catenin in skin. Cell. 2001;104:605–17.
47. Sato T, Fujita N, Yamada A, Ooshio T, Okamoto R, Irie K, Takai Y. Regulation of the assembly and adhesion activity of E-cadherin by nectin and afadin for the formation of adherens junctions in Madin-Darby canine kidney cells. J Biol Chem. 2006;281:5288–99.
48. Adams CL, Chen YT, Smith SJ, Nelson WJ. Mechanisms of epithelial cell-cell adhesion and cell compaction revealed by high-resolution tracking of E-cadherin-green fluorescent protein. J Cell Biol. 1998;142:1105–19.
49. Chen YT, Stewart DB, Nelson WJ. Coupling assembly of the E-cadherin/beta-catenin complex to efficient endoplasmic reticulum exit and basal-lateral membrane targeting of E-cadherin in polarized MDCK cells. J Cell Biol. 1999;144:687–99.
50. Aberle H, Butz S, Stappert J, Weissig H, Kemler R, Hoschuetzky H. Assembly of the cadherin-catenin complex in vitro with recombinant proteins. J Cell Sci. 1994;107(Pt 12):3655–63.
51. Huber O, Krohn M, Kemler R. A specific domain in alpha-catenin mediates binding to beta-catenin or plakoglobin. J Cell Sci. 1997;110(Pt 15):1759–65.
52. Vasioukhin V, Bauer C, Yin M, Fuchs E. Directed actin polymerization is the driving force for epithelial cell-cell adhesion. Cell. 2000;100:209–19.
53. Gumbiner BM. Regulation of cadherin adhesive activity. J Cell Biol. 2000;148:399–404.
54. Daniel JM, Reynolds AB. Tyrosine phosphorylation and cadherin/catenin function. BioEssays. 1997;19:883–91.
55. Behrens J, Vakaet L, Friis R, Winterhager E, van Roy F, Mareel MM, Birchmeier W. Loss of epithelial differentiation and gain of invasiveness correlates with tyrosine phosphorylation of the E-cadherin/beta-catenin complex in cells transformed with a temperature-sensitive v-SRC gene. J Cell Biol. 1993;120:757–66.
56. Matsuyoshi N, Hamaguchi M, Taniguchi S, Nagafuchi A, Tsukita S, Takeichi M. Cadherin-mediated cell-cell adhesion is perturbed by v-src tyrosine phosphorylation in metastatic fibroblasts. J Cell Biol. 1992;118:703–14.
57. Shibamoto S, Hayakawa M, Takeuchi K, Hori T, Oku N, Miyazawa K, Kitamura N, Takeichi M, Ito F. Tyrosine phosphorylation of beta-catenin and plakoglobin enhanced by hepatocyte growth factor and epidermal growth factor in human carcinoma cells. Cell Adhes Commun. 1994;1:295–305.
58. Kaibuchi K, Kuroda S, Fukata M, Nakagawa M. Regulation of cadherin-mediated cell-cell adhesion by the Rho family GTPases. Curr Opin Cell Biol. 1999;11:591–6.
59. Hall A. Rho GTPases and the actin cytoskeleton. Science. 1998;279:509–14.
60. Garrod DR, Merritt AJ, Nie Z. Desmosomal cadherins. Curr Opin Cell Biol. 2002;14:537–45.
61. Laxmidevi LB, Angadi PV, Pillai RK, Chandreshekar C. Aberrant beta-catenin expression in the histologic differentiation of oral squamous cell carcinoma and verrucous carcinoma: an immunohistochemical study. J Oral Sci. 2010;52:633–40.
62. Nelson WJ, Nusse R. Convergence of Wnt, beta-catenin, and cadherin pathways. Science. 2004;303:1483–7.
63. Wheelock MJ, Johnson KR. Cadherin-mediated cellular signaling. Curr Opin Cell Biol. 2003;15:509–14.
64. Thiery JP. Epithelial-mesenchymal transitions in tumour progression. Nat Rev Cancer. 2002;2:442–54.
65. Moh MC, Shen S. The roles of cell adhesion molecules in tumor suppression and cell migration: a new paradox. Cell Adhes Migr. 2009;3:334–6.
66. Shinmura K, Kohno T, Takahashi M, Sasaki A, Ochiai A, Guilford P, Hunter A, Reeve AE, Sugimura H, Yamaguchi N, Yokota J. Familial gastric cancer: clinicopathological characteristics, RER phenotype and germline p53 and E-cadherin mutations. Carcinogenesis. 1999;20:1127–31.

67. Wijnhoven BP, Dinjens WN, Pignatelli M. E-cadherin-catenin cell-cell adhesion complex and human cancer. Br J Surg. 2000;87:992–1005.
68. Fanjul-Fernandez M, Quesada V, Cabanillas R, Cadinanos J, Fontanil T, Obaya A, Ramsay AJ, Llorente JL, Astudillo A, Cal S, Lopez-Otin C. Cell-cell adhesion genes CTNNA2 and CTNNA3 are tumour suppressors frequently mutated in laryngeal carcinomas. Nat Commun. 2013;4:2531.
69. Culkins CC, Setzer SV. Spotting desmosomes: the first 100 years. J Invest Dermatol. 2007;127:E2–3.
70. Delva E, Tucker DK, Kowalczyk AP. The desmosome. Cold Spring Harb Perspect Biol. 2009;1:a002543.
71. Garrod DR, Berika MY, Bardsley WF, Holmes D, Tabernero L. Hyper-adhesion in desmosomes: its regulation in wound healing and possible relationship to cadherin crystal structure. J Cell Sci. 2005;118:5743–54.
72. Kimura TE, Merritt AJ, Garrod DR. Calcium-independent desmosomes of keratinocytes are hyper-adhesive. J Invest Dermatol. 2007;127:775–81.
73. Teh MT, Parkinson EK, Thurlow JK, Liu F, Fortune F, Wan H. A molecular study of desmosomes identifies a desmoglein isoform switch in head and neck squamous cell carcinoma. J Oral Pathol Med. 2011;40:67–76.
74. Brennan D, Hu Y, Joubeh S, Choi YW, Whitaker-Menezes D, O'Brien T, Uitto J, Rodeck U, Mahoney MG. Suprabasal Dsg2 expression in transgenic mouse skin confers a hyperproliferative and apoptosis-resistant phenotype to keratinocytes. J Cell Sci. 2007;120:758–71.
75. Harmon RM, Simpson CL, Johnson JL, Koetsier JL, Dubash AD, Najor NA, Sarig O, Sprecher E, Green KJ. Desmoglein-1/Erbin interaction suppresses ERK activation to support epidermal differentiation. J Clin Invest. 2013;123:1556–70.
76. Brown L, Wan H. Desmoglein 3: a help or a hindrance in cancer progression? Cancers. 2015;7:266–86.
77. Brown L, Waseem A, Cruz IN, Szary J, Gunic E, Mannan T, Unadkat M, Yang M, Valderrama F, O'Toole EA, Wan H. Desmoglein 3 promotes cancer cell migration and invasion by regulating activator protein 1 and protein kinase C-dependent-Ezrin activation. Oncogene. 2014;33:2363–74.
78. Tsang SM, Brown L, Gadmor H, Gammon L, Fortune F, Wheeler A, Wan H. Desmoglein 3 acting as an upstream regulator of Rho GTPases, Rac-1/Cdc42 in the regulation of actin organisation and dynamics. Exp Cell Res. 2012;318:2269–83.
79. Tsang SM, Brown L, Lin K, Liu L, Piper K, O'Toole EA, Grose R, Hart IR, Garrod DR, Fortune F, Wan H. Non-junctional human desmoglein 3 acts as an upstream regulator of Src in E-cadherin adhesion, a pathway possibly involved in the pathogenesis of pemphigus vulgaris. J Pathol. 2012;227:81–93.
80. Tsang SM, Liu L, Teh MT, Wheeler A, Grose R, Hart IR, Garrod DR, Fortune F, Wan H. Desmoglein 3, via an interaction with E-cadherin, is associated with activation of Src. PLoS One. 2010;5:e14211.
81. Wan H, Lin K, Tsang SM, Uttagomol J. Evidence for Dsg3 in regulating Src signaling by competing with it for binding to caveolin-1. Data Brief. 2016;6:124–34.
82. Zhurinsky J, Shtutman M, Ben-Ze'ev A. Differential mechanisms of LEF/TCF family-dependent transcriptional activation by beta-catenin and plakoglobin. Mol Cell Biol. 2000;20:4238–52.
83. North AJ, Bardsley WG, Hyam J, Bornslaeger EA, Cordingley HC, Trinnaman B, Hatzfeld M, Green KJ, Magee AI, Garrod DR. Molecular map of the desmosomal plaque. J Cell Sci. 1999;112(Pt 23):4325–36.
84. Chen J, Nekrasova OE, Patel DM, Klessner JL, Godsel LM, Koetsier JL, Amargo EV, Desai BV, Green KJ. The C-terminal unique region of desmoglein 2 inhibits its internalization via tail-tail interactions. J Cell Biol. 2012;199:699–711.
85. Watt FM, Mattey DL, Garrod DR. Calcium-induced reorganization of desmosomal components in cultured human keratinocytes. J Cell Biol. 1984;99:2211–5.
86. Chitaev NA, Leube RE, Troyanovsky RB, Eshkind LG, Franke WW, Troyanovsky SM. The binding of plakoglobin to desmosomal cadherins: patterns of binding sites and topogenic potential. J Cell Biol. 1996;133:359–69.
87. de Bruin A, Caldelari R, Williamson L, Suter MM, Hunziker T, Wyder M, Müller EJ. Plakoglobin-dependent disruption of the desmosomal plaque in pemphigus vulgaris. Exp Dermatol. 2007;16:468–75.
88. Muller EJ, Hunziker T, Suter MM. Keratin intermediate filament retraction is linked to plakoglobin-dependent signaling in pemphigus vulgaris. J Am Acad Dermatol. 2007;56:890–1.
89. Andl CD, Stanley JR. Central role of the plakoglobin-binding domain for desmoglein 3 incorporation into desmosomes. J Invest Dermatol. 2001;117:1068–74.
90. Miravet S, Piedra J, Miro F, Itarte E, García de Herreros A, Duñach M. The transcriptional factor Tcf-4 contains different binding sites for beta-catenin and plakoglobin. J Biol Chem. 2002;277:1884–91.
91. Yin T, Green KJ. Regulation of desmosome assembly and adhesion. Semin Cell Dev Biol. 2004;15:665–77.
92. Hatzfeld M. Plakophilins: multifunctional proteins or just regulators of desmosomal adhesion? Biochim Biophys Acta. 2007;1773:69–77.
93. Kottke MD, Delva E, Kowalczyk AP. The desmosome: cell science lessons from human diseases. J Cell Sci. 2006;119:797–806.
94. McGrath JA, McMillan JR, Shemanko CS, Runswick SK, Leigh IM, Lane EB, Garrod DR, Eady RA. Mutations in the plakophilin 1 gene result in ectodermal dysplasia/skin fragility syndrome. Nat Genet. 1997;17:240–4.
95. Mertens C, Hofmann I, Wang Z, Teichmann M, Sepehri CS, Schnolzer M, Franke WW. Nuclear particles containing RNA polymerase III complexes associated with the junctional plaque protein plakophilin 2. Proc Natl Acad Sci U S A. 2001;98:7795–800.

96. Chen X, Bonne S, Hatzfeld M, van Roy F, Green KJ. Protein binding and functional characterization of plakophilin 2. Evidence for its diverse roles in desmosomes and beta -catenin signaling. J Biol Chem. 2002;277:10512–22.
97. Green KJ, Simpson CL. Desmosomes: new perspectives on a classic. J Invest Dermatol. 2007;127:2499–515.
98. Acehan D, Petzold C, Gumper I, Sabatini DD, Muller EJ, Cowin P, Stokes DL. Plakoglobin is required for effective intermediate filament anchorage to desmosomes. J Invest Dermatol. 2008;128:2665–75.
99. Kowalczyk AP, Bornslaeger EA, Norvell SM, Palka HL, Green KJ. Desmosomes: intercellular adhesive junctions specialized for attachment of intermediate filaments. Int Rev Cytol. 1999;185:237–302.
100. Kowalczyk AP, Hatzfeld M, Bornslaeger EA, Kopp DS, Borgwardt JE, Corcoran CM, Settler A, Green KJ. The head domain of plakophilin-1 binds to desmoplakin and enhances its recruitment to desmosomes. Implications for cutaneous disease. J Biol Chem. 1999;274:18145–8.
101. Cabral RM, Wan H, Cole CL, Abrams DJ, Kelsell DP, South AP. Identification and characterization of DSPIa, a novel isoform of human desmoplakin. Cell Tissue Res. 2010;341:121–9.
102. Bornslaeger EA, Corcoran CM, Stappenbeck TS, Green KJ. Breaking the connection: displacement of the desmosomal plaque protein desmoplakin from cell-cell interfaces disrupts anchorage of intermediate filament bundles and alters intercellular junction assembly. J Cell Biol. 1996;134:985–1001.
103. Nekrasova O, Green KJ. Desmosome assembly and dynamics. Trends Cell Biol. 2013;23:537–46.
104. Burdett ID, Sullivan KH. Desmosome assembly in MDCK cells: transport of precursors to the cell surface occurs by two phases of vesicular traffic and involves major changes in centrosome and Golgi location during a Ca(2+) shift. Exp Cell Res. 2002;276:296–309.
105. Andersen NJ, Yeaman C. Sec3-containing exocyst complex is required for desmosome assembly in mammalian epithelial cells. Mol Biol Cell. 2010;21:152–64.
106. Godsel LM, Hsieh SN, Amargo EV, Bass AE, Pascoe-McGillicuddy LT, Huen AC, Thorne ME, Gaudry CA, Park JK, Myung K, Goldman RD, Chew TL, Green KJ. Desmoplakin assembly dynamics in four dimensions: multiple phases differentially regulated by intermediate filaments and actin. J Cell Biol. 2005;171:1045–59.
107. Grossmann KS, Grund C, Huelsken J, Behrend M, Erdmann B, Franke WW, Birchmeier W. Requirement of plakophilin 2 for heart morphogenesis and cardiac junction formation. J Cell Biol. 2004;167:149–60.
108. Green KJ, Getsios S, Troyanovsky S, Godsel LM. Intercellular junction assembly, dynamics, and homeostasis. Cold Spring Harb Perspect Biol. 2010;2:a000125.
109. Aoyama Y, Yamamoto Y, Yamaguchi F, Kitajima Y. Low to high Ca^{2+}-switch causes phosphorylation and association of desmocollin 3 with plakoglobin and desmoglein 3 in cultured keratinocytes. Exp Dermatol. 2009;18:404–8.
110. Sheu HM, Kitajima Y, Yaoita H. Involvement of protein kinase C in translocation of desmoplakins from cytosol to plasma membrane during desmosome formation in human squamous cell carcinoma cells grown in low to normal calcium concentration. Exp Cell Res. 1989;185:176–90.
111. van Hengel J, Gohon L, Bruyneel E, Vermeulen S, Cornelissen M, Mareel M, von Roy F. Protein kinase C activation upregulates intercellular adhesion of alpha-catenin-negative human colon cancer cell variants via induction of desmosomes. J Cell Biol. 1997;137:1103–16.
112. Bass-Zubek AE, Hobbs RP, Amargo EV, Garcia NJ, Hsieh SN, Chen X, Wahl JK III, Denning MF, Green KJ. Plakophilin 2: a critical scaffold for PKC alpha that regulates intercellular junction assembly. J Cell Biol. 2008;181:605–13.
113. Pasdar M, Krzeminski KA, Nelson WJ. Regulation of desmosome assembly in MDCK epithelial cells: coordination of membrane core and cytoplasmic plaque domain assembly at the plasma membrane. J Cell Biol. 1991;113:645–55.
114. Pasdar M, Nelson WJ. Regulation of desmosome assembly in epithelial cells: kinetics of synthesis, transport, and stabilization of desmoglein I, a major protein of the membrane core domain. J Cell Biol. 1989;109:163–77.
115. Amagai M, Ishii K, Hashimoto T, Gamou S, Shimizu N, Nishikawa T. Conformational epitopes of pemphigus antigens (Dsg1 and Dsg3) are calcium dependent and glycosylation independent. J Invest Dermatol. 1995;105:243–7.
116. Rotzer V, Hartlieb E, Vielmuth F, Gliem M, Spindler V, Waschke J. E-cadherin and Src associate with extradesmosomal Dsg3 and modulate desmosome assembly and adhesion. Cell Mol Life Sci. 2015;72:4885–97.
117. Awad MM, Dalal D, Tichnell C, James C, Tucker A, Abraham T, Spevak PJ, Calkins H, Judge DP. Recessive arrhythmogenic right ventricular dysplasia due to novel cryptic splice mutation in PKP2. Hum Mutat. 2006;27:1157.
118. Bolling MC, Jonkman MF. Skin and heart: une liaison dangereuse. Exp Dermatol. 2009;18:658–68.
119. Simpson MA, Mansour S, Ahnood D, Kalidas K, Patton MA, McKenna WJ, Behr ER, Crosby AH. Homozygous mutation of desmocollin-2 in arrhythmogenic right ventricular cardiomyopathy with mild palmoplantar keratoderma and woolly hair. Cardiology. 2009;113:28–34.
120. Kljuic A, Bazzi H, Sundberg JP, Martinez-Mir A, O'Shaughnessy R, Mahoney MG, Levy M, Montagutelli X, Ahmad W, Aita VM, Gordon D, Uitto J, Whiting D, Ott J, Fischer S, Gilliam TC,

Jahoda CA, Morris RJ, Panteleyev AA, Nguyen VT, Christiano AM. Desmoglein 4 in hair follicle differentiation and epidermal adhesion: evidence from inherited hypotrichosis and acquired pemphigus vulgaris. Cell. 2003;113:249–60.

121. Brooke MA, Nitoiu D, Kelsell DP. Cell-cell connectivity: desmosomes and disease. J Pathol. 2012;226:158–71.
122. Broussard JA, Getsios S, Green KJ. Desmosome regulation and signaling in disease. Cell Tissue Res. 2015;360:501–12.
123. Thomason HA, Scothern A, McHarg S, Garrod DR. Desmosomes: adhesive strength and signalling in health and disease. Biochem J. 2010;429:419–33.
124. Amagai M, Matsuyoshi N, Wang ZH, Andl C, Stanley JR. Toxin in bullous impetigo and staphylococcal scalded-skin syndrome targets desmoglein 1. Nat Med. 2000;6:1275–7.
125. Anhalt GJ. Paraneoplastic pemphigus. J Investig Dermatol Symp Proc. 2004;9:29–33.
126. Stanley JR, Amagai M. Pemphigus, bullous impetigo, and the staphylococcal scalded-skin syndrome. N Engl J Med. 2006;355:1800–10.
127. Berkowitz P, Hu P, Liu Z, Diaz LA, Enghild JJ, Chua MP, Rubenstein DS. Desmosome signaling. Inhibition of p38MAPK prevents pemphigus vulgaris IgG-induced cytoskeleton reorganization. J Biol Chem. 2005;280:23778–84.
128. Gliem M, Heupel WM, Spindler V, Harms GS, Waschke J. Actin reorganization contributes to loss of cell adhesion in pemphigus vulgaris. Am J Physiol Cell Physiol. 2010;299:C606–13.
129. Kitajima Y. 150(th) anniversary series: desmosomes and autoimmune disease, perspective of dynamic desmosome remodeling and its impairments in pemphigus. Cell Commun Adhes. 2014;21:269–80.
130. Chidgey M, Dawson C. Desmosomes: a role in cancer? *Br*. J Cancer. 2007;96:1783–7.
131. Dusek RL, Attardi LD. Desmosomes: new perpetrators in tumour suppression. Nat Rev Cancer. 2011;11:317–23.
132. Brennan D, Mahoney MG. Increased expression of Dsg2 in malignant skin carcinomas: a tissue-microarray based study. Cell Adhes Migr. 2009;3:148–54.
133. Breuninger S, Reidenbach S, Sauer CG, Strobel P, Pfitzenmaier J, Trojan L, Hofmann I. Desmosomal plakophilins in the prostate and prostatic adenocarcinomas: implications for diagnosis and tumor progression. Am J Pathol. 2010;176:2509–19.
134. Chen YJ, Chang JT, Lee L, Wang HM, Liao CT, Chiu CC, Chen PJ, Cheng AJ. DSG3 is overexpressed in head neck cancer and is a potential molecular target for inhibition of oncogenesis. Oncogene. 2007;26:467–76.

Immunology of the Oral Mucosa

4

Lesley Ann Bergmeier

4.1 Introduction

The oral mucosa is defined as a mucous membrane that is continuous with the skin at the lips, and more importantly continuous with the pharyngeal mucosa and the gastrointestinal mucosa. While the oral mucosa shares many features with the skin and gastrointestinal mucosa it has many unique features that enable this sophisticated tissue to act as a gatekeeper controlling the effects of both inhaled and ingested antigens and the levels of inflammation and immune responses that are permitted in a normal healthy oral cavity. The structure, function and barrier of the oral mucosa have been described in detail in Chap. 1 (Cruchley & Bergmeier). In Chap. 7 (Saloom & Carpenter) the contribution of the saliva and gingival crevicular fluid to oral mucosal homeostasis is described and Chap. 5 (Allam & Novak) elaborates on the homeostasis of the mucosa.

Most invaders access the body via external surfaces and the oral mucosa is exposed to a huge antigenic challenge in the form of ingested food and the microbes that make up the commensal oral flora. It has been estimated that >1000 kg of nutrients will pass through the adult gut per year and more than 700 different species colonise the oral cavity. Microbes, necrotic cells and hypoxia initiate inflammatory responses which, depending on the duration and severity, may result in clinical or pathological manifestations.

The purpose of this chapter is to introduce the concept of the **oral immune system**, to describe the functions of immune cells within the oral cavity and to place this highly sophisticated network in the context of health and disease.

The main structural features of the oral mucosa are the oral epithelium, lamina propria and submucosa. The oral epithelium is described as a stratified squamous (in some areas keratinised) epithelium and contains multiple layers of cells with different morphologies arranged into discrete layers. The oral mucosa undergoes two distinct patterns of maturation resulting in the keratinised epithelium of the hard pallet and the gingivae and the non-keratinised epithelium of the sublingual and buccal mucosa [1]. The distinction of these tissues is important in understanding the differential immune responses that are possible within the oral cavity.

The major lymphoid organs are the tonsils and adenoids, making up **Waldeyer's ring** and there are numerous lymph nodes draining the head and neck that contribute to the immune function of the oral immune system (Fig. 4.1).

Mammals have evolved a sophisticated innate and adaptive immune system that integrates this network of tissues, cells and effector molecules and protects the body from disease by recognition of potential pathogens or diseased tissues (Fig. 4.2).

L.A. Bergmeier
Centre for Oral Immunology and Regenerative Medicine, Institute of Dentistry, Queen Mary School of Medicine and Dentistry, London, UK
e-mail: l.a.bergmeier@qmul.ac.uk

L.A. Bergmeier (ed.), *Oral Mucosa in Health and Disease*,
https://doi.org/10.1007/978-3-319-56065-6_4

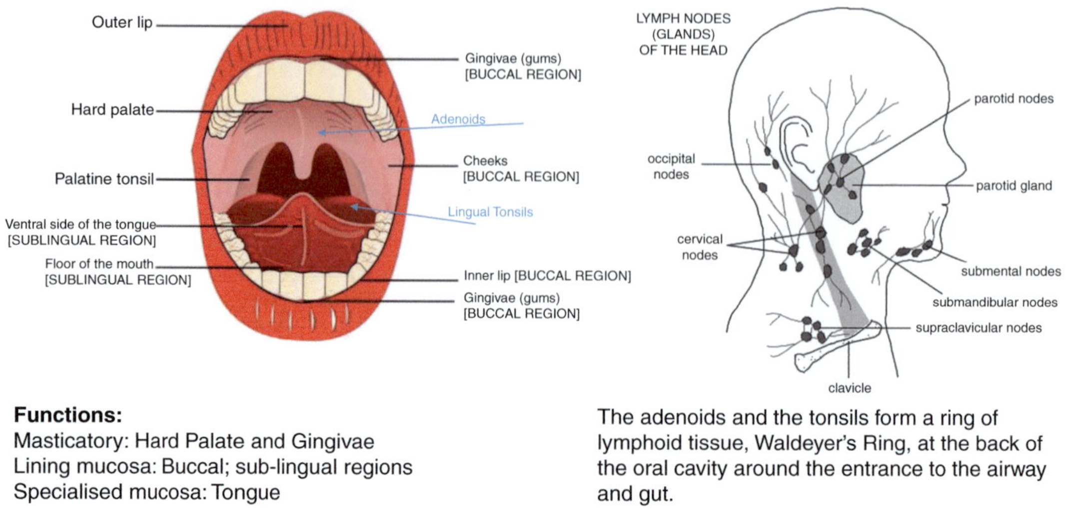

Fig. 4.1 The anatomy of the oral cavity and the draining lymph nodes (Adapted from [1] with permission of the Publishers)

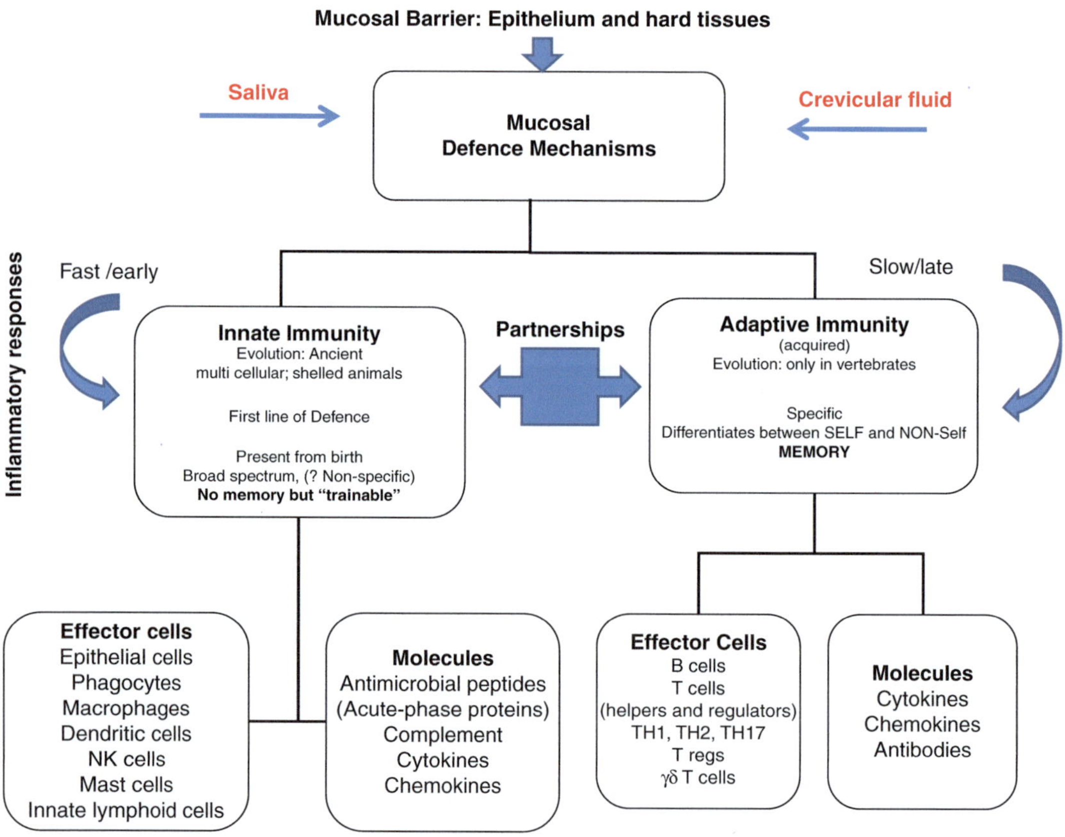

Fig. 4.2 The oral mucosal barrier. The partnerships of innate and adaptive responses. Cells and molecules in Innate and Adaptive immune responses interact with the epithelial tissues providing cells and molecules that contribute to protection but can also contribute to disease when the system becomes dysregulated

Discrimination between self, "non-self" and damaged or altered self is key to the immune responses in health and disease. The exquisite specificity of the adaptive immune response depends on the ability of T and B cells to rearrange the chains of their antigen receptors in order to recognise unique peptide antigens as they are presented to them on the MHC molecules of antigen-presenting cells. The molecular mechanisms of this "generation of diversity" are beyond the scope of this chapter but are the basis of the specificity of the adaptive immune response.

The epithelium is a physical barrier that prevents access of pathogenic organisms into underlying tissues. The presence of mucous and the desquamate nature of the epithelium also ensure that entrapped or adherent microorganisms are propelled by swallowing and peristalsis into the gut and destroyed.

The immune system is traditionally divided into the **systemic immune response**, an enclosed system protecting the systemic tissues, and includes the blood circulation and lymphatic drainage through which cells migrate to areas of inflammation and/or injury to effect destruction of invading organisms or to initiate repair mechanisms (Smith: Chap. 6).

The oral cavity is part of the **mucosal immune system** and is continuous in, but distinct from, the gut. The oral mucosa has a total surface area of about 0.8 m^2 compared with about 400 m^2 in the gut and about 2 m^2 of skin.

The **mucosal immune system** is open to the environment and constantly bathed in fluid (saliva). Antigenic material entering the oral cavity will be partially broken down by the masticatory process and by salivary enzymes but has the potential of inducing immune responses and early work on the development of vaccines against dental caries showed that ingestion of *Strep. mutans* gave rise to the induction of IgA antibodies not only in the saliva but also in tears. This introduced the concept of a **common mucosal immune system** where induction of immune responses at one mucosal surface gave rise to a widely disseminated immune response into other mucosal tissues [2, 3]. Later the compartmentalisation of the common mucosal immune system was redefined in the context of vaccine development [4] and it was established that mucosal administration of antigens induces specific antibodies in external secretions but a specific unresponsiveness in the systemic tissues, known as oral tolerance [5].

While it is convenient to regard the systemic and mucosal immune systems as separate entities it is clear from vaccine studies and the work on oral tolerance that there is continual crosstalk between these systems. In the oral cavity, this is highlighted by the exudation of serum components through the gingival sulcus and the active secretion of antibodies and effector molecules from the salivary glands into the oral cavity.

4.2 Inflammation and the Immune Response: Defending the Barricades

Inflammation is a protective response designed to rid the body of invading microorganisms that cause tissue injury, but also to deal with the consequences of injury such as necrosis. The inflammatory process is driven and controlled by cytokines such as interleukin-1β (IL-1β), IL-6 and tumour necrosis factor α (TNFα). Phagocytic cells such as neutrophils and macrophages (as well as mast cells and eosinophils) are recruited into inflamed tissues where they can destroy invading microorganisms.

Leucocytes can home rapidly to sites of infection or tissue damage while plasma proteins can diffuse into tissues [6].

Inflammation can be acute or chronic depending on the stimulus. Acute inflammation is rapid and usually of short duration, and is regarded as part of the **innate immune response** and characterised by neutrophil infiltration (Fig. 4.3).

Inflammation is usually terminated once the injurious agent is eliminated and anti-inflammatory mechanisms are activated that limit the response, preventing tissue damage and initiating tissue repair. Chronic inflammation is characterised by mononuclear cell infiltration such as macrophages,

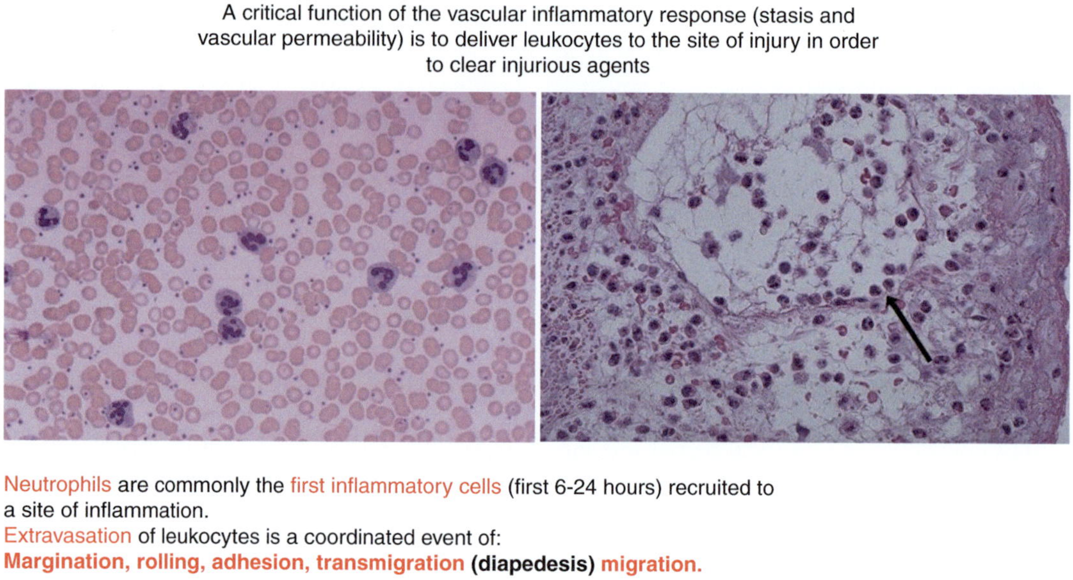

Fig. 4.3 Sequentail migration of Neutrophils and Leucocytes into sites of inflammation

lymphocytes and plasma cells. It can follow acute inflammation or arise de novo and represents a failure of resolution of the acute response resulting in fibrosis and tissue destruction. Chronic inflammatory responses are known to be associated with disease such as rheumatoid arthritis, atherosclerosis as well as life-threatening hypersensitivity reactions. Inflammation has been recognised as a key component of tumour progression where cytokines play an influential role in the development of malignant phenotypes in oral (and other) cancers. The overlap between host-pathogen interactions in the context of periodontal disease and cancer was recently reviewed [7].

Innate responses are present from birth, are a first line of defence and respond rapidly (Fig. 4.2). There are many soluble factors that are secreted by epithelial cells or secreted into the fluids that bathe the oral cavity and have a variety of functions such as the antimicrobial properties of defensins, histatins, lysozyme and lactoferrin. Some of these molecules can interact in a synergistic manner with secretory IgA (SIgA: Table 4.1). High-molecular-weight agglutinins are able to form heterotypic associations and in particular gp340 in human saliva can bind HIV-1 gp120 and prevent infection [8], whereas the same glycoprotein in genital tract secretions promotes infection of target cells (CD4⁺T cells) at normally sub-infectious doses [9].

Table 4.1 Antimicrobial properties of saliva

Antimicrobial agent	Activity
Secretory IgA	Major antibody in saliva. Inhibits adherence; Agglutinates bacteria; Virus neutralisation
Lactoferrin	Bacteriostatic-Iron binding
Lysozyme	Destroys the outer membrane of bacteria (effective against *S. mutans*)
Agglutinins	Glycoproteins, mucins; fibronectin; Histatins; Proline rich proteins, B-2 Microglobulin
Myeloperoxidase system	Bactericidal in the presence of thiocyanate/halide H_2O_2
Salivary peroxidase system	Enzyme-thiocyanate-halide H_2O_2
Complement	C3 in GCF
Leucocytes	Cells in gingival sulcus: >98% Neutrophils, but 50% not Phagocytic.

In the last decade, our understanding of the innate immune has undergone a paradigm shift with the identification of several different cell types that orchestrate the innate immune response and interface with the adaptive immune response. As one paper put it "Innate lymphoid cells-how did we miss them"? [10].

Innate cells protect against invading pathogens by the recognition of a wide range of pathogen-associated molecular patterns (PAMPS) or damage-associated molecular patterns (DAMPS) in diseased or damaged cells and induce an inflammatory response to which cells of the **adaptive** immune response are also recruited. Genetically encoded **pattern recognition receptors** (PRRs) present on epithelial cells activate the innate immune system which in turn "educates" adaptive cells which migrate to the draining lymph nodes where the induction of a robust immune response is initiated. Polymorphisms in these molecules (TLRs, NLRs and RIGI receptors) are known to be associated with inflammatory mucosal disease [11]. PRRs include the Toll-like receptors which recognise a wide range of bacterial and viral antigens and have been shown to exhibit splice variants in diseases such as Behçet's disease which might result in aberrant signalling and induction of chronic inflammation characteristic of this disease [12]. In oral lichen planus (OPL) TLR signalling has also been shown to be defective [13].

The **adaptive** immune response by contrast is slower, occurs later and is dependent on information provided by the innate immune system for full activation. This usually occurs through presentation of antigenic material, usually in the form of peptides that are processed by digestion within **antigen-presenting cells** (APCs) and expressed on their surface bound to either MHC class I or class II molecules.

APCs acting at the interface of the innate and adaptive responses include Langerhans cells (LCs) and dendritic cells (DCs), macrophages and other phagocytic cells such as neutrophils. DCs and Langerhans cells in the mucosal epithelium extend "fingerlike" projections that penetrate from the lamina propria right through the epithelial layer and "sample" substances in the lumen of the oral cavity [14, 15].

As we have learned more about the different immune responses in the mucosal system the relative thickness of the different mucosal compartments has become relevant to vaccine design and to the potential for infection. This has been especially true for HIV where the very thin rectal mucosa is easily disrupted while the oral mucosa has shown little or no infection and infection through the vagina/cervical route is often because of ongoing inflammation which brings target cells ($CD4^+$ T cells) to the mucosa (Fig. 4.4) [16].

APCs control the adaptive immune response and maintain homeostasis [17, 18]. The phenotype of these cells is different depending on whether they are isolated from buccal, sublingual or gingival mucosa while their function is affected by risk factors for oral pathologies such as age, alcohol consumption and smoking [19–21] resulting in changes to proteins that are transcribed (the transcriptome) and by definition alter the homeostasis of the oral mucosa.

Following activation, clones of antigen-specific effector cells migrate from the lymph

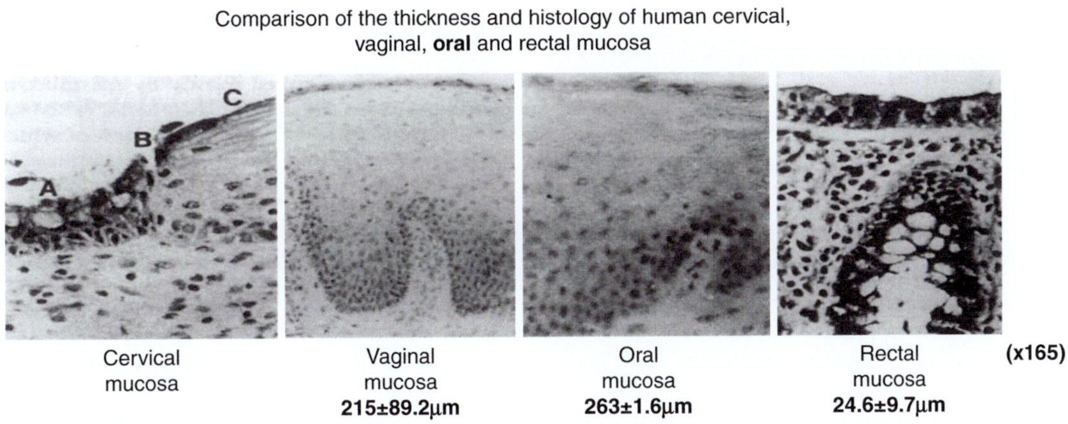

Fig. 4.4 Relative thickness of different mucosal epithelia

node to the site of infection/inflammation and carry out effector functions such as secretion of specific antibody (B cells) for the clearance of bacterial infections; induction of cytotoxic CD8+ T cells which will kill viral (or intracellular bacteria) infected cells or damaged or dying cells; cytotoxicity occurs either directly (through the action of CD8+ T cells) or with the help of antibodies (through the action of NK cells in antibody-dependent cellular cytotoxicity—ADCC). NK cells circulate and migrate to tissues where they carry out both effector and regulatory functions through the secretion of cytokines. In Behçet's disease there is a depletion in the circulating NK cells which have probably migrated to tissues where autoinflammatory reactions are driving pathology [22].

4.3 Interface of Innate and Adaptive Immunity

In recent years, as new cytokines and effector functions have been recognised in both T and B cells, there has been a blurring of the strict division between the innate and adaptive immune responses [23]. B cells with innate-like functions are found at mucosal epithelial barriers and can make "natural antibodies" that recognise bacterial carbohydrates and phospholipids. These cells express TLRs that can be activated by many bacterial antigens. The antibodies tend to be of low affinity but can certainly protect the host against bacterial pathogens early in infection [24].

A subset of T cells has been described with a very restricted T cell receptor that has limited ability to rearrange its β-chains and therefore has restricted specificity. These cells have a cytotoxic capacity and have been designated invariant natural killer T cells (iNKT) and recognise lipids in the context of unusual MHC molecule, CD1d [25].

Mucosal associated invariant T cells (MAIT) are another type of innate cell restricted in their ability to rearrange their T cell receptors and therefore have restricted specificities [26]. These cells appear to recognise an MHC-related molecule, MR1, which binds to riboflavin metabolites produced by bacteria and fungi. To date there is little or no literature documenting their distribution in human oral mucosal tissues as most studies have concentrated on the gut or in mouse models of human diseases.

In addition to conventional T cells bearing the αβ T cell receptor, a second type of T cell was recognised in the 1980s which carries the **γδ** T cell receptor, recognises non-protein antigens and homes to mucosal tissues [27–29]. These cells have been associated with lesions in oral mucosal diseases such as Behçet's disease [30–32]. The expansion of γδ T cells is driven by non-peptide antigens such as phosphoantigens, many of which are generated by oral commensal microorganisms and indeed by heat-shock proteins [33–35]. Levels of the γδ T cell mRNA were shown to decrease in mucosal sites but increase in lymphoid tissues after experimental oral infection with Simian immunodeficiency virus (SIV) [36].

These atypical cells are multifunctional with characteristics of T cells, NK cells and antigen-presenting cells. In vitro studies have suggested that activated γδT cells expressed a repertoire of antigen presentation and co-stimulatory molecules and that the antigen-presenting phenotypes could prime αβ T cells to induce strong adaptive response [37, 38]. These cells are also capable of interacting with dendritic cells (DCs), directly regulating their function and mutually promoting maturation. Activated γδ T cells can produce high levels of cytokines which identify their effector roles in immune response and additionally provide a regulatory role. IL-17 is produced by γδ T cells and contributes to inflammation in the oral mucosa in response to candida infection [39].

These innate immune functions of T and B cells expand the repertoire of responses beyond the peptide responses associated with fully activated adaptive immune cells to include carbohydrates, lipids and phosphoantigens generated by microorganisms or cells under stress.

A recent systematic characterisation of the immune cell network at the gingival barrier in a large cohort of healthy individuals indicated a predominant number of T cells, minimal B cells and large numbers of granulocytes/neutrophils and a sophisticated network of antigen-presenting cells. A small number of **innate lymphoid cells** were also present [40]. However, in six untreated periodontitis patients, displaying severe bone loss and visible inflammation, there was a significant increase in IL-17 producing CD4$^+$ T helper cells.

Innate lymphoid cells are a relatively recent discovery and are key to barrier defence in skin, intestine and airways [41, 42]. These cells lack antigen-specific receptors which are characterised by both conventional T cells and B cells, and by the unconventional γδ T cells. To date three subsets have been identified: ILC-1, -2 and -3 [41]. These cells have a dual function of initiating innate responses but also maintaining homeostasis through inflammation and tissue repair mechanisms. They are important in pathogen clearance, lymphoid organogenesis and tissue remodelling. There are distinct differences in the transcription factors and cytokines that are used and produced in the three subsets. ILC1 uses the transcription factor, T-bet, and secretes IFN-γ. This class of cells also includes NK cells.

ILC-2 cells secrete IL-5 and IL-13 through RORα activation, and express the GATA3 transcription factor, while ILC3 utilises the RORγt transcription factor and secretes pro-inflammatory cytokines such as IL-17 (and/or IL-22). There have been several recent studies proposing immune-regulatory mechanisms for ILCs in different tissues under different inflammatory pressures [43, 44]. To date there is very little in the literature examining the role of ILCs in the oral mucosa but a recent paper on surgical repair of cleft lip identified a significantly higher level of ILC2 cells than the other ILCs suggesting that their regulatory function might contribute to the low levels of surgical site infection [45, 46].

4.4 Antibody-Mediated Protection (B Cell Compartment)

The fluids that bathe the oral mucosa have two sources: direct secretion from the salivary glands (saliva) or as a serum exudate from the circulation into the gingival crevice (GCF).

Antibodies produced following immunisation or introduced by passive immunisation can be very effective against extracellular organisms and their products (toxins). Antibodies can block viruses and bacteria from entering and infecting host cells and can also mediate killing of pathogens. The ability of antibodies to neutralise toxins can prevent the damaging effects of infections such as *Diphtheria* or *Clostridium.* The type of antibody induced is important as IgG is primarily effective in blood whereas secretory IgA (SIgA) is the principal antibody associated with mucosal surfaces.

Vaccine strategies against dental caries have a long history with the *S. mutans* antigen I/II demonstrated to be effective in animal studies [47, 48]. Natural IgG antibodies to *S. mutans* were shown in serum and were associated with low or no caries [49] while passage of immunoglobulins from the serum to the GCF had previously been demonstrated in the rhesus macaque [50, 51].

Passive application of monoclonal antibodies specific for antigen I/II was successfully used in preventing recolonisation by *Strep. mutans* in the 1990s [52–54]. Monoclonal antibodies to periodontal pathogens had more limited success [55–57]. This is probably due to causality of periodontal disease being more associated with a pathological host response rather than a direct bacterial affect [58, 59].

Secretory IgA from salivary glands provides protection against adhesion of both invasive bacteria and viruses. The heterotypic functions of SIgA with lysozyme, etc. provide additional protection at the oral mucosa with antibody complexes anchored to the epithelium and antibody actively secreted into the lumen and also present intracellularly and in lamina propria (reviewed in [60, 61]).

In the oral cavity, IgG is present in the gingival crevicular fluid as a serum transudate, while IgA is actively secreted from the salivary glands (Fig. 4.5). Both these classes of antibody are important in both prevention of disease, e.g. caries, and in some cases as part of a pathogenic process such as IgG anti-Dsg3 in pemphigus vulgaris (see Chap. 10).

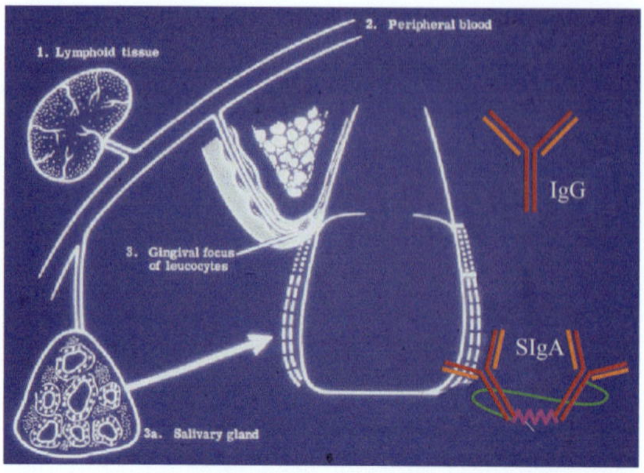

Fig. 4.5 The contribution of antibodies to the protection of the gingival marginal mucosa and tooth surface (reproduced with permission)

4.5 Cell-Mediated Immune Protection (T Cell Compartment)

The cell-mediated immune response is characterised by two distinct activities: **cytotoxic T cells** ($CD8^+$), able to kill cells infected with viruses or other intracellular pathogens including some bacterial, fungal and protozoan organisms, and **$CD4^+$ T helper cells** which are important activators of other T cells but also activate cells of the innate immune system, such as polymorphonuclear leucocytes (PMNs), natural killer cells and NKT cells, by their ability to secrete a wide range of cytokines, chemokines and growth factors. The help provided by $CD4^+$ T cells can provide maturation and survival signals for innate immune cells that prolong the innate response.

$CD4^+$ T helper cells are also key to the induction and maturation of antibody-secreting cells, the induction of class switching to the class of antibody most appropriate to the response required, and their ability to generate memory cells which migrate back to the draining lymph nodes.

The cytokines secreted by T helper cells are grouped according to the effector cells which they induce and the two basic categories were first identified by Mosmann et al. in 1986 [62] . The T helper 1 (Th1) cytokines (IL-2, IFNγ, TNFα) activate macrophages, induce B cells to class switch to IgG1or IgG3 and supress Th2 responses, while Th2 cytokines (IL-4, -5, -6, -10 and -13) activate B cells; induce class switching to IgG2, IgA or IgE; and supress the Th1 response. In recent years, more subsets of T helper cells have been identified and Th17 cells have been recognised as important contributors to the host response to periodontal pathogens and are characterised by the secretion of IL-17. The pro-inflammatory properties of IL-17 are important in the context of early immune responses to pathogens and here tissue-resident γδ T cells are important as they are a significant source of IL-17 [38]. However, continued signalling and IL-17 production are associated with autoimmune disease and cancer progression [63, 64].

This subset is related to Th1 but its induction is dependent on IL-21 and TGFβ. Another subset of CD4 helper cells are the T regulatory cells (T_{regs}) which play an important role in maintaining self-tolerance. They are also important in contributing to the limitation of immune responses—in other words when to shut down a response when that is no longer required (Fig. 4.6: [65]).

Dendritic cells were shown to polarise effector responses towards TH1 (inflammatory) or TH2 (anti-inflammatory) cytokine secretion depending on the type of endotoxin encountered [66–69]. The effect of *P. gingivalis* LPS on DCs demonstrated that suboptimal maturation of DCs occurs with the result that IL-10 (an immunosuppressive/TH2 cytokine) tends to be induced (reviewed in [70]). Dental pulp DCs have been shown to migrate to regional lymph nodes where they induce adaptive immune responses to cariogenic organisms [71]. Immature DCs are thought to maintain a tolerogenic environment in the oral mucosa; thus it is often thought of as an immune-privileged site [72–76]. However, mature DCs can drive potent immune responses.

Memory cells are induced which recirculate back to the draining lymph nodes where they will wait, "armed" for the next exposure to their "cognate" antigen.

Activated "effector" T cells are capable of secreting substances such as cytokines and enzymes that will passively diffuse into the local environment. These potentially destructive effector mechanisms require powerful regulation, and the immune system has evolved intricate feedback loops that limit the duration of responses, thus avoiding the potential for **bystander damage** resulting from prolonged inflammatory responses. This balance

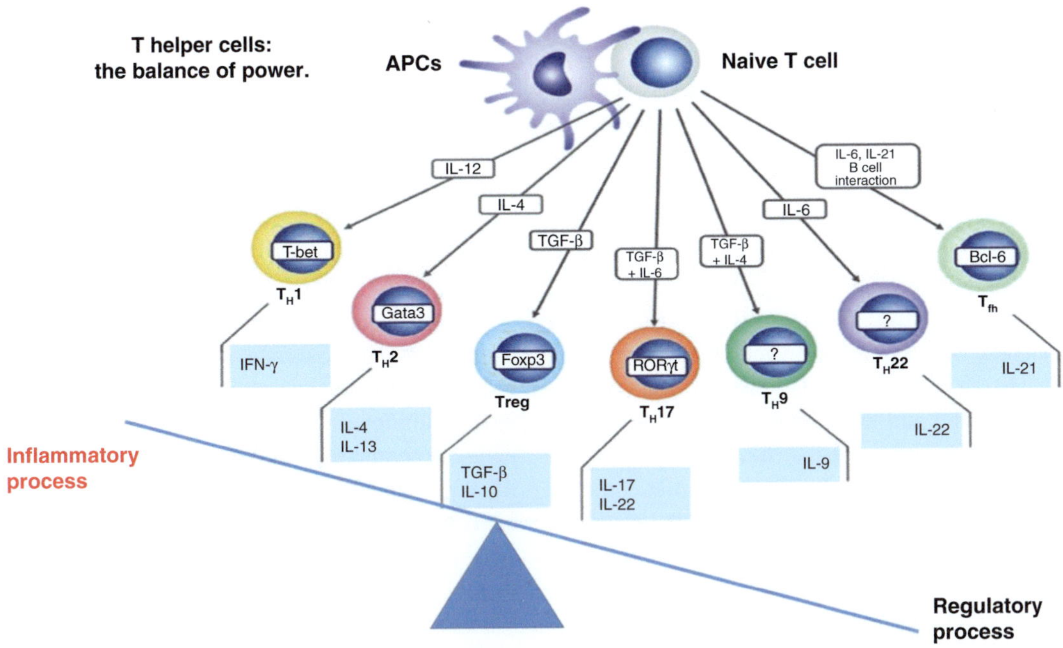

Fig. 4.6 The diversity of T helper cell responses. Key to regulation of inflammation and effector mechanisms of the immune response. Signalling pathways induced by APC/T cell interactions dictate T helper cell effector functions anad downstream pathogenesis or regulation. Cytokines and chemokines are important effector molecules in the oral mucosa influencing recruitment of other cells into the tissue and orchestrating the balance between inflammation anad regulation as well as tissue breakdown and repair (adapted from [87] with permission)

between inflammation and regulation is dictated by the induction of specific effector T cells responding to unique signalling and transcription pathways resulting in the production of cytokines able to either regulate the inflammation or drive specific functions (Fig. 4.6); for example production of IL-4 and IL-13 by Th2 cells supports antibody production while Th17 cells secrete IL-17 which has been associated with autoimmune diseases and also the exacerbation of periodontal disease, once dysbiosis has been established. Regulatory molecules such as the suppressor of cytokine signalling family (SOCS) have also been shown to be upregulated in conditions such as Behçet's and Sjögren's syndromes [77, 78].

Mucosal disease is frequently caused by immune deficiency or dysregulation. Alterations in T cell activity are frequently associated with disease. A classic example is the loss of T helper cells in HIV predisposing patients to opportunistic infections such as candidiasis and necrotising periodontitis, characteristic of AIDS. Patients with hyper IgE syndrome have mutations in the STAT3 transcription factors controlling the development of Th17 cells [79, 80]. Although Th17 responses have been shown to be important in responses to fungi, it has been suggested that uncontrolled Th17 responses lead to chronic inflammation and autoimmunity [81–84]. T helper cells play a key role in recruiting neutrophils and osteoclasts into periodontal lesions [85, 86]. In OLP an intense lymphocyte infiltration into the lamina propria results in the destruction of the basal layer of the epithelium and is associated with T cells, which are the major contributors to this inflammation.

4.6 Inductive and Effector Sites in the Oral Cavity

The tissues of the mucosal immune system have been segregated into areas where responses are *induced* and those areas to which the resulting *effectors cells* migrate to carry out their functions.

Mucosal **inductive sites** include the nasal associated lymphoid tissues (NALT) (which drain into the back of the oral cavity) and the Peyer's patches of the gut (GALT), for the induction of mucosal secretory IgA (SIgA) antibody responses. Homing of memory and/or activated T and B cells from NALT to nasal passages and oral cavity occurs, and is part of the compartmentalisation of the immune system [4]. In the oral cavity, inductive sites consist of the buccal mucosa, salivary glands and Waldeyer's ring consisting of the adenoids (unpaired nasopharyngeal tonsils) and the paired palatine and lingual tonsils. Human tonsils have deep-branched antigen-retaining crypts with a reticular endothelium which contains M cells, a highly developed antigen capture cell which is vital for the induction of B cell diversity and memory. About 50% of the cells present in lymphoid follicles contain germinal centres where immune responses are induced. Palatine tonsils contain a significant sub-epithelial population that may be crucial in the production of antibody specific for inhaled antigens taken up by M cells. These cells act as a portal to the outside environment delivering antigens to the lymphoid cells in the lamina propria for the induction of antigen-specific immune responses.

Effector sites for oral mucosal immune responses include the epithelium, lamina propria and salivary glands. There are also scattered intraepithelial lymphoid cells throughout the mucosa [76].

Epithelial cells, macrophages and dendritic cells all secrete a variety of cytokines that have profound effects on the recruitment of T cells to the oral mucosa and their ultimate effector function [87].

4.7 Tolerance in the Oral Mucosa

Oral tolerance via the **gut mucosa** has a long history in inducing a state of specific immunological unresponsiveness. Our ability to tolerate the vast array of food antigens is dependent on this phenomenon and has been exploited in experimental animals to induce tolerance to many antigens associated with autoimmune disease. The potential for oral tolerance to ameliorate allergic or autoimmune disease is an important area of research and has been successfully used in clinical trials (reviewed by Sun et al. [88]). Oral tolerance has been used to ameliorate the uveitis in Behçet's patients by oral administration of heat-shock protein peptides along with the mucosal adjuvant cholera toxin B subunit [89].

However, more recently the **oral mucosa** has been investigated as a site of tolerisation [76, 90]. Exploiting the tolerogenic function of the APCs in the sublingual mucosa has considerable therapeutic potential especially regarding allergy [75, 91–93]. The mechanisms involve the induction of T regulatory cells which secrete IL-10 and TGF-β. Along with inhibitory ligands expressed in the Tregs, such as CTLA-4, these cytokines limit the T helper cell responses that would normally drive immune responses to antigens applied to the mucosal surfaces. It has been suggested that intra-oral administration of peptides, prior to challenge with allergens, could limit T cell proliferation in the oral-pharyngeal lymph nodes in a mouse model of tree pollen allergy [94]. The evidence is building that this route of oral tolerance may indeed be superior to that of gut oral tolerance (Fig. 4.7).

4.8 Immunology in the Dental Clinic

The oral manifestations of both local and systemic disease make the oral cavity an important part of the immune system and it has frequently been suggested that the oral cavity can act as a mirror reflecting the homeostasis—or lack of it—that dictates health and disease. Table 4.2 is a reminder of the types of disease that are frequently seen in the mouth and the contribution of infection and dysregulation is further explored in Chap. 5 (Allam) and Chap. 10 (Bergmeier).

The complex nature of the interactions within the oral cavity is common to those throughout the entire immune system and we are beginning to unravel the connections that allow this normally quiescent tissue to become dysregulated and to exhibit both local and systemic diseases (Fig. 4.8).

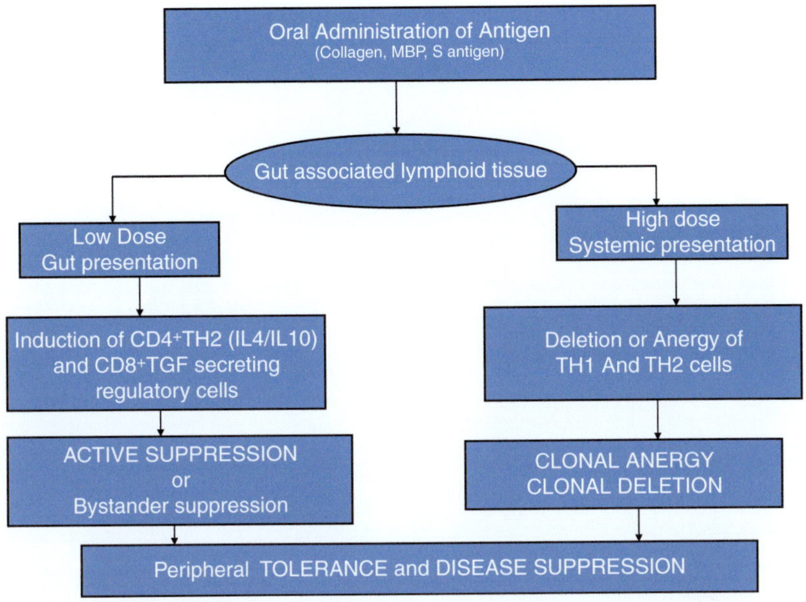

Fig. 4.7 Mechanisms of oral tolerance

Table 4.2 Immunology in the dental clinic. A range of organisms give rise to significant oral disease. Oral mucosal disease may also have systemic involvment and systemic disease have oral manifestations. The immune response is involved in either resolving disease manifestations or in exacerbating symptoms due to immune dysregulation

Infections	Oral mucosal disease	Systemic diseases with oral manifestations
• Bacteria • Carles • Periodcntal Disease • Fungi • Candidiasis • Viruses • Heroes viruses (HPV, HSV] • CMV, EBV • Enteroviruses • Measles • Scarlet fever • (HIV)	• Behcet's Disease* • Recurrent Aphthous stomatitis • Lichen Planus* • Pempriigus/Pemphigoid* • Erythema multiforrne*	• Celiac disease • Crohn's disease • Ulcerative colitis • Food allergy and oral tolerance • Sjögrens Syndrome • HIV

Immune responses to infective agents are important in resolution of infections but when dysregulated can give rise to changes in homeostatic environments that allow for disease development with or without systemic involvement. * Systemic involvment of diseases with oral origins

The term "interactome" has been coined to emphasise the interconnectedness of the mechanisms within biological systems and several studies have been carried out on the interaction of the salivary proteome and the microbiome in the oral cavity. The tools for studying these interactions are highly relevant to the interface between the immune responses in the oral mucosa and the ecosystem of the oral cavity [95–97].

While this has not been an exhaustive re-examination of the immunology of the oral cavity it is hoped that the new knowledge of the functional capacity of the different immune processes in the oral cavity repositions this unique tissue at the forefront of understanding the *gatekeeper* and *housekeeper* properties of the oral mucosa.

This mucosa is not just the entrance to the gastrointestinal tract and there is much still to learn about the immune functions which differ considerably from gut. Assumptions based on gut biology may not apply to the oral mucosa.

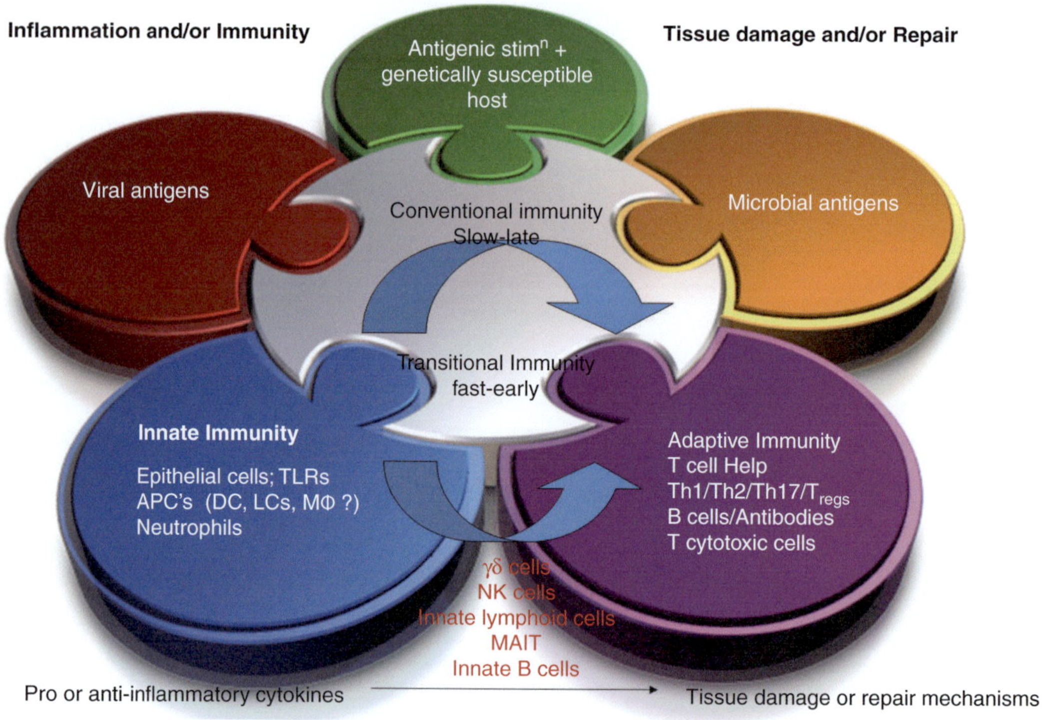

Fig. 4.8 The "interactome" of cells and molecules of the oral cavity

References

1. Kraan H, Vrieling H, Czerkinsky C, Jiskoot W, Kersten G, Amorij JP. Buccal and sublingual vaccine delivery. J Control Release. 2014;190:580–92.
2. Mestecky J, McGhee JR, Arnold RR, Michalek SM, Prince SJ, Babb JL. Selective induction of an immune response in human external secretions by ingestion of bacterial antigen. J Clin Invest. 1978a;61(3):731–7.
3. Mestecky J, McGhee JR, Michalek SM, Arnold RR, Crago SS, Babb JL. Concept of the local and common mucosal immune response. Adv Exp Med Biol. 1978b;107:185–92.
4. Holmgren J, Czerkinsky C. Mucosal immunity and vaccines. Nat Med. 2005;11(4 Suppl):S45–53.
5. Mestecky J, Moldoveanu Z, Elson CO. Immune response versus mucosal tolerance to mucosally administered antigens. Vaccine. 2005;23(15):1800–3.
6. Ley K, Laudanna C, Cybulsky MI, Nourshargh S. Getting to the site of inflammation: the leukocyte adhesion cascade updated. Nat Rev Immunol. 2007;7(9):678–89.
7. Sahingur SE, Yeudall WA. Chemokine function in periodontal disease and oral cavity cancer. Front Immunol. 2015;6:214.
8. Wu Z, Lee S, Abrams W, Weissman D, Malamud D. The N-terminal SRCR-SID domain of gp-340 interacts with HIV type 1 gp120 sequences and inhibits viral infection. AIDS Res Hum Retrovir. 2006;22(6):508–15.
9. Stoddard E, Cannon G, Ni H, Kariko K, Capodici J, Malamud D, Weissman D. gp340 expressed on human genital epithelia binds HIV-1 envelope protein and facilitates viral transmission. J Immunol. 2007;179(5):3126–32.
10. Walker JA, Barlow JL, McKenzie AN. Innate lymphoid cells--how did we miss them? Nat Rev Immunol. 2013;13(2):75–87.
11. Lavelle EC, Murphy C, O'Neill LAJ, Creagh EM. The role of TLRs, NLRs, and RLRs in mucosal innate immunity and homeostasis. Mucosal Immunol. 2010;3(1):17–28.
12. Seoudi N, Bergmeier LA, Hagi-Pavli E, Bibby D, Curtis MA, Fortune F. The role of TLR2 and 4 in Behçet's disease pathogenesis. Innate Immun. 2014;20(4):412–22.
13. Sinon SH, Rich AM, Parachuru VP, Firth FA, Milne T, Seymour GJ. Downregulation of toll-like receptor-mediated signalling pathways in oral lichen planus. J Oral Pathol Med. 2016;45(1):28–34.
14. Allen F, Tong AA, Huang AY. Unique transcompartmental bridge: antigen-presenting cells sampling across endothelial and mucosal barriers. Front Immunol. 2016;7:231.
15. Hussain LA, Lehner T. Comparative investigation of Langerhans' cells and potential receptors for HIV in oral, genitourinary and rectal epithelia. Immunology. 1995;85(3):475–84.

16. Lehner T, Bergmeier L, Wang Y, Tao L, Mitchell E. A rational basis for mucosal vaccination against HIV infection. Immunol Rev. 1999;170:183–96.
17. Hovav A-H. Dendritic cells of the oral mucosa. Mucosal Immunol. 2014;7(1):27–37.
18. Palm NW, Medzhitov R. Pattern recognition receptors and control of adaptive immunity. Immunol Rev. 2009;227(1):221–33.
19. Boyle JO, Gümüs ZH, Kacker A, Choksi VL, Bocker JM, Zhou XK, Yantiss RK, Hughes DB, Du B, Judson BL, Subbaramaiah K, Dannenberg AJ. Effects of cigarette smoke on the human oral mucosal transcriptome. Cancer Prev Res (Phila). 2010;3(3):266–78.
20. Cruchley AT, Williams DM, Farthing PM, Lesch CA, Squier CA. Regional variation in Langerhans cell distribution and density in normal human oral mucosa determined using monoclonal antibodies against CD1, HLADR, HLADQ and HLADP. J Oral Pathol Med. 1989;18(9):510–6.
21. Cruchley AT, Williams DM, Farthing PM, Speight PM, Lesch CA, Squier CA. Langerhans cell density in normal human oral mucosa and skin: relationship to age, smoking and alcohol consumption. J Oral Pathol Med. 1994;23(2):55–9.
22. Hasan MS, Ryan PL, Bergmeier LA, Fortune F. Circulating NK cells and their subsets in Behcet's disease. Clin Exp Immunol. 2017;188(2):311–22.
23. Lanier LL. Shades of grey--the blurring view of innate and adaptive immunity. Nat Rev Immunol. 2013;13(2):73–4.
24. Cerutti A, Cols M, Puga I. Marginal zone B cells: virtues of innate-like antibody-producing lymphocytes. Nat Rev Immunol. 2013;13(2):118–32.
25. Brennan PJ, Brigl M, Brenner MB. Invariant natural killer T cells: an innate activation scheme linked to diverse effector functions. Nat Rev Immunol. 2013;13(2):101–17.
26. Gold MC, Lewinsohn DM. Co-dependents: MR1-restricted MAIT cells and their antimicrobial function. Nat Rev Microbiol. 2013;11(1):14–9.
27. Itohara S, Farr AG, Lafaille JJ, Bonneville M, Takagaki Y, Haas W, Tonegawa S. Homing of a gamma delta thymocyte subset with homogeneous T-cell receptors to mucosal epithelia. Nature. 1990;343(6260):754–7.
28. McCarthy NE, Bashir Z, Vossenkamper A, Hedin CR, Giles EM, Bhattacharjee S, Brown SG, Sanders TJ, Whelan K, MacDonald TT, Lindsay JO, Stagg AJ. Proinflammatory Vdelta2+ T cells populate the human intestinal mucosa and enhance IFN-gamma production by colonic alphabeta T cells. J Immunol. 2013;191(5):2752–63.
29. McCarthy NE, Hedin CR, Sanders TJ, Amon P, Hoti I, Ayada I, Baji V, Giles EM, Wildemann M, Bashir Z, Whelan K, Sanderson I, Lindsay JO, Stagg AJ. Azathioprine therapy selectively ablates human Vdelta2(+) T cells in Crohn's disease. J Clin Invest. 2015;125(8):3215–25.
30. Freysdottir J, Hussain L, Farmer I, Lau S-H, Fortune F. Diversity of gammadelta T cells in patients with Behcet's disease is indicative of polyclonal activation. Oral Dis. 2006;12(3):271–7.
31. Hasan A, Fortune F, Wilson A, Warr K, Shinnick T, Mizushima Y, van der Zee R, Stanford MR, Sanderson J, Lehner T. Role of gamma delta T cells in pathogenesis and diagnosis of Behcet's disease. Lancet. 1996;347(9004):789–94.
32. Hasan MS, Bergmeier LA, Petrushkin H, Fortune F. Gamma delta (gammadelta) T cells and their involvement in Behcet's disease. J Immunol Res. 2015;2015:705831.
33. Bank I, Duvdevani M, Livneh A. Expansion of gammadelta T-cells in Behcet's disease: role of disease activity and microbial flora in oral ulcers. J Lab Clin Med. 2003;141(1):33–40.
34. Eberl M, Hintz M, Reichenberg A, Kollas AK, Wiesner J, Jomaa H. Microbial isoprenoid biosynthesis and human gammadelta T cell activation. FEBS Lett. 2003;544(1-3):4–10.
35. Lehner T, Mitchell E, Bergmeier L, Singh M, Spallek R, Cranage M, Hall G, Dennis M, Villinger F, Wang Y. The role of gammadelta T cells in generating antiviral factors and beta-chemokines in protection against mucosal simian immunodeficiency virus infection. Eur J Immunol. 2000;30(8):2245–56.
36. Kosub DA, Durudas A, Lehrman G, Milush JM, Cano CA, Jain MK, Sodora DL. Gamma/delta T cell mRNA levels decrease at mucosal sites and increase at lymphoid sites following an oral SIV infection of macaques. Curr HIV Res. 2008;6(6):520–30.
37. Su D, Shen M, Li X, Sun L. Roles of γδ T cells in the pathogenesis of autoimmune diseases. Clin Dev Immunol. 2013;2013:985753.
38. Vantourout P, Hayday A. Six-of-the-best: unique contributions of gammadelta T cells to immunology. Nat Rev Immunol. 2013;13(2):88–100.
39. Conti HR, Bruno VM, Childs EE, Daugherty S, Hunter JP, Mengesha BG, Saevig DL, Hendricks MR, Coleman BM, Brane L, Solis N, Cruz JA, Verma AH, Garg AV, Hise AG, Richardson JP, Naglik JR, Filler SG, Kolls JK, Sinha S, Gaffen SL. IL-17 receptor signaling in oral epithelial cells is critical for protection against oropharyngeal candidiasis. Cell Host Microbe. 2016;20(5):606–17.
40. Dutzan N, Konkel JE, Greenwell-Wild T, Moutsopoulos NM. Characterization of the human immune cell network at the gingival barrier. Mucosal Immunol. 2016;9(5):1163–72.
41. Artis D, Spits H. The biology of innate lymphoid cells. Nature. 2015;517(7534):293–301.
42. Tait Wojno ED, Artis D. Innate lymphoid cells: balancing immunity, inflammation, and tissue repair in the intestine. Cell Host Microbe. 2012;12(4):445–57.
43. Hazenberg MD, Spits H. Human innate lymphoid cells. Blood. 2014;124(5):700–9.
44. McKenzie AN, Spits H, Eberl G. Innate lymphoid cells in inflammation and immunity. Immunity. 2014;41(3):366–74.
45. Klose CS, Artis D. Innate lymphoid cells as regulators of immunity, inflammation and tissue homeostasis. Nat Immunol. 2016;17(7):765–74.
46. Simmerman E, Qin X, Marshall B, Perry L, Cai L, Wang T, Yu J, Akbari O, Baban B. Innate lymphoid

cells: a paradigm for low SSI in cleft lip repair. J Surg Res. 2016;205(2):312–7.
47. Russell MW, Bergmeier LA, Zanders ED, Lehner T. Protein antigens of Streptococcus Mutans: purification and properties of a double antigen and its protease-resistant component. Infect Immun. 1980a;28(2):486–93.
48. Russell MW, Zanders ED, Bergmeier LA, Lehner T. Affinity purification and characterization of protease-susceptible antigen I of Streptococcus Mutans. Infect Immun. 1980b;29(3):999–1006.
49. Challacombe SJ, Bergmeier LA, Czerkinsky C, Rees AS. Natural antibodies in man to Streptococcus Mutans: specificity and quantification. Immunology. 1984;52(1):143–50.
50. Challacombe SJ, Russell MW, Hawkes J. Passage of intact IgG from plasma to the oral cavity via crevicular fluid. Clin Exp Immunol. 1978a; 34(3):417–22.
51. Challacombe SJ, Russell MW, Hawkes JE, Bergmeier LA, Lehner T. Passage of immunoglobulins from plasma to the oral cavity in rhesus monkeys. Immunology. 1978b;35(6):923–31.
52. Lehner T, Ma JK, Kelly CG. A mechanism of passive immunization with monoclonal antibodies to a 185,000 M(r) streptococcal antigen. Adv Exp Med Biol. 1992;327:151–63.
53. Ma JK, Hunjan M, Smith R, Kelly C, Lehner T. An investigation into the mechanism of protection by local passive immunization with monoclonal antibodies against Streptococcus Mutans. Infect Immun. 1990;58(10):3407–14.
54. Ma JK, Lehner T. Prevention of colonization of Streptococcus Mutans by topical application of monoclonal antibodies in human subjects. Arch Oral Biol. 1990;35(Suppl):115S–22S.
55. Booth V, Ashley FP, Lehner T. Passive immunization with monoclonal antibodies against Porphyromonas gingivalis in patients with periodontitis. Infect Immun. 1996;64(2):422–7.
56. Booth V, Lehner T. Characterization of the Porphyromonas gingivalis antigen recognized by a monoclonal antibody which prevents colonization by the organism. J Periodontal Res. 1997;32(1 Pt 1):54–60.
57. Kelly CG, Booth V, Kendal H, Slaney JM, Curtis MA, Lehner T. The relationship between colonization and haemagglutination inhibiting and B cell epitopes of Porphyromonas gingivalis. Clin Exp Immunol. 1997;110(2):285–91.
58. Curtis MA, Zenobia C, Darveau RP. The relationship of the oral microbiotia to periodontal health and disease. Cell Host Microbe. 2011;10(4):302–6.
59. Hajishengallis G. Periodontitis: from microbial immune subversion to systemic inflammation. Nat Rev Immunol. 2015;15(1):30–44.
60. Brandtzaeg P. Secretory immunity with special reference to the oral cavity. J Oral Microbiol. 2013;5.
61. Woof JM, Mestecky J. Mucosal immunoglobulins. Immunol Rev. 2005;206:64–82.
62. Mosmann TR, Cherwinski H, Bond MW, Giedlin MA, Coffman RL. Two types of murine helper T cell clone. I. Definition according to profiles of lymphokine activities and secreted proteins. J Immunol. 1986;136(7):2348–57.
63. Amatya N, Garg AV, Gaffen SL. IL-17 signaling: the yin and the Yang. Trends Immunol. 2017;38(5):310–22.
64. Wadia P, Atre N, Pradhan T, Mistry R, Chiplunkar S. Heat shock protein induced TCR gammadelta gene rearrangements in patients with oral cancer. Oral Oncol. 2005;41(2):175–82.
65. Wu R-Q, Zhang D-F, Tu E, Chen Q-M, Chen W. The mucosal immune system in the oral cavity-an orchestra of T cell diversity. Int J Oral Sci. 2014a;6(3):125–32.
66. Jotwani R, Pulendran B, Agrawal S, Cutler CW. Human dendritic cells respond to Porphyromonas gingivalis LPS by promoting a Th2 effector response in vitro. Eur J Immunol. 2003;33(11):2980–6.
67. Pulendran B, Kumar P, Cutler CW, Mohamadzadeh M, Van Dyke T, Banchereau J. Lipopolysaccharides from distinct pathogens induce different classes of immune responses in vivo. J Immunol. 2001;167(9):5067–76.
68. Pulendran B, Lingappa J, Kennedy MK, Smith J, Teepe M, Rudensky A, Maliszewski CR, Maraskovsky E. Developmental pathways of dendritic cells in vivo: distinct function, phenotype, and localization of dendritic cell subsets in FLT3 ligand-treated mice. J Immunol. 1997;159(5):2222–31.
69. Pulendran B, Smith JL, Caspary G, Brasel K, Pettit D, Maraskovsky E, Maliszewski CR. Distinct dendritic cell subsets differentially regulate the class of immune response in vivo. Proc Natl Acad Sci U S A. 1999;96(3):1036–41.
70. Cutler CW, Jotwani R. Dendritic cells at the oral mucosal interface. J Dent Res. 2006;85(8):678–89.
71. Bhingare AC, Ohno T, Tomura M, Zhang C, Aramaki O, Otsuki M, Tagami J, Azuma M. Dental pulp dendritic cells migrate to regional lymph nodes. J Dent Res. 2014;93(3):288–93.
72. Allam JP, Peng WM, Appel T, Wenghoefer M, Niederhagen B, Bieber T, Bergé S, Novak N. Toll-like receptor 4 ligation enforces tolerogenic properties of oral mucosal Langerhans cells. J Allergy Clin Immunol. 2008a;121(2):368–374.e361.
73. Allam JP, Stojanovski G, Friedrichs N, Peng W, Bieber T, Wenzel J, Novak N. Distribution of Langerhans cells and mast cells within the human oral mucosa: new application sites of allergens in sublingual immunotherapy? Allergy. 2008b;63(6):720–7.
74. Novak N, Bieber T, Allam J-P. Immunological mechanisms of sublingual allergen-specific immunotherapy. Allergy. 2011;66(6):733–9.
75. Novak N, Gros E, Bieber T, Allam J-P. Human skin and oral mucosal dendritic cells as 'good guys' and 'bad guys' in allergic immune responses. Clin Exp Immunol. 2010;161(1):28–33.
76. Novak N, Haberstok J, Bieber T, Allam JP. The immune privilege of the oral mucosa. Trends Mol Med. 2008;14(5):191–8.

77. Hamedi M, Bergmeier LA, Hagi-Pavli E, Vartoukian SR, Fortune F. Differential expression of suppressor of cytokine signalling proteins in Behçet's disease. Scand J Immunol. 2014;80(5):369–76.
78. Vartoukian SR, Tilakaratne WM, Seoudi N, Bombardieri M, Bergmeier L, Tappuni AR, Fortune F. Dysregulation of the suppressor of cytokine signalling 3-signal transducer and activator of transcription-3 pathway in the aetiopathogenesis of Sjogren's syndrome. Clin Exp Immunol. 2014;177(3):618–29.
79. Conti HR, Peterson AC, Brane L, Huppler AR, Hernández-Santos N, Whibley N, Garg AV, Simpson-Abelson MR, Gibson GA, Mamo AJ, Osborne LC, Bishu S, Ghilardi N, Siebenlist U, Watkins SC, Artis D, McGeachy MJ, Gaffen SL. Oral-resident natural Th17 cells and γδ T cells control opportunistic Candida Albicans infections. J Exp Med. 2014;211(10):2075–84.
80. Conti HR, Shen F, Nayyar N, Stocum E, Sun JN, Lindemann MJ, Ho AW, Hai JH, Yu JJ, Jung JW, Filler SG, Masso-Welch P, Edgerton M, Gaffen SL. Th17 cells and IL-17 receptor signaling are essential for mucosal host defense against oral candidiasis. J Exp Med. 2009;206(2):299–311.
81. Bai J, Lin M, Zeng X, Zhang Y, Wang Z, Shen J, Jiang L, Gao F, Chen Q. Association of polymorphisms in the human IFN-gamma and IL-4 gene with oral lichen planus: a study in an ethnic Chinese cohort. J Interf Cytokine Res. 2008;28(6):351–8.
82. Romani L. Immunity to fungal infections. Nat Rev Immunol. 2011;11(4):275–88.
83. Zhang X, Jin J, Peng X, Ramgolam VS, Markovic-Plese S. Simvastatin inhibits IL-17 secretion by targeting multiple IL-17-regulatory cytokines and by inhibiting the expression of IL-17 transcription factor RORC in CD4+ lymphocytes. J Immunol. 2008a;180(10):6988–96.
84. Zhang Y, Lin M, Zhang S, Wang Z, Jiang L, Shen J, Bai J, Gao F, Zhou M, Chen Q. NF kappaB-dependent cytokines in saliva and serum from patients with oral lichen planus: a study in an ethnic Chinese population. Cytokine. 2008b;41(2):144–9.
85. Cochran DL. Inflammation and bone loss in periodontal disease. J Periodontol. 2008;79(8 Suppl):1569–76.
86. Gemmell E, Yamazaki K, Seymour GJ. Destructive periodontitis lesions are determined by the nature of the lymphocytic response. Crit Rev Oral Biol Med. 2002;13(1):17–34.
87. Wu RQ, Zhang DF, Tu E, Chen QM, Chen W. The mucosal immune system in the oral cavity-an orchestra of T cell diversity. Int J Oral Sci. 2014b;6(3):125–32.
88. Sun JB, Czerkinsky C, Holmgren J. Mucosally induced immunological tolerance, regulatory T cells and the adjuvant effect by cholera toxin B subunit. Scand J Immunol. 2010;71(1):1–11.
89. Stanford M, Whittall T, Bergmeier LA, Lindblad M, Lundin S, Shinnick T, Mizushima Y, Holmgren J, Lehner T. Oral tolerization with peptide 336-351 linked to cholera toxin B subunit in preventing relapses of uveitis in Behcet's disease. Clin Exp Immunol. 2004;137(1):201–8.
90. Novak N, Allam J-P. Mucosal dendritic cells in allergy and immunotherapy. Allergy. 2011;66(Suppl 95):22–4.
91. Mascarell L, Saint-Lu N, Moussu H, Zimmer A, Louise A, Lone Y, Ladant D, Leclerc C, Tourdot S, Van Overtvelt L, Moingeon P. Oral macrophage-like cells play a key role in tolerance induction following sublingual immunotherapy of asthmatic mice. Mucosal Immunol. 2011a;4(6):638–47.
92. Mascarell L, Zimmer A, Van Overtvelt L, Tourdot S, Moingeon P. Induction of allergen-specific tolerance via mucosal routes. Curr Top Microbiol Immunol. 2011b;352:85–105.
93. Moingeon P, Mascarell L. Induction of tolerance via the sublingual route: mechanisms and applications. Clin Dev Immunol. 2012;2012:623474.
94. Yoshitomi T, Nakagami Y, Hirahara K, Taniguchi Y, Sakaguchi M, Yamashita M. Intraoral administration of a T-cell epitope peptide induces immunological tolerance in cry j 2-sensitized mice. J Pept Sci. 2007;13(8):499–503.
95. Coelho ED, Arrais JP, Matos S, Pereira C, Rosa N, Correia MJ, Barros M, Oliveira JL. Computational prediction of the human-microbial oral interactome. BMC Syst Biol. 2014;8(1):24.
96. Douglas CWI. Bacterial-protein interactions in the oral cavity. Adv Dent Res. 1994;8(2):254–62.
97. Salvucci E. Microbiome, holobiont and the net of life. Crit Rev Microbiol. 2016;42(3):485–94.

5 Mucosal Homeostasis of the Oral Mucosa

Jean-Pierre Allam and Natalija Novak

5.1 Introduction

The oral mucosa contains a stratified squamous epithelium and resembles in great part the architecture of its epidermal counterpart [1]. However, the most striking difference to epidermal epithelium is the absence of a prominent granular layer, which serves as a physical barrier in the skin, and its absence in turn leads to an enhanced permeability in oral mucosa [2]. Thus, contact with several foreign antigens such as bacterial products from local mucosal microbiota or nutrition components from food is likely to be quite frequent in oral mucosal epithelium. In this regard local homeostasis is critical in order to inhibit immunological reactions towards commensal microbiota or harmless food proteins but also to prevent pathogens from invading the tissue. As the oral mucosa also represents the entry point to the gastrointestinal tract (GIT), where tolerance induction predominates to maintain mucosal homeostasis, it is more than likely that corresponding pro-tolerogenic mechanisms take place in the oral mucosa.

J.-P. Allam (✉) • N. Novak
Department of Dermatology and Allergy, University of Bonn, Bonn, Germany
e-mail: jean-pierre.allam@ukb.uni-bonn.de

L.A. Bergmeier (ed.), *Oral Mucosa in Health and Disease*,
https://doi.org/10.1007/978-3-319-56065-6_5

5.2 Oral Mucosal Architecture

The oral cavity and its structures are covered by distinct type of mucosal stratified squamous epithelium depending on anatomical and functional characteristics in different regions [1]. In this respect the three main types are lining mucosa, lingual mucosa and masticatory mucosa including dentogingival mucosa [1]. Masticatory mucosa and lingual mucosa are orthokeratinized lacking a prominent granular layer covering regions exposed to strong shear forces such as tongue, attached gingiva and hard palatum. Lining mucosa contains a non-keratinized epithelium, which lines the remaining part of the oral cavity. Common features of the different types of oral mucosal epithelium are a high vascularization and permeability as well as undistinguishable papillary and reticular dermis by which they differ to skin epithelium [1, 2]. Despite high permeability, the oral mucosal epithelium still represents the first barrier preventing pathogens from invasion. Recent published data suggests that oral epithelial cells participate in controlling the oral microbiota by producing pro-inflammatory as well as antimicrobial molecules [3, 4]. However, in steady-state conditions the oral mucosa also harbours several immune cells such as Langerhans cells (LCs) located within the epithelium itself but also other leucocytes such as T-lymphocytes (T-cells) distributed within dermal compartment of oral mucosa (Fig. 5.1) [2, 5]. LCs belong to the group

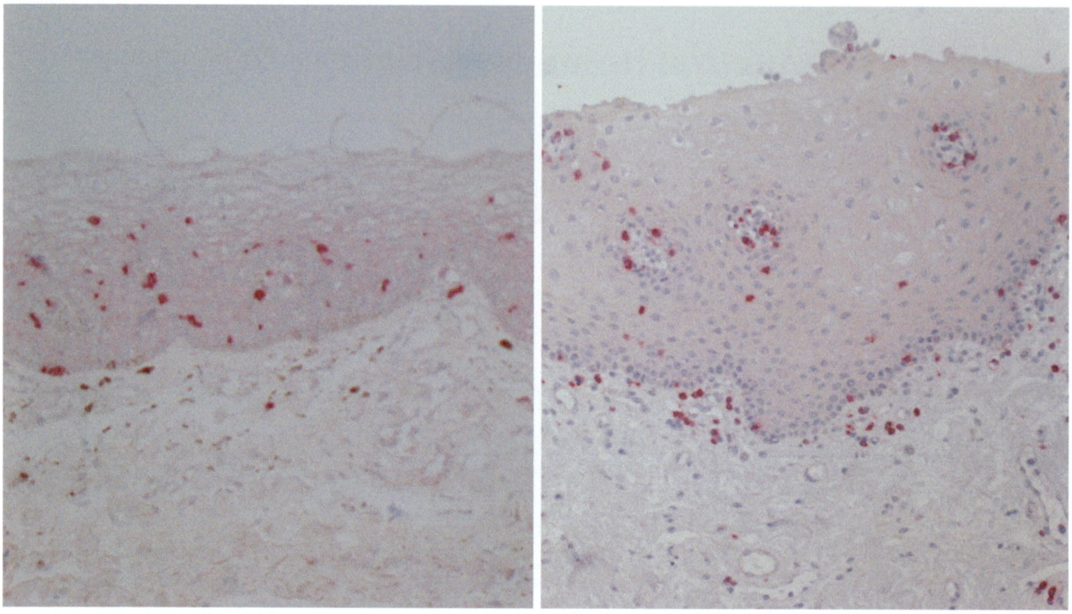

Fig. 5.1 DCs are located within the suprabasal layer of the oral mucosal epithelium as well as in the dermal compartment (left). $CD3^+$ T cells are distributed within the dermal compartment and the rete ridges of the oral mucosa (right)

of antigen-presenting cells (APCs) and are characterized by the expression CD1a and C-type lectin Langerin (CD207) as well as several co-stimulatory molecules and MHC-I and MHC-II for antigen presentation [6]. Similar to LCs in epidermal skin they are located within the suprabasal layer of the epithelium [2]. However, recent studies have shown that most oral mucosal epithelium sites contain higher numbers of LCs than epidermal epithelium [2]. LCs in the oral epithelium are responsible for continuous antigen uptake and they are key players in preventing entry of harmful pathogens on the one hand. On the other hand, they need to avoid immune reactions towards harmless foreign substances or particles from the microbiota or from nutrition [1, 7]. Although great scientific progress has been made within recent years in elucidating induction and maintenance of tolerance in the GIT, only little is known about the immune system of the oral mucosa, which represents much more than a simple transition zone between skin and GIT. In view of a microbiota comprising over 500 different bacterial species of harmless commensal and pathogenic microbes as well as exposure to antigens from food proteins, it is more than likely that local oral mucosal homeostasis is maintained by a sophisticated network of local epithelial cells and immune cells leading to a balance of active pro-tolerogenic anti-inflammatory and pro-inflammatory mechanisms.

5.3 Mucosal Homeostasis in GIT and Oral Mucosa

Chase introduced and defined the term "oral tolerance" in 1946 as a stage of active inhibition of an immune response to an antigen previously presented through the oral route to the immune system [8]. While the term "oral route" refers to the gastrointestinal mucosa, the term "active inhibition" implies systemic protection in particular [8]. However, mechanisms of oral tolerance also contribute to mucosal homeostasis. The latter is achieved by (1) the epithelial barrier in the sense of a mechanical protection and as a "first line of defence" with primarily innate immune mechanisms; (2) immune exclusion with the production of secretory immunoglobulin (Ig) A or IgM by plasma cells; and (3) immune suppression through T-cell anergy, T-cell depletion and induction of

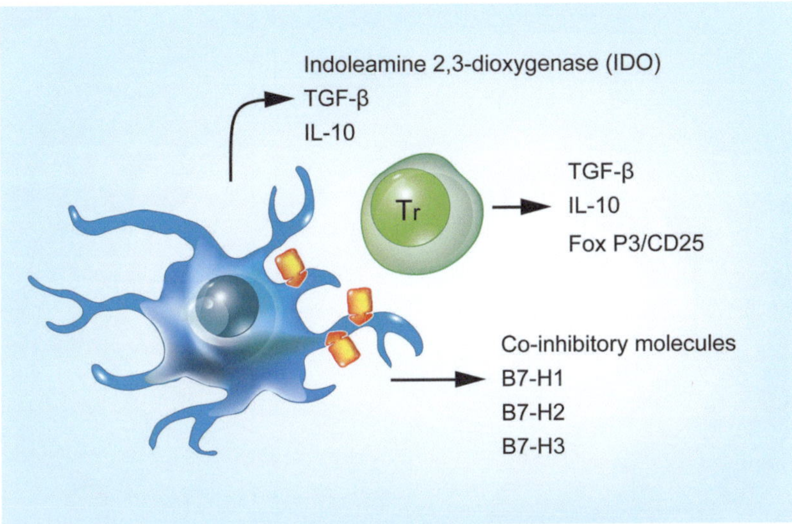

Fig. 5.2 DCs induce regulatory T cells

regulatory T-cells (Treg) [9]. Immune suppression by T-cells is induced by DCs via several mechanisms. It has been shown that especially co-inhibitory molecules such as B7-H1, B7-H2 and B7-H3 on DCs mediate a pro-tolerogenic T-cell response. Furthermore, several soluble factors also produced by DCs such as IL-10, TGF-β or indoleamine-2,3-dioxygenase (IDO) force T-cells towards tolerogenic lineage [10]. Classic Tregs express CD4 and CD25, in addition to the transcription factor Forkhead box p 3 (Foxp3), and are able to mediate immune suppression by means of cell–cell contact or IL-10 and TGF-β production (Fig. 5.2) [11, 12]. Currently, the induction of Tregs is considered to be the key mechanism of oral tolerance [7].

5.3.1 Epithelial Barrier in GIT and Oral Mucosa

Intestinal epithelial cells (IEC) build the epithelial barrier to the lumen in the GIT where tight junctions provide epithelial integrity to prevent pathogens and commensal bacteria from invading [13]. Moreover, IEC produce antimicrobial peptides such as defensins or cathelicidin, which contribute to regulation of microbiota [14]. Further on, IEC express several different pattern recognition receptors (PRR), such as Toll-like receptors (TLR) [15]. These innate immune receptors sense bacterial substances and upon activation trigger a pro-inflammatory immune response [16]. It has been shown that IEC predominantly express intracellular TLR like TLR3, TLR7, TLR8 and TLR9, which can only be activated once bacteria have penetrated IEC. TLR5, on the other hand, is expressed only in the basolateral zone of IEC and can also only be activated upon bacterial invasion [14]. Primarily pathogenic bacteria activate these processes, which lead to a local pro-inflammatory immune response, whereas commensal bacteria cause indirect inhibition of TLR leading to prevention of pro-inflammatory responses [14]. Moreover, IEC carry only few luminal extracellular TLR, mostly TLR2 and TLR4 [17, 18]. It has been shown that ligation of TLR4 by lipopolysaccharide (LPS) in IEC causes interleukin (IL)-1 receptor-associated kinase-1-dependent loss in activation of the pro-inflammatory responses in mice, which has been considered as "endotoxin tolerance" [19].

Comparable TLR expression profiles have been demonstrated in human and mouse oral mucosal epithelial cells [20]. Furthermore, similar mechanisms to the GIT have been described in the oral mucosal epithelium. In this regard it has been shown that TLR4 is involved in the protection against invasion and cell injury caused by *Candida albicans* in mice [20]. Especially, *growth arrest specific 6* (GAS6) has been

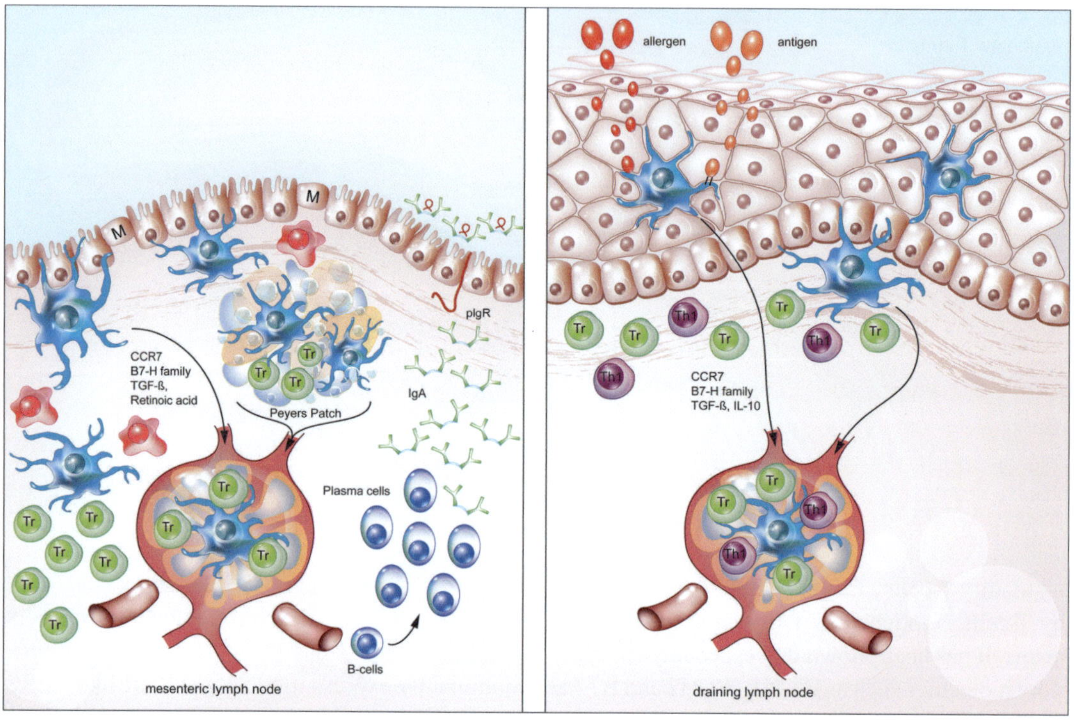

Fig. 5.3 DCs sample antigens from the gut to induce regulatory T cells. Antigens are also passed through M cells and then processed in the Peyer's patch or in mesenteric lymph node to induce regulatory T cells. Plasma cells produce IgA which is transported as secretory IgA into the gut lumen to perform immune exclusion (left). In the oral mucosa antigens are processed by intraepithelial DCs which induce regulatory T cells (right)

suggested to be critical in regulating microbiota in the oral mucosa in mice [21]. GAS6 is expressed only in the outer layer of the oral mucosal epithelium and is induced by local microbiota. It is known as a ligand of the TYRO3-AXL-MERTK (TAM) receptor family and together they play a crucial role in the resolution of inflammation [22]. In turn, knocking out GAS6 in the mouse oral mucosa leads to an upregulation of pro-inflammatory activity and expansion of anaerobic bacteria [21]. As TAM receptors are induced by TLR it is likely that the described mechanism is strongly dependent on TLR such as TLR2 and TLR4 which have been demonstrated to be expressed by oral mucosal epithelial cells [20, 21]. Moreover, in line with IEC in the GIT oral mucosal epithelial cells also produce antimicrobial peptides such as defensins and LL-37 participating in regulation of microbiota [23]. In some oral diseases such as chronic periodontitis these antimicrobial peptides are deregulated allowing uncontrolled growth of pathogenic bacteria (Fig. 5.3) [24, 25].

5.3.2 Tolerance Induction and Immune Exclusion in GIT and Oral Mucosa

Most mucosal tissues contain a specialized immune network composed of inductive and effector sites of which the latter include the lamina propria mucosae (LP), the stroma of exocrine glands and surface epithelia. Inductive sites consist of mucosa-associated lymphoid tissue (MALT) as well as local and regional draining lymph nodes [26–28]. The histological architecture of MALT is similar to the structure of lymph nodes, although MALT lacks afferent lymphatics. In respect of anatomical site of mucosal tissue MALT includes nasopharynx- or nose-associated lymphoid tissue (NALT), bronchus-associated lymphoid tissue (BALT) and genital-associated lymphoid tissue (GENALT) and gut-associated lymphoid tissue (GALT) [28]. The latter consists of Peyer's patches (PP) and isolated lymphoid follicles [7]. In the GIT antigens are captured and processed directly from mucosal luminal side through a specialized follicle-associated epithelium (FAE) containing so-called microfold or membrane (M) cells and dendritic cells (DCs). These cells deliver antigens to APCs in PP or isolated lymphoid follicles, which are able to stimulate naïve B to produce IgA and to mediate T-cell anergy, T-cell depletion and induction of Treg [9, 26–28]. Therefore, GALT is considered to be critical in oral tolerance to soluble antigens [7]. However, there is mounting evidence supporting oral tolerance in the absence of PP and that DCs in the LP of effector sites are crucial for inducing tolerance. In this context it has been shown in mice that CD11c + DC in the LP bind antigen 30–60 min after feeding and that DCs expressing integrin chain α_E (CD103) in the LP are able to induce Tregs and therefore are critical for oral tolerance [7]. Apart from inducing tolerance effector sites also contribute to the formation of secretory IgA (sIgA) whereas antigen-specific B cells are induced in the GALT [9]. After their activation in the GALT B cells migrate via mesenteric lymph node to the thoracic duct where they enter the bloodstream to circulate back to the LP for maturation into IgA-producing plasma cells. In turn, plasma cells in the LP produce dimeric IgA consisting of two monomeric IgA linked to each other by the so-called joining (j) chain [9, 26, 27]. After binding to polymeric Ig receptor located at basolateral zone of the IEC it is secreted to the gut lumen as sIgA for binding of respective antigen preventing contact and resorption of antigen with the GIT epithelium [9, 26, 27]. This process is referred to as immune exclusion (Fig. 5.3).

By contrast, less is known about the structures of oral mucosal immune system, which rather corresponds to effector sites by lacking MALT [2]. In the past some authors consider oral mucosa as part of cranial-, oral- and nasal-associated tissue (CONALT) containing oropharyngeal and nasopharyngeal tissue with tonsils as well as Waldeyer's ring and cervical lymph nodes [29, 30]. Other authors used to prefer the term cranial-, oral- and nasal-associated lymph nodes (CONALN) to emphasize the absence of lymphoid structures or sampling of antigens directly from mucosal surfaces through M cells and to acknowledge the induction of immune responses within local and regional lymph nodes [9]. However, several key mechanisms have been discovered which both initiate and perpetuate tolerance in the oral mucosa. These take place concomitantly within the mucosal epithelium, oral lamina propria mucosa and salivary glands. Resident DCs can be found throughout the whole oral mucosal epithelium and are composed of myeloid DCs from the Langerhans cell (LC) subtype expressing CD1a and the LC-specific lectin Langerin/CD207. Only in respect to expression of costimulatory molecules such as B7.1/CD80 and B7.2/CD86 and other myeloid markers such as CD11b they resemble DC in gut MALT [31, 32]. Next to oral mucosal epithelium LCs, lamina propria DCs (LPDC) expressing CD11c, CD11b and major histocompatibility class (MHC) II have been described in oral mucosa as well [32]. Although the oral cavity represents a small surface area, numbers of resident LCs differ depending on the oral mucosal region. Highest numbers of LCs are located in the vestibulum, bucca, hard palatum and lingua, while in the gingiva and sublingual region lower numbers of LCs were detected [2]. It is most likely that these DC populations are involved in tolerance induction as suppression of T cell-mediated allergic immune reactions to nickel has been shown in individuals wearing mucosal nickel containing braces at an early age [33].

Oral mucosal DCs express several specific receptors and sense the environment for invading pathogens in order to induce an effective defence. In this context mucosal DCs express TLR2 and TLR4 (Fig. 5.3) [34]. These receptors could be involved in maintenance of mucosal homeostasis by downregulation of pattern recognition receptors after multiple stimulations by microbial components. For instance, TLR2 and TLR4 are downregulated from the surface of antigen-presenting cells after repetitive stimulation with LPS from *Porphyromonas gingivalis* in vitro [35]. Concomitantly, production of proinflammatory cytokines in response to these stimuli decreases [35]. Moreover, it has been shown that activation of TLR4 on oral mucosal DCs induced up-regulation of immunosuppressive IL-10 production, which is required for the induction of regulatory T lymphocytes [36, 37]. In turn, upon TLR4 ligation, oral LCs induce FoxP3 expressing and IL-10 as well as TGF-β-producing Tregs [34]. Similar results have been obtained using proteins known to be strong environmental allergens. In mouse models and human system the challenge with these allergens of oral mucosal DCs leads to the induction of IL-10 and TGF-β-producing Tregs [38, 39]. Apart from DCs, other studies focused on the oral mucosal presence of T cells during a steady-state situation. In this regard, a constant infiltration of T cells could be demonstrated in non-inflammatory oral mucosa. Because of their cytokine-producing profile these T cells appear to be mainly Treg (IL-10 and TGF-β) next to Th1 (IFN-γ) and Th17 (IL-17) cells. IL-4-producing T cells could not be detected [5]. While Tregs are more likely involved in regulation of excessive immune response Th17 have been shown to contribute to the prevention of fungal infection (Fig. 5.3) [40].

Altogether, resident oral mucosal DCs are kept in an immature state in the presence of commensals, while DCs newly recruited to the mucosa in the case of invasion of pathogenic organisms induce an active immune response [41]. It is suggested that antigen presentation by oral mucosal DCs takes part mainly in regional draining lymph nodes [42]. However, under certain pathologic conditions such as chronic periodontitis DCs and B and T lymphocyte-rich infiltrates are detectable within the gingival lamina propria, suggesting that local antigen presentation apart from antigen presentation in regional draining lymph nodes might play an important role in mucosal homeostasis induction [30].

Another important part in respect to maintenance of oral mucosal homeostasis is the release of sIgA produced by B plasma cells within the salivary glands such as the parotid, submandibular as well as numerous minor salivary glands distributed throughout the whole oral mucosa which produce over 1000 ml saliva per day [1]. IgA is the most prominent antibody class lining the oral mucosal surfaces and thereby contributes to immune exclusion of antigens [26, 27].

5.4 Concluding Remarks

Taken together, several immune mechanisms have been described in the oral mucosa. However, to date most scientific effort has been made investigating pathologic mechanisms in inflammatory diseases of the oral mucosa, but not much is known about physiologic pathways of mucosal homeostasis at this site. Therefore, future investigations should focus on the physiology of this underestimated mucosal tissue to improve our knowledge about natural ways of mucosal homeostasis.

References

1. Novak N, Haberstok J, Bieber T, Allam JP. The immune privilege of the oral mucosa. Trends Mol Med. 2008;14(5):191–8.
2. Allam JP, Stojanovski G, Friedrichs N, Peng W, Bieber T, Wenzel J, Novak N. Distribution of Langerhans cells and mast cells within the human oral mucosa: new application sites of allergens in sublingual immunotherapy? Allergy. 2008b;63(6):720–7.
3. Eskan MA, Rose BG, Benakanakere MR, Zeng Q, Fujioka D, Martin MH, Lee MJ, Kinane DF. TLR4 and S1P receptors cooperate to enhance inflammatory cytokine production in human gingival epithelial cells. Eur J Immunol. 2008;38(4):1138–47.
4. Greer A, Zenobia C, Darveau RP. Defensins and LL-37: a review of function in the gingival epithelium. Periodontol 2000. 2013;63(1):67–79.
5. Allam JP, Novak N, Duan Y, Winter J, Stojanovski G, Fronhoffs F, Wenghoefer M, Bieber T, Peng WM. Tolerogenic T cells, Th1/Th17 cytokines and TLR2/TLR4 expressing dendritic cells predominate the microenvironment within distinct oral mucosal sites. Allergy. 2011;66(4):532–9.
6. Steinman RM. Dendritic cells: understanding immunogenicity. Eur J Immunol. 2007;37(Suppl 1):S53–60.
7. Pabst O, Mowat AM. Oral tolerance to food protein. Mucosal Immunol. 2012;5(3):232–9.
8. Chehade M, Mayer L. Oral tolerance and its relation to food hypersensitivities. J Allergy Clin Immunol. 2005;115(1):3–12; quiz 13.
9. Brandtzaeg P, Pabst R. Let's go mucosal: communication on slippery ground. Trends Immunol. 2004;25(11):570–7.

10. Rescigno M. Dendritic cells in oral tolerance in the gut. Cell Microbiol. 2011;13(9):1312–8.
11. Jonuleit H, Schmitt E. The regulatory T cell family: distinct subsets and their interrelations. J Immunol. 2003;171(12):6323–7.
12. Wambre E. Effect of allergen-specific immunotherapy on CD4+ T cells. Curr Opin Allergy Clin Immunol. 2015;15(6):581–7.
13. Shen L, Turner JR. Role of epithelial cells in initiation and propagation of intestinal inflammation. Eliminating the static: tight junction dynamics exposed. Am J Physiol Gastrointest Liver Physiol. 2006;290(4):G577–82.
14. Artis D. Epithelial-cell recognition of commensal bacteria and maintenance of immune homeostasis in the gut. Nat Rev Immunol. 2008;8(6):411–20.
15. Brandtzaeg P. Food allergy: separating the science from the mythology. Nat Rev Gastroenterol Hepatol. 2010;7(7):380–400.
16. Rakoff-Nahoum S, Hao L, Medzhitov R. Role of toll-like receptors in spontaneous commensal-dependent colitis. Immunity. 2006;25(2):319–29.
17. Abreu MT, Vora P, Faure E, Thomas LS, Arnold ET, Arditi M. Decreased expression of toll-like receptor-4 and MD-2 correlates with intestinal epithelial cell protection against dysregulated proinflammatory gene expression in response to bacterial lipopolysaccharide. J Immunol. 2001;167(3):1609–16.
18. Melmed G, Thomas LS, Lee N, Tesfay SY, Lukasek K, Michelsen KS, Zhou Y, Hu B, Arditi M, Abreu MT. Human intestinal epithelial cells are broadly unresponsive to toll-like receptor 2-dependent bacterial ligands: implications for host-microbial interactions in the gut. J Immunol. 2003;170(3):1406–15.
19. Lotz M, Gutle D, Walther S, Menard S, Bogdan C, Hornef MW. Postnatal acquisition of endotoxin tolerance in intestinal epithelial cells. J Exp Med. 2006;203(4):973–84.
20. Weindl G, Naglik JR, Kaesler S, Biedermann T, Hube B, Korting HC, Schaller M. Human epithelial cells establish direct antifungal defense through TLR4-mediated signaling. J Clin Invest. 2007;117(12):3664–72.
21. Nassar M, Tabib Y, Capucha T, Mizraji G, Nir T, Pevsner-Fischer M, Zilberman-Schapira G, Heyman O, Nussbaum G, Bercovier H, Wilensky A, Elinav E, Burstyn-Cohen T, Hovav AH. GAS6 is a key homeostatic immunological regulator of host-commensal interactions in the oral mucosa. Proc Natl Acad Sci U S A. 2017;114(3):E337–46.
22. Rothlin CV, Carrera-Silva EA, Bosurgi L, Ghosh S. TAM receptor signaling in immune homeostasis. Annu Rev Immunol. 2015;33:355–91.
23. da Silva BR, de Freitas VA, Nascimento-Neto LG, Carneiro VA, Arruda FV, de Aguiar AS, Cavada BS, Teixeira EH. Antimicrobial peptide control of pathogenic microorganisms of the oral cavity: a review of the literature. Peptides. 2012;36(2):315–21.
24. Dommisch H, Acil Y, Dunsche A, Winter J, Jepsen S. Differential gene expression of human beta-defensins (hBD-1, −2, −3) in inflammatory gingival diseases. Oral Microbiol Immunol. 2005;20(3):186–90.
25. Li X, Duan D, Yang J, Wang P, Han B, Zhao L, Jepsen S, Dommisch H, Winter J, Xu Y. The expression of human beta-defensins (hBD-1, hBD-2, hBD-3, hBD-4) in gingival epithelia. Arch Oral Biol. 2016;66:15–21.
26. Brandtzaeg P. 'ABC' of mucosal immunology. Nestle Nutr Workshop Ser Pediatr Program. 2009a;64:23–38; discussion 38-43, 251-257.
27. Brandtzaeg P. Mucosal immunity: induction, dissemination, and effector functions. Scand J Immunol. 2009b;70(6):505–15.
28. Wang M, Gao Z, Zhang Z, Pan L, Zhang Y. Roles of M cells in infection and mucosal vaccines. Hum Vaccin Immunother. 2014;10(12):3544–51.
29. Csencsits KL, Jutila MA, Pascual DW. Mucosal addressin expression and binding-interactions with naive lymphocytes vary among the cranial, oral, and nasal-associated lymphoid tissues. Eur J Immunol. 2002;32(11):3029–39.
30. Cutler CW, Jotwani R. Dendritic cells at the oral mucosal interface. J Dent Res. 2006;85(8):678–89.
31. Allam JP, Novak N, Fuchs C, Asen S, Berge S, Appel T, Geiger E, Kochan JP, Bieber T. Characterization of dendritic cells from human oral mucosa: a new Langerhans' cell type with high constitutive FcepsilonRI expression. J Allergy Clin Immunol. 2003;112(1):141–8.
32. Iwasaki A. Mucosal dendritic cells. Annu Rev Immunol. 2007;25:381–418.
33. Van Hoogstraten IM, Andersen KE, Von Blomberg BM, Boden D, Bruynzeel DP, Burrows D, Camarasa JG, Dooms-Goossens A, Kraal G, Lahti A, et al. Reduced frequency of nickel allergy upon oral nickel contact at an early age. Clin Exp Immunol. 1991;85(3):441–5.
34. Allam JP, Peng WM, Appel T, Wenghoefer M, Niederhagen B, Bieber T, Berge S, Novak N. Toll-like receptor 4 ligation enforces tolerogenic properties of oral mucosal Langerhans cells. J Allergy Clin Immunol. 2008a;121(2):368–374 e361.
35. Muthukuru M, Jotwani R, Cutler CW. Oral mucosal endotoxin tolerance induction in chronic periodontitis. Infect Immun. 2005;73(2):687–94.
36. Hawrylowicz CM, O'Garra A. Potential role of interleukin-10-secreting regulatory T cells in allergy and asthma. Nat Rev Immunol. 2005;5(4):271–83.
37. Levings MK, Gregori S, Tresoldi E, Cazzaniga S, Bonini C, Roncarolo MG. Differentiation of Tr1 cells by immature dendritic cells requires IL-10 but not CD25+CD4+ Tr cells. Blood. 2005;105(3):1162–9.
38. Allam JP, Wurtzen PA, Reinartz M, Winter J, Vrtala S, Chen KW, Valenta R, Wenghoefer M, Appel T, Gros E, Niederhagen B, Bieber T, Lund K, Novak N. Phl p 5 resorption in human oral mucosa leads to dose-dependent and time-dependent allergen binding

by oral mucosal Langerhans cells, attenuates their maturation, and enhances their migratory and TGF-beta 1 and IL-10-producing properties. J Allergy Clin Immunol. 2010;126(3):638–U635.
39. Mascarell L, Lombardi V, Louise A, Saint-Lu N, Chabre H, Moussu H, Betbeder D, Balazuc AM, Van Overtvelt L, Moingeon P. Oral dendritic cells mediate antigen-specific tolerance by stimulating TH1 and regulatory CD4+ T cells. J Allergy Clin Immunol. 2008;122(3):603–609 e605.
40. Conti HR, Gaffen SL. IL-17-mediated immunity to the opportunistic fungal pathogen Candida Albicans. J Immunol. 2015;195(3):780–8.
41. Iliev ID, Matteoli G, Rescigno M. The yin and yang of intestinal epithelial cells in controlling dendritic cell function. J Exp Med. 2007;204(10):2253–7.
42. Samsom JN. Regulation of antigen-specific regulatory T-cell induction via nasal and oral mucosa. Crit Rev Immunol. 2004;24(3):157–77.

Wound Healing in the Oral Mucosa

6

Patricio C. Smith and Constanza Martínez

6.1 Introduction

Oral mucosa wound healing comprises a series of sequential responses that allow the closure of ruptures in this tissue. This process is of critical importance to prevent the invasion of microorganisms or other agents into tissues avoiding the establishment of chronic inflammation. Since the oral mucosa is continually exposed to traumatic and infectious challenges, this tissue has developed evolutionary strategies to circumvent this adverse environment. Therefore, the oral mucosa has several advantages in terms of the efficiency of the wound healing response. Wound healing may also play an important role during the cell and tissue reactions that occur during the development of chronic inflammatory diseases and cancer. Therefore, knowledge on the mechanisms that regulate wound healing is essential for the comprehension of pathological events in this tissue.

After tissue injury, different biological mechanisms become immediately activated to reconstitute the damaged tissues. Diverse cell types are sequentially recruited and activated to take action in this process. Importantly, these cell types must also be eliminated or silenced for the normal evolution of the wound healing process. The cells engaged during wound healing include components of the immunological system (neutrophils, monocytes, lymphocytes, and dendritic cells), as well as endothelial cells, keratinocytes, and fibroblasts [1]. Cell activation involves in several cases the regulation of the expression numerous genes [2] that control cell proliferation, differentiation, and migration [1]. The extracellular matrix represents another important tissue component involved in wound healing [3]. This is because cells must secrete and organize several molecules including glycoproteins like collagens and fibronectin as well as proteoglycans and matricellular proteins [4]. This event is of outmost importance to permit cell migration and differentiation and finally to restore the damaged tissues. In addition, the level of tension perceived by the cells during cell migration and matrix organization is an important source of information for cells that modify gene expression, proliferation, migration, and differentiation [4]. Therefore, successful wound healing and tissue regeneration require both normally responding cells and a healthy ECM.

P.C. Smith (✉) • C. Martínez
Faculty of Medicine, Pontificia Universidad Católica de Chile, Santiago, Chile
e-mail: psmithf@uc.cl

L.A. Bergmeier (ed.), *Oral Mucosa in Health and Disease*,
https://doi.org/10.1007/978-3-319-56065-6_6

6.2 Wound Healing in the Oral Mucosa: General Mechanisms

The wound healing response involves three overlapping and distinct stages: (1) coagulation-inflammation, (2) new tissue formation, and (3) remodeling [1]. Figure 6.1 illustrates the main stages of the wound healing response along with the main cell types and functions involved. Figure 6.2 shows the main histological features of the coagulation-inflammation and new tissue formation phase. During these events, cells experience important modifications in gene expression driven by soluble mediators and cell-matrix interactions.

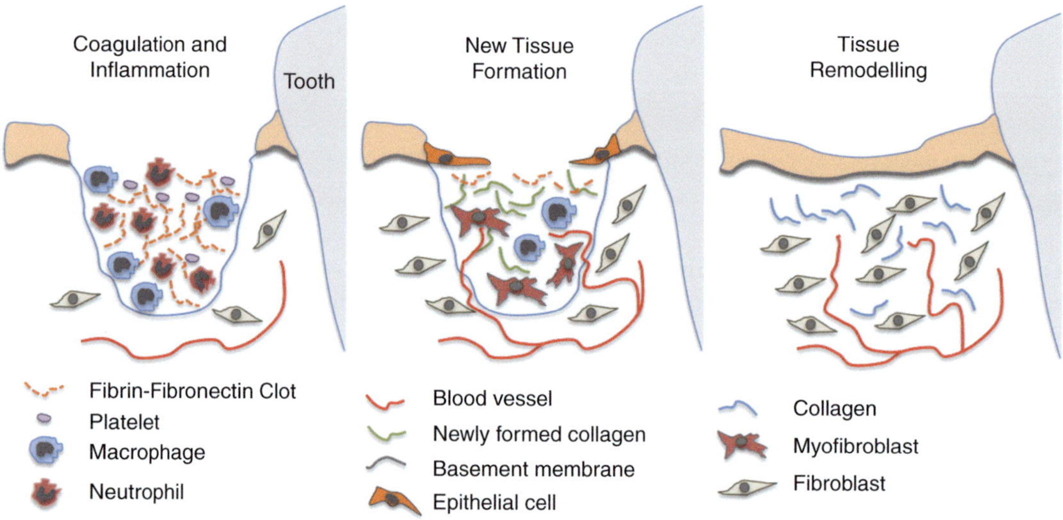

Fig. 6.1 Wound healing stages in the oral mucosa. The three main phases of the wound healing process are represented along with the main cellular components involved

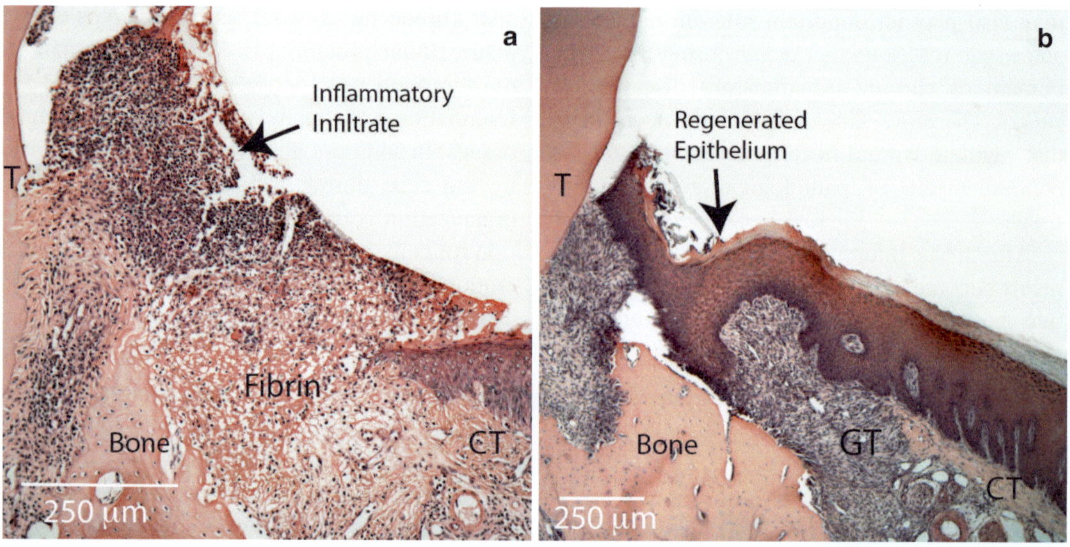

Fig. 6.2 Morphological features of wound healing phases. Representative histological sections obtained from 2 (**a**) and 7 (**b**) day-old gingival wounds performed in rats. Tissues were stained with eosin and hematoxylin. Figure in (**a**) highlights the inflammatory phase of wound healing. Note the inflammatory infiltrate at the tip of the fibrin-fibronectin clot. (**b**) shows the new tissue formation phase. Note the regenerated epithelium and the granulation tissue (GT) that has filled the wound defect. CT = connective tissue. T = tooth

6.3 The Coagulation Process and Its Role in Tissue Repair

Immediately after injury, components of the coagulation cascade are activated in order to prevent excessive bleeding. In association with this, inflammatory pathways and cells of the immune system are recruited to remove cell debris and prevent the proliferation and invasion of infecting agents [5] (see Fig. 6.1). The formation of a platelet plug embedded in a fibrin-fibronectin matrix is critically important in this phase. Platelets play a critical role in wound healing through the release of growth factors, cytokines, and chemokines present in their granules that promote cell migration and proliferation. This response represents one of the first steps that drive the healing of tissues. Patients with disorders in platelet adhesion or reduced platelet numbers may show an altered coagulation process that will delay healing [6]. Interestingly, classic studies have evaluated skin wound healing in mice treated with an antiplatelet serum that induces thrombocytopenia [7]. Under these experimental conditions, injuries are characterized by increased numbers of infiltrating inflammatory cells. Nevertheless, wounds heal normally, suggesting that platelets are mostly involved in hemostasia [7]. In apparent contradiction to these studies, it has been proposed that factors released from platelets promote the proliferation and migration of the cells that will be involved in wound healing. These factors include platelet-derived growth factor, transforming growth factor-β1, and fibroblast growth factor 2 among others. This concept has led to the application of a diverse array of autologous platelet-derived products with the purpose of promoting wound healing in different surgical procedures. Although the concept behind this innovation is interesting, these techniques still require further studies and development [8].

6.4 Role of Inflammation in Wound Healing

Neutrophils are critically important cells that contribute to eliminate infection and are recruited to the wound attracted by components of the complement system, molecules derived from platelets, and factors derived from bacteria [9]. Neutrophil migration typically increases up to 2 days after wounding (Figs. 6.1 and 6.2a). However, it then declines in the absence of infection. Interestingly, neutrophils are eliminated via phagocytic engulfment by macrophages [10]. Neutrophils control bacteria through several mechanisms that include the secretion of antimicrobial peptides, generation of reactive oxygen species, organization of extracellular traps containing DNA and histones, and phagocytosis. Although recruited at later stages, macrophages are also attracted to the wound and play important roles in immune defense and in the development of the granulation tissue and angiogenesis [11]. Although it appears that inflammation is important to prevent wound infection, studies have proposed that the presence of neutrophils and macrophages is not essential for tissue repair. As an example, skin wounds performed in mice depleted of neutrophils heal faster when compared to control animals [5]. In addition, tissue repair studies performed in mice deficient in macrophages and neutrophils have demonstrated an accelerated wound healing response when compared to their wild-type littermates [12].

The connective tissue of the oral mucosa contains resident macrophages (see Fig. 6.1). However, the precise role played by these cells in wound healing is still not well defined. Macrophages originated from circulating monocytes arrive to the wound site where they play several important functions. These cells are recruited at the injury site several days after the appearance of neutrophils. It has been reported that macrophages populating the wound may belong to the M1 (inflammatory) or M2 (alternatively activated or reparative macrophage) subpopulations [13]. The M1 phenotype (pro-inflammatory) is the predominant subpopulation at the initial stages of wound healing and M2 macrophages predominate at later time points [13]. Moreover, lack of the M2 phenotype has been associated with delayed wound healing [14]. Studies in skin and lung have identified that two different chemokines are important to attract macrophages at the wound at early and late stages. The first macrophages that arrive to the wound are

CCR2-expressing cells. Later these cells are replaced by the CX3CR-1-expressing monocytes. These two different waves of macrophages contribute to the initial inflammatory response and then to the resolution of this process [15].

In conclusion, there is general agreement that neutrophils are critically important to prevent infection during wound healing. However, they do not seem to be important in the absence of microbial insult. Moreover, prolonged inflammation may delay wound healing establishing the conditions for abnormal wound repair or tissue fibrosis [16, 17]. Macrophages play a critical role during the inflammatory phase, resolution of inflammation, and initiation of the new tissue formation.

6.5 The New Tissue Formation Phase: An Essential Role in Tissue Regeneration

The formation of new tissue phase corresponds to the second stage of wound repair and occurs between 2 and 10 days after injury [1]. This phase involves the migration and proliferation of epithelial cells, activation of myofibroblasts, and proliferation of new capillaries into the newly formed tissue (see Figs. 6.1 and 6.2b).

6.6 Epithelial Closure

A striking step in this stage is the migration of keratinocytes over the wound bed [1] (Fig. 6.2b). Classic studies identified that after wounding, epithelial cells change their morphology from a polarized cell into a more elongated and migratory cell that recapitulates a mesenchymal phenotype [18]. Cell proliferation is observed between 48 and 72 h after wounding in the basal epithelial cells adjacent to the wound that provide new cells for tissue healing [18]. Interestingly, it has been observed that migrating keratinocytes populating the wound bed do not divide [18]. A critically important issue in this event is the regulation of cell-matrix interactions that control cell migration. In normal unwounded tissue, basal epithelial cells interact with the intact basement membrane through integrin receptors. However, after tissue injury, keratinocytes must migrate over a new cellular environment enriched in matrix components like type I collagen, adhesive molecules like fibronectin, and polymerized fibrin [19, 20]. Two different models have been proposed to explain the migration of keratinocytes over the wound. One of the models proposes that basal keratinocytes creep over the wound provisional matrix as a sheet. On the contrary, it has been proposed that suprabasal keratinocytes leap over the basal keratinocytes and attach to the wound matrix forming a new migration front. Classic studies have identified that the activation of specific integrin receptors, specifically collagen and fibronectin-binding integrins, controls cell migration and expression of proteinases that help the progression of cells in the wound environment including matrix metalloproteinases and plasminogen activators [21, 22]. Proteolytic events are fundamental for proper epithelial cell migration and were described by seminal studies that identified how cells are able to attach and degrade collagen in a controlled manner. In human skin keratinocytes, cell interaction with type I collagen occurs through the $\alpha2\beta1$ integrin that stimulates the expression of matrix metalloproteinase-1 (MMP) that degrades collagen and allows the directional migration of cells [23]. Besides matrix metalloproteinase-1, migrating keratinocytes express a wide array of MMPs and TIMPs (endogenous tissue inhibitors of MMPs) that modulate MMP activity and control matrix degradation in the wound [24]. In this regard, an appropriate balance between extracellular matrix synthesis and degradation is important for the normal evolution and healing of the wound. Keratinocyte migration is also associated with the dissolution of the hemidesmosomal adhesions that intervene in their interaction with the basement membrane [25]. Moreover, the migration of epithelial cells is also associated with modifications in their intercellular connections characterized by a decrease in the lateral desmosomes and connexins [26]. From a topographical point of view, keratinocytes migrate at the initial stages through the provisional matrix and later travel in contact with the forming granulation tissue. In the skin, epidermal stem cells, located in the basal layer of the epidermis and in dermal appendages, are mobilized and recruited to augment the number of cells involved in the repair process [27]. Since the oral mucosa has no appendages, the main source for wound keratinocytes is the basal layer of the oral epithelium. In addition, the hypoxic environment of the wound further stimulates the migration of epithelial cells [28]. Several factors control epithelial cell migration including the release of soluble mediators and the activation of an electric circuit [29]. Soluble mediators comprise cytokines, growth factors, and chemokines released in an autocrine manner [30, 31]. In particular, soluble ligands for the epidermal growth factor receptor (EGFR) play a critical role in the regulation of keratinocyte migration [32]. Once the migrating epithelial cells confront after covering the wound, the keratinocytes start forming hemidesmosomal adhesions to the basal lamina. Interestingly, both keratinocytes and connective tissue cells contribute to the regeneration of basement membrane. After the reconstitution of this structure, keratinocytes restart a normal tissue phenotype. At the end of this phase the integrin αvβ6 increases in its expression. Importantly, this integrin regulates keratinocyte proliferation and synthesis of the newly formed extracellular matrix through activation of TGF-β1 [19].

6.7 Healing of the Connective Tissue

Repair of the connective tissue involves two main stages of the wound healing process that includes the formation of granulation tissue and the tissue-remodeling phase. Although these two phases are observed in all the organs and tissues analyzed, the timing of the healing events may vary according to several factors that include the location and in particular the size of the wound. Accordingly, wounds that heal through primary intention (optimal approximation of wound edges) heal faster when compared to secondary intention wounds (characterized abundant granulation tissue). Connective tissue wound healing may result in complete tissue regeneration or suboptimal healing (nonhealing chronic wounds). Nonhealing chronic wounds are characterized by an abundant outgrowth of poorly organized extracellular matrix that does not restore the structure and function of the tissue [33].

Cell proliferation is observed in the connective tissue as early as 2 days after wounding (Häkkinen et al., 2011). After injury, several cell types residing in the oral mucosa are activated including fibroblasts, and stromal progenitor stem cells among others [1, 34]. In addition to the local fibroblasts and progenitor cells, a specific circulating cell population originated from the bone marrow known as fibrocytes may also migrate into the wounded tissue [35]. Nevertheless, the potential contribution of these cell populations to the restoration of tissues is still far from being understood. A complex array of signals may initiate the regeneration of connective tissues. These include molecules released from the circulation, from infiltrating inflammatory cells, and from cells residing in the tissue (fibroblasts, epithelial cells, nerve endings, salivary gland cells, and vasculature-associated cells). Besides the signals guided by soluble mediators, wounds alter the mechanical stiffness of the tissue. Tissue stiffness is sensed by residing fibroblasts through their integrin receptors [33]. Therefore, cell activation may also involve mechanical perturbations of the tissue [33, 36]. Seminal studies have identified that connective tissue cells residing close to the blood vessels and in the connective tissue near the wound site are primarily engaged in cell proliferation [37]. The cells that populate the wound receive diverse signals that will drive cell proliferation, migration, and differentiation. This signaling may derive from the cellular niche (environment) in which the cells interplay, the distinct extracellular matrix proteins present in the wound, growth factors, cytokines, and mechanical cues sensed by the cells. Moreover, at least in the skin it has been identified that cells derived from the deep or superficial connective tissue have a distinct phenotype that may result in different healing outcomes [38].

After migrating, fibroblasts within the wound proliferate expanding the number of cells available for tissue regeneration. Several growth factors, actively secreted or stored within the wound matrix, may stimulate both the secretion of matrix components and the proliferation of wound fibroblasts. These growth factors may include fibroblast growth factor-2, insulin-like growth factor-1, connective tissue growth factor, platelet-derived growth factor, and transforming growth factor-β1 [39]. During this period, deposition of collagen is regulated, for instance, by macrophages that stimulate fibroblast activity through the secretion of transforming growth factor-β1 at early time points within the wound [40].

During wound healing, a specific phenotype of mesenchymal cells known as myofibroblast is transiently differentiated [41] (see Figs. 6.1 and 6.3). Myofibroblasts play a critical role during wound healing by secreting and organizing matrix components through the remodeling of the newly formed tissues [41]. As previously suggested, the origin of myofibroblasts is still not well defined since several cell types may contribute to the growth of this cell population. Myofibroblasts may derive from resident tissue fibroblasts and mesenchymal stem cells residing in the lamina propria of the oral mucosa, pericytes, fibrocytes derived from the bone marrow that arrive through the circulation, and even epithelial cells that may be transformed by the growth factors released in the wound environment [41]. At least three local events are important to drive myofibroblastic differentiation. These include the presence of biologically active transforming growth factor-β1 released from the extracellular matrix, increased

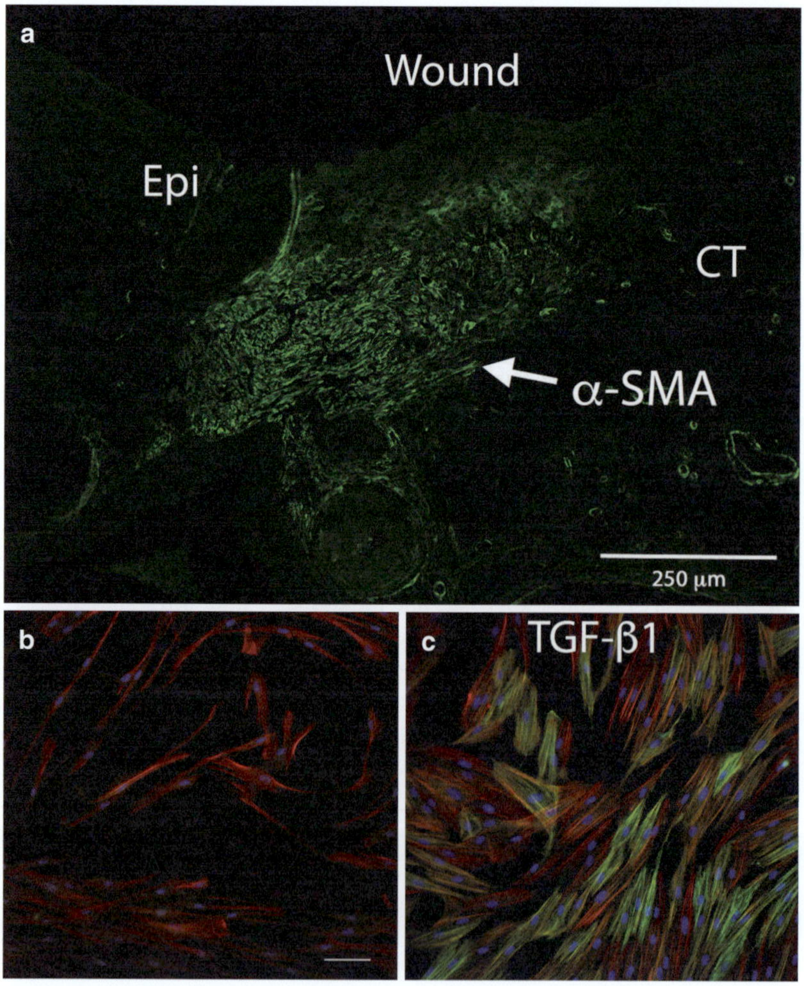

Fig. 6.3 Distribution of myofibroblasts during wound healing. Histological section of a 5-day-old wound performed in the palatal mucosa of mice. Green immunofluorescence staining shows the abundant distribution of the myofibroblast marker α-smooth muscle actin (α-SMA). Epi = epithelium. CT = Connective tissue (**a**). Cell culture of human gingival fibroblasts in both control conditions (**b**) or stimulated with 5 ng/mL transforming growth factor-beta1 (TGF-β1) (**c**). Actin cytoskeleton (red), α-SMA (green), cell nuclei (blue). Magnification bar equals 50 microns

levels of stiffness perceived by the cells, and formation of a specialized extracellular matrix enriched in adhesive proteins like the ED-A spliced form of fibronectin [42, 43]. Several features characterize myofibroblasts. These include the de novo expression of the actin isoform alpha smooth muscle actin (α-SMA), increased contractile capability of the cells, and reinforcement of cell matrix adhesions that allow active remodeling of the extracellular matrix [42]. Figure 6.2a shows the abundant distribution of the myofibroblast marker α-SMA in a 5-day-old palatal wound. Moreover, increased levels of α-SMA are shown in human gingival fibroblasts that have been stimulated with TGF-β1 in vitro (Fig. 6.2b, c). These images highlight the prominent expression of the myofibroblast marker α-SMA during wound healing.

As previously indicated, wound stiffness is an important factor that modulates the activity of fibroblasts. It has been described that increased stiffness is an important factor that promotes the differentiation of myofibroblasts [44]. Wound tissue stiffness experiments a gradual increase during healing due to deposition and cross-linking of collagen [45–47]. Although increased stiffness may stimulate the differentiation of myofibroblasts necessary for normal wound healing, prolonged rigidity of the matrix may also promote scarring and fibrosis [44].

Finally, proteolytic enzymes like matrix metalloproteinases may also play a role in granulation tissue differentiation. This effect was identified in mice deficient in matrix metalloproteinase-13 (MMP-13) that showed defective skin wound healing characterized by delayed granulation tissue and myofibroblastic differentiation [48]. In addition, granulation tissue is also defective in animals treated with an inhibitor of matrix metalloproteinases [49]. These studies strongly suggest that proteolytic events are important for the release of matrix components or for the activation/inactivation of growth factors and cytokines involved in granulation tissue development. Clearly further studies are needed to characterize these events.

Angiogenesis corresponds to an essential response developed as part of the granulation tissue formation process. Blood vessels are needed to restore the nutrition for the new cells that will populate that damaged tissues. To this end, angiogenesis is orchestrated by the formation of new blood vessels from the preexisting vasculature. Critical factors that stimulate angiogenesis include the hypoxic environment of the wound as well as inflammatory mediators (tumor necrosis factor-α, interleukin-1β, interferon-γ, interleukin-8) and growth factors (vascular endothelial growth factor, platelet-derived growth factor, fibroblast growth factor-2) [50]. Both endothelial and mural cells (vascular smooth muscle cells) are involved in this response [51, 52]. For the growth of new blood vessels capillaries grow in association with fibroblasts and macrophages that replace the temporary fibrin matrix during the granulation tissue phase. Blood vessels develop rapidly after wounding by sprouting from preexisting capillaries through the incorporation of endothelial precursors and by differentiation of circulating monocytes into endothelial cells [53].

6.8 Tissue Remodeling

The extracellular matrix deposited during the new tissue formation phase is primitive and unorganized and resembles the organization of the early connective tissue observed during the fetal stages of development [1] (see Fig. 6.1). It is enriched in glycosaminoglycans like hyaluronic acid and contains increased levels of fibronectin, matricellular proteins, and type III collagen [1]. It is important to consider that the cells involved in the new tissue formation phase including myofibroblasts, macrophages, and endothelial cells are eliminated during the remodeling phase [54, 55]. Myofibroblasts undergo apoptosis and are replaced by fibroblasts with a reduced capacity to secrete extracellular matrix components. During this stage, downregulation of the inflammatory response is also important to reduce the development of scar tissue [56].

The duration of the remodeling phase is highly inconstant and will be modified by the size of the wound and whether the injury has healed by primary or secondary intention. However, this phase starts at approximately 2 weeks after injury and may last for 1 year or more [1]. During this stage, all the biological responses activated after injury are downregulated and conclude. One of the important transformations detected during the tissue remodeling phase is the substitution of the new-formed extracellular matrix deposited in the wound. During the new tissue formation phase type III collagen is the main structural protein secreted. However, type III collagen is resorbed and replaced by type I collagen fibers [1]. Collagen fiber degradation is probably executed by members of the matrix metalloproteinase (MMP) family of proteinases that adequate the amount of collagen present in the wound [57]. Besides the degradation and synthesis of new collagen fibers, the extracellular matrix must be organized in order to restore the functional demands of the tissue. To this end, fibroblasts adhere to the collagen fibers through integrins as well as other proteins including the discoidin domain receptors [58]. At the intracellular level, integrins are connected with the actin cytoskeleton through several proteins that contribute to the organization and signaling of focal adhesions [59]. Active contraction of the actin-myosin complex allows the deformation of collagen fibers at the extracellular level [60, 61]. Therefore, cell contraction and remodeling permit the reorientation of the collagen fibers that will finally constitute a complex and mature tissue. Another important change detected during the remodeling phase is the gradual increase in the cross-linking of the wound collagen: this is exerted by several enzymes that include lysyl oxidases, lysyl hydroxylases, and transglutaminases that increase the stability and strength of the collagen network [33].

6.9 Privileged Wound Healing in the Oral Mucosa

Although cutaneous and oral mucosal wounds progress through the same phases, oral mucosal wound healing is characterized by an accelerated rate of tissue healing with minimal scar formation [62–65]. A similar wound healing phenotype has been described in fetal wounds that are characterized by the rapid resolution of lesions with minimal fibrosis [66, 67]. Scar tissue and fibrosis have been associated with several factors that include hypoxia, an increased inflammatory response, abnormal angiogenesis, and the persistence of myofibroblasts in the wound environment [68]. Therefore, studies have focused on whether some of these particular aspects of the wound healing process might be different in the oral mucosa. These features include a less robust inflammatory response, saliva in the oral environment, a different pattern of growth factors, a more restricted angiogenic response, distinct fibroblast subpopulations, and an increased capacity of the connective tissue cells to remodel the extracellular matrix [62, 65, 69–72]. The specific features of the wound healing response in the oral mucosa are summarized in Table 6.1. All these characteristics may contribute to the privileged wound healing phenotype in the oral mucosa.

Table 6.1 Factors explaining privileged wound healing response in the oral mucosa

Specific feature	References
Decreased inflammatory phase during oral mucosal wound healing	[64, 65]
Negative modulation of inflammation by gingival mesenchymal stem cells	[73, 74]
Increased matrix remodeling activity of gingival fibroblasts	[69, 71, 75]
Differential expression of growth factors in oral mucosal wounds	[65, 70]
Restricted angiogenesis in oral wounds	[72]
Presence of growth factors and salivary proteins in saliva	[76, 77, 78]

6.10 Involvement of Mesenchymal Stem Cells During Oral Mucosa Wound Healing

Healing of the oral mucosa is a complex process orchestrated by different subpopulations of resident and infiltrating cells. As previously indicated, oral mucosa wound healing is characterized by a faster response and by the ability to heal with minimal scar tissue formation [79]. Interestingly, these properties may be in part attributed to the presence of mesenchymal stem cells present in the lamina propria of the oral mucosa [80–82]. Oral mucosa-derived mesenchymal stem cells have similar characteristics to bone marrow mesenchymal stem cells, displaying a high self-renewal ability and multipotent differentiation capacity evaluated both in vitro and in vivo [81, 83]. In vitro, these cell populations have been characterized following the criteria established by the International Society for Cellular Therapy [84], along with their properties to differentiate into multiple lineages and to generate connective tissue-like structures after transplantation into immunocompromised mice [82, 85]. In addition, other cell markers displayed by these cells include STRO-1, CD146, CD166, SSEA-4, CD271, Nanog, Sox-2, and Oct-4 [86, 87]. Table 6.2 describes the main phenotypic cell markers and differentiation potential of human gingival mesenchymal stem cells. It is interesting to consider that from the developmental point of view, oral mucosa cells are predominantly derived from the neural crest, a cell population with stem cell-like properties emerging from the dorsal neural plate border during embryonic gastrulation [87, 91]. A recent study reported the isolation of human gingival stem cells that displayed neural crest-related markers and showed high neural lineage differentiation ability, enabling gingival tissue as a potential source of cells with versatility for diverse tissue-regenerative modalities [81, 91]. The fast wound healing response in the oral mucosa has been explained by a milder inflammatory response when compared to skin wound healing [64]. Reduced levels of inflammatory cytokines like interleukin-6 and tumor necrosis factor-α and decreased levels of transforming growth factor-β1 have also been reported in oral mucosa wounds [80, 82]. Interestingly, recent studies have proposed that gingival mesenchymal stem cells might improve wound healing by reducing the inflammatory response [79]. These anti-inflammatory properties have been highlighted by studies that identified that human gingival mesenchymal stem cells may dampen the inflammatory response in experimental models of colitis and [73] and collagen-induced arthritis [74].

Table 6.2 Phenotypic cell markers and differentiation potential of human gingival mesenchymal stem cells

Origin	Positive markers	Negative markers	Differentiation potential	References
Healthy gingival Mesenchymal stem cells	CD73 CD90 CD105 CD166 CD146 CD271 CD 29 STRO-1 Nanog Sox2 OCT4 SSEA-4 Nestin	CD14 CD19 CD34 CD45 CD117 SSEA-1 SSEA-3	Adipose Cartilage Bone Connective tissue-like structures (in vivo)	[81, 83, 85, 87–90]; Jin et al., 2010
Inflamed Gingival MSC	CD44 CD73 CD90 CD105 CD166	CD14 CD34 CD45	Adipose Cartilage Bone Connective tissue-like structures (in vivo)	[85]
Gingival neural crest-derived stem cells	Nestin Snail1 Twist 1 Pax3 Sox9 FoxD3	Non-reported	Neuronal Glial	[81]

6.11 Factors Affecting Wound Healing in the Oral Mucosa

Although the oral mucosa has remarkable wound healing capabilities, these functions may be altered by important diseases and conditions including aging, tobacco smoking, and diabetes.

Aging involves as a complex biological process characterized by a decrease in cell and tissue function [92]. Several biological mechanisms have been proposed to explain aging. Among these, cellular senescence refers to a complex cellular program that derives in a permanent proliferative arrest that may affect several cell types involved in wound healing like fibroblasts and keratinocytes along with important changes in gene expression that alter cell and tissue function [92]. Experimental animal studies have identified that aging affects the wound healing response in the periodontal ligament, cementum, and bone [93]. In addition, recent studies have determined that wound healing in the oral mucosa is delayed due to defects in the migration of gingival epithelial cells and differentiation of myofibroblasts among other cellular functions ([94]; Smith et al., 2015).

Tobacco smoking is an important environmental factor that negatively affects wound healing in the oral mucosa. This detrimental response has been extensively studied in the context of the gingival tissues where the response to periodontal therapy including surgical and nonsurgical procedures is profoundly affected [95]. Cigarette smoke may induce the secretion of inflammatory mediators and tissue remodeling enzymes by gingival fibroblasts [96], migration and differentiation of gingival myofibroblasts [97], and collagen production and remodeling as well [98, 99]. Cigarette smoke also has a strong impact on inflammation that will impact tissue healing at several levels [100].

Diabetes mellitus corresponds to a complex group of diseases characterized by high blood glucose levels [101]. Diabetes has a strong impact on oral health increasing the prevalence of caries, periodontal diseases, and traumatic ulcers [102, 103]. Several factors may explain the effects of diabetes on delayed wound healing. Increased levels of glucose cause cellular stress generating an increased inflammatory response and the generation of advanced glycation end products that have a strong impact on cell function [104]. Diabetes strongly affects wound healing in the oral mucosa causing a delay in epithelial cell migration, decreased connective tissue regeneration, and increased inflammation [105]. Interestingly, recent studies have identified that the transcription factor FOXO1 stimulates oral wound healing in normoglycemic mice and is also critically important for delayed wound healing in diabetic mice [106–108].

6.12 Concluding Remarks

Wound healing in the oral mucosa is a complex process that aims the restitution of this important barrier. Several cell types including resident and infiltrating cells are involved. These cell populations actively interact with a rapidly evolving extracellular matrix that controls cell behavior and ultimately determine the evolution of this process. Given the increased functional demand of this tissue and the continuous exposure to bacterial biofilms, the oral mucosa has developed a refined system of wound healing that involves a controlled inflammatory response and specialized cells that are able to resolve breaches in the oral mucosa with high efficiency. However, important diseases and conditions like diabetes and aging may affect this wound healing capability. Therefore, it is critically important to increase our knowledge on the protective mechanisms of the oral mucosa in both health and disease.

References

1. Gurtner GC, Werner S, Barrandon Y, Longaker MT. Wound repair and regeneration. Nature. 2008;453:314–21.
2. Iyer VR, Eisen MB, Ross DT, Schuler G, Moore T, Lee JC, Trent JM, Staudt LM, Hudson J Jr, Boguski MS, Lashkari D, Shalon D, Botstein D, Brown PO. The transcriptional program in the response of human fibroblasts to serum. Science. 1999;283:83–7.
3. Greiling D, Clark RA. Fibronectin provides a conduit for fibroblast transmigration from collagenous stroma into fibrin clot provisional matrix. J Cell Sci. 1997;110:861–70.

4. Frantz C, Stewart KM, Weaver VM. The extracellular matrix at a glance. J Cell Sci. 2010;123:4195–200.
5. Dovi JV, He LK, DiPietro LA. Accelerated wound closure in neutrophil-depleted mice. J Leukoc Biol. 2003;73:448–55.
6. Nurden AT, Freson K, Seligsohn U. Inherited platelet disorders. Haemophilia. 2012;18(Suppl. 4):154–60.
7. Szpaderska AM, Egozi EI, Gamelli RL, DiPietro LA. The effect of thrombocytopenia on dermal wound healing. J Invest Dermatol. 2003a;120:1130–7.
8. Martínez CE, Smith PC, Palma Alvarado VA. The influence of platelet-derived products on angiogenesis and tissue repair: a concise update. Front Physiol. 2015;6:290.
9. Kolaczkowska E, Kubes P. Neutrophil recruitment and function in health and inflammation. Nat Rev Immunol. 2013;13:159–75.
10. Almzaiel AJ, Billington R, Smerdon G, Moody AJ. Hyperbaric oxygen enhances neutrophil apoptosis and their clearance by monocyte-derived macrophages. Biochem Cell Biol. 2015;93:405–16.
11. Davies LC, Jenkins SJ, Allen JE, Taylor PR. Tissue-resident macrophages. Nat Immunol. 2013;14:986–95.
12. Martin P, D'Souza D, Martin J, Grose R, Cooper L, Maki R, McKercher SR. Wound healing in the PU.1 null mouse—tissue repair is not dependent on inflammatory cells. Curr Biol. 2003;13:1122–8.
13. Novak ML, Koh TJ. Macrophage phenotypes during tissue repair. J Leukoc Biol. 2013;93:875–81.
14. Klinkert K, Whelan D, Clover AJ, Leblond AL, Kumar AH, Caplice NM. Selective M2 macrophage depletion leads to prolonged inflammation in surgical wounds. Eur Surg Res. 2017;58:109–20.
15. Minutti CM, Knipper JA, Allen JE, Zaiss DM. Tissue-specific contribution of macrophages to wound healing. Semin Cell Dev Biol. 2017;61:3–11.
16. Arancibia R, Oyarzún A, Silva D, Tobar N, Martínez J, Smith PC. Tumor necrosis factor-α inhibits transforming growth factor-β-stimulated myofibroblastic differentiation and extracellular matrix production in human gingival fibroblasts. J Periodontol. 2013;84:683–93.
17. Pacios S, Kang J, Galicia J, Gluck K, Patel H, Ovaydi-Mandel A, Petrov S, Alawi F, Graves DT. Diabetes aggravates periodontitis by limiting repair through enhanced inflammation. FASEB J. 2012;26:1423–30.
18. Odland G, Ross R. Human wound repair I. Epidermal regeneration. J Cell Biol. 1968;39:135–68.
19. Larjava H, Koivisto L, Häkkinen L, Heino J. Epithelial integrins with special reference to oral epithelia. J Dent Res. 2011;90:1367–76.
20. Sudbeck BD, Pilcher BK, Welgus HG, Parks WC. Induction and repression of collagenase-1 by keratinocytes is controlled by distinct components of different extracellular matrix compartments. J Biol Chem. 1997;272:22103–10.
21. Garlick JA, Parks WC, Welgus HG, Taichman LB. Re-epithelialization of human oral keratinocytes *in vitro*. J Dent Res. 1996;75:912–8.
22. Netzel-Arnett S, Mitola DJ, Yamada SS, Chrysovergis K, Holmbeck K, Birkedal-Hansen H, Bugge TH. Collagen dissolution by keratinocytes requires cell surface plasminogen activation and matrix metalloproteinase activity. J Biol Chem. 2002;277:45154–61.
23. Dumin JA, Dickeson SK, Stricker TP, Bhattacharyya-Pakrasi M, Roby JD, Santoro SA, Parks WC. Procollagenase-1 (matrix metalloproteinase-1) binds the alpha(2)beta(1) integrin upon release from keratinocytes migrating on type I collagen. J Biol Chem. 2001;276:29368–74.
24. Martins VL, Caley M, O'Toole EA. Matrix metalloproteinases and epidermal wound repair. Cell Tissue Res. 2013;351:255–68.
25. Margadant C, Charafeddine RA, Sonnenberg A. Unique and redundant functions of integrins in the epidermis. FASEB J. 2010;24:4133–52.
26. Scott CA, Tattersall D, O'Toole EA, Kelsell DP. Connexins in epidermal homeostasis and skin disease. Biochim Biophys Acta. 2012;1818:1952–61.
27. Morasso MI, Tomic-Canic M. Epidermal stem cells: the cradle of epidermal determination, differentiation and wound healing. Biol Cell. 2005;97:173–83.
28. Lokmic Z, Musyoka J, Hewitson TD, Darby IA. Hypoxia and hypoxia signaling in tissue repair and fibrosis. Int Rev Cell Mol Biol. 2012;296:139–85.
29. Zhao M. Electrical fields in wound healing – an overriding signal that directs cell migration. Semin Cell Dev Biol. 2009;20:674–82.
30. Kroeze KL, Boink MA, Sampat-Sardjoepersad SC, Waaijman T, Scheper RJ, Gibbs S. Autocrine regulation of re-epithelialization after wounding by chemokine receptors CCR1, CCR10, CXCR1, CXCR2, and CXCR3. J Invest Dermatol. 2012;132:216–25.
31. Myers SR, Leigh IM, Navsaria H. Epidermal repair results from activation of follicular and epidermal progenitor keratinocytes mediated by a growth factor cascade. Wound Repair Regen. 2007;15:693–701.
32. Barrientos S, Stojadinovic O, Golinko MS, Brem H, Tomic-Canic M. Growth factors and cytokines in wound healing. Wound Repair Regen. 2008;16:585–601.
33. Coelho NM, McCulloch CA. Contribution of collagen adhesion receptors to tissue fibrosis. Cell Tissue Res. 2016;365:521–38.
34. Pretheeban T, Lemos DR, Paylor B, Zhang RH, Rossi FM. Role of stem/progenitor cells in reparative disorders. Fibrogenesis Tissue Repair. 2012;5:20.
35. Kao HK, Chen B, Murphy GF, Li Q, Orgill DP, Guo L. Peripheral blood fibrocytes: enhancement of wound healing by cell proliferation, re-epithelialization, contraction, and angiogenesis. Ann Surg. 2011;254:1066–74.
36. Wong VW, Longaker MT, Gurtner GC. Soft tissue mechanotransduction in wound healing and fibrosis. Semin Cell Dev Biol. 2012;23:981–6.
37. McCulloch CA, Melcher AH. Cell density and cell generation in the periodontal ligament of mice. Am J Anat. 1983;167:43–58.

38. Wang J, Dodd C, Shankowsky HA, Scott PG, Tredget EE. Deep dermal fibroblasts contribute to hypertrophic scarring. Lab Investig. 2008;88:1278–90.
39. Werner S, Grose R. Regulation of wound healing by growth factors and cytokines. Physiol Rev. 2003;83:835–70.
40. Rodero MP, Legrand JM, Bou-Gharios G, Khosrotehrani K. Wound-associated macrophages control collagen 1α2 transcription during the early stages of skin wound healing. Exp Dermatol. 2013;22:143–5.
41. Hinz B, Phan SH, Thannickal VJ, Prunotto M, Desmoulière A, Varga J, De Wever O, Mareel M, Gabbiani G. Recent developments in myofibroblast biology: paradigms for connective tissue remodeling. Am J Pathol. 2012;180:1340–55.
42. Hinz B, Phan SH, Thannickal VJ, Galli A, Bochaton-Piallat ML, Gabbiani G. The myofibroblast: one function, multiple origins. Am J Pathol. 2007;170:1807–16.
43. Tomasek JJ, Gabbiani G, Hinz B, Chaponnier C, Brown RA. Myofibroblasts and mechano-regulation of connective tissue remodelling. Nat Rev Mol Cell Biol. 2002;3:349–63.
44. van Putten S, Shafieyan Y, Hinz B. Mechanical control of cardiac myofibroblasts. J Mol Cell Cardiol. 2016;93:133–42.
45. Chiron S, Tomczak C, Duperray A, Lainé J, Bonne G, Eder A, Hansen A, Eschenhagen T, Verdier C, Coirault C. Complex interactions between human myoblasts and the surrounding 3D fibrin-based matrix. PLoS One. 2012;7:e36173.
46. Hinz B. The myofibroblast: paradigm for a mechanically active cell. J Biomech. 2010;43:146–55.
47. Ogawa R. Mechanobiology of scarring. Wound Repair Regen. 2011;19:s2–9.
48. Toriseva M, Laato M, Carpén O, Ruohonen ST, Savontaus E, Inada M, Krane SM, Kähäri VM. MMP-13 regulates growth of wound granulation tissue and modulates gene expression signatures involved in inflammation, proteolysis, and cell viability. PLoS One. 2012;7(8):e42596.
49. Mirastschijski U, Haaksma CJ, Tomasek JJ, Agren MS. Matrix metalloproteinase inhibitor GM 6001 attenuates keratinocyte migration, contraction and myofibroblast formation in skin wounds. Exp Cell Res. 2004;299:465–75.
50. Presta M, Dell'Era P, Mitola S, Moroni E, Ronca R, Rusnati M. Fibroblast growth factor/fibroblast growth factor receptor system in angiogenesis. Cytokine Growth Factor Rev. 2005;16:159–78.
51. Arroyo AG, Iruela-Arispe ML. Extracellular matrix, inflammation, and the angiogenic response. Cardiovasc Res. 2010;86:226–35.
52. DiPietro LA. Angiogenesis and wound repair: when enough is enough. J Leukoc Biol. 2016;100:979–84.
53. Carmeliet P. Angiogenesis in health and disease. Nat Med. 2003;9:653–60.
54. Desmoulière A, Redard M, Darby I, Gabbiani G. Apoptosis mediates the decrease in cellularity during the transition between granulation tissue and scar. Am J Pathol. 1995;146:56–66.
55. Zhang X, Kohli M, Zhou Q, Graves DT, Amar S. Short- and long-term effects of IL-1 and TNF antagonists on periodontal wound healing. J Immunol. 2004;173:3514–23.
56. Mak K, Manji A, Gallant-Behm C, Wiebe C, Hart DA, Larjava H, Häkkinen L. Scarless healing of oral mucosa is characterized by faster resolution of inflammation and control of myofibroblast action compared to skin wounds in the red Duroc pig model. J Dermatol Sci. 2009;56:168–80.
57. Ravanti L, Häkkinen L, Larjava H, Saarialho-Kere U, Foschi M, Han J, Kähäri VM. Transforming growth factor-beta induces collagenase-3 expression by human gingival fibroblasts via p38 mitogen-activated protein kinase. J Biol Chem. 1999;274:37292–373300.
58. Staudinger LA, Spano SJ, Lee W, Coelho N, Rajshankar D, Bendeck MP, Moriarty T, McCulloch CA. Interactions between the discoidin domain receptor 1 and β1 integrin regulate attachment to collagen. Biol Open. 2013;2:1148–59.
59. Segal G, Lee W, Arora PD, McKee M, Downey G, McCulloch CA. Involvement of actin filaments and integrins in the binding step in collagen phagocytosis by human fibroblasts. J Cell Sci. 2001;114:119–29.
60. Conrad PA, Giuliano KA, Fisher G, Collins K, Matsudaira PT, Taylor DL. Relative distribution of actin, myosin I, and myosin II during the wound healing response of fibroblasts. J Cell Biol. 1993;120:1381–91.
61. Grinnell F. Fibroblast biology in three-dimensional collagen matrices. Trends Cell Biol. 2003;13:264–9.
62. Häkkinen L, Uitto VJ, Larjava H. Cell biology of gingival wound healing. Periodontol. 2000;24:127–52.
63. Sciubba JJ, Waterhouse JP, Meyer J. A fine structural comparison of the healing of incisional wounds of mucosa and skin. J Oral Pathol. 1978;7(4):214–27.
64. Szpaderska AM, Zuckerman JD, DiPietro LA. Differential injury responses in oral mucosal and cutaneous wounds. J Dent Res. 2003b;82:621–6.
65. Wong JW, Gallant-Behm C, Wiebe C, Mak K, Hart DA, Larjava H, Häkkinen L. Wound healing in oral mucosa results in reduced scar formation as compared with skin: evidence from the red Duroc pig model and humans. Wound Repair Regen. 2009;17:717–29.
66. Ferguson MW, O'Kane S. Scar-free healing: from embryonic mechanisms to adult therapeutic intervention. Philos Trans R Soc Lond Ser B Biol Sci. 2004;359:839–50.
67. Redd MJ, Cooper L, Wood W, Stramer B, Martin P. Wound healing and inflammation: embryos reveal the way to perfect repair. Philos Trans R Soc Lond Ser B Biol Sci. 2004;359:777–84.
68. Häkkinen L, Larjava H, Koivisto L. Granulation tissue formation and remodeling. In: Larjava H, editor. Oral wound healing. Cell biology and clinical management. New Jersey: Wiley-Blackwell; 2012. p. 125–73.

69. McKeown ST, Barnes JJ, Hyland PL, Lundy FT, Fray MJ, Irwin CR. Matrix metalloproteinase-3 differences in oral and skin fibroblasts. J Dent Res. 2007;86:457–62.
70. Schrementi ME, Ferreira AM, Zender C, DiPietro LA. Site-specific production of TGF-beta in oral mucosal and cutaneous wounds. Wound Repair Regen. 2008;16:80–6.
71. Shannon DB, McKeown ST, Lundy FT, Irwin CR. Phenotypic differences between oral and skin fibroblasts in wound contraction and growth factor expression. Wound Repair Regen. 2006;14:172–8.
72. Szpaderska AM, Walsh CG, Steinberg MJ, DiPietro LA. Distinct patterns of angiogenesis in oral and skin wounds. J Dent Res. 2005;84:309–14.
73. Zhang Q, Shi S, Liu Y, Uyanne J, Shi Y, Shi S, et al. Mesenchymal stem cells derived from human gingiva are capable of immunomodulatory functions and ameliorate inflammation-related tissue destruction in experimental colitis. J Immunol. 2009;183:7787–98.
74. Chen M, Su W, Lin X, Guo Z, Wang J, Zhang Q, Brand D, Ryffel B, Huang J, Liu Z, He X, Le AD, Zheng SG. Adoptive transfer of human gingiva-derived mesenchymal stem cells ameliorates collagen-induced arthritis via suppression of Th1 and Th17 cells and enhancement of regulatory T cell differentiation. Arthritis Rheum. 2013;65:1181–93.
75. Stephens P, Davies KJ, Occleston N, Pleass RD, Kon C, Daniels J, Khaw PT, Thomas DW. Skin and oral fibroblasts exhibit phenotypic differences in extracellular matrix reorganization and matrix metalloproteinase activity. Br J Dermatol. 2001;144:229–37.
76. Royce LS, Baum BJ. Physiologic levels of salivary epidermal growth factor stimulate migration of an oral epithelial cell line. Biochim Biophys Acta. 1991;1092:401–3.
77. Jahovic N, Güzel E, Arbak S, Yeğen BC. The healing-promoting effect of saliva on skin burn is mediated by epidermal growth factor (EGF): role of the neutrophils. Burns. 2004;30:531–8.
78. Oudhoff MJ, Bolscher JG, Nazmi K, Kalay H, van 't Hof W, Amerongen AV, Veerman EC. Histatins are the major wound-closure stimulating factors in human saliva as identified in a cell culture assay. FASEB J. 2008;22:3805–12.
79. Leavitt T, Hu MS, Marshall CD, Barnes LA, Lorenz HP, Longaker MT. Scarless wound healing: finding the right cells and signals. Cell Tissue Res. 2016;365:483–93.
80. Fournier BP, Larjava H, Häkkinen L. Gingiva as a source of stem cells with therapeutic potential. Stem Cells Dev. 2013;22:3157–77.
81. Fournier BP, Loison-Robert LS, Ferre FC, Owen GR, Larjava H, Häkkinen L. Characterization of human gingival neural crest-derived stem cells in monolayer and neurosphere cultures. Eur Cell Mater. 2016;31:40–58.
82. Zhang QZ, Nguyen AL, Yu WH, Le AD. Human oral mucosa and gingiva: a unique reservoir for mesenchymal stem cells. J Dent Res. 2012;91:1011–8.
83. Fournier BP, Ferre FC, Couty L, Lataillade JJ, Gourven M, Naveau A, Coulomb B, Lafont A, Gogly B. Multipotent progenitor cells in gingival connective tissue. Tissue Eng Part A. 2010;16:2891–9.
84. Dominici M, Le Blanc K, Mueller I, Slaper-Cortenbach I, Marini F, Krause D, Deans R, Keating A, Prockop DJ, Horwitz E. Minimal criteria for defining multipotent mesenchymal stromal cells. The International Society for Cellular Therapy position statement. Cytotherapy. 2006;8:315–7.
85. Ge S, Mrozik KM, Menicanin D, Gronthos S, Bartold PM. Isolation and characterization of mesenchymal stem cell-like cells from healthy and inflamed gingival tissue: potential use for clinical therapy. Regen Med. 2012;7:819–32.
86. Jin SH, Lee JE, Yun JH, Kim I, Ko Y, Park JB. Isolation and characterization of human mesenchymal stem cells from gingival connective tissue. J Periodontal Res. 2015;50:461–7.
87. Tang L, Li N, Xie H, Jin Y. Characterization of mesenchymal stem cells from human normal and hyperplastic gingiva. J Cell Physiol. 2011;226:832–42.
88. Otabe K, Muneta T, Kawashima N, Suda H, Tsuji K, Sekiya I. Comparison of gingiva, dental pulp, and periodontal ligament cells from the standpoint of Mesenchymal stem cell properties. Cell Med. 2012;4:13–21.
89. Marynka-Kalmani K, Treves S, Yafee M, Rachima H, Gafni Y, Cohen MA, Pitaru S. The lamina propria of adult human oral mucosa harbors a novel stem cell population. Stem Cells. 2010;28:984–95.
90. Ferré FC, Larjava H, Loison-Robert LS, Berbar T, Owen GR, Berdal A, Chérifi H, Gogly B, Häkkinen L, Fournier BP. Formation of cartilage and synovial tissue by human gingival stem cells. Stem Cells Dev. 2014;23:2895–907.
91. Xu X, Chen C, Akiyama K, Chai Y, Le AD, Wang Z, Shi S. Gingivae contain neural-crest- and mesoderm-derived mesenchymal stem cells. J Dent Res. 2013;92:825–32.
92. Sousounis K, Baddour JA, Tsonis PA. Aging and regeneration in vertebrates. Curr Top Dev Biol. 2014;108:217–46.
93. Benatti BB, Neto JB, Casati MZ, Sallum EA, Sallum AW, Nociti FH Jr. Periodontal healing may be affected by aging: a histologic study in rats. J Periodontal Res. 2006;41:329–33.
94. Cáceres M, Oyarzun A, Smith PC. Defective wound-healing in aging gingival tissue. J Dent Res. 2014;93:691–7.
95. Wan CP, Leung WK, Wong MC, Wong RM, Wan P, Lo EC, Corbet EF. Effects of smoking on healing response to non-surgical periodontal therapy: a multilevel modelling analysis. J Clin Periodontol. 2009;36:229–39.
96. González R, Arancibia R, Cáceres M, Martínez J, Smith PC. Cigarette smoke condensate stimulates urokinase production through the generation of reactive oxygen species and activation of the mitogen activated protein kinase pathways in human gingival fibroblasts. J Periodontal Res. 2009;44:386–94.

97. Silva D, Cáceres M, Arancibia R, Martínez C, Martínez J, Smith PC. Effects of cigarette smoke and nicotine on cell viability, migration and myofibroblastic differentiation. J Periodontal Res. 2012;47:559–607.
98. Takeuchi H, Kubota S, Murakashi E, Zhou Y, Endo K, Ng PS, Takigawa M, Numabe Y. Nicotine-induced CCN2: from smoking to periodontal fibrosis. J Dent Res. 2010;89:34–9.
99. Zhang W, Fang M, Song F, Windsor LJ. Effects of cigarette smoke condensate and nicotine on human gingival fibroblast-mediated collagen degradation. J Periodontol. 2011;82:1071–9.
100. Lee J, Taneja V, Vassallo R. Cigarette smoking and inflammation: cellular and molecular mechanisms. J Dent Res. 2012;91:142–9.
101. American Diabetes Association. Position paper. Diagnosis and classification of diabetes mellitus. Diabetes Care. 2006;29:S43–8.
102. Guggenheimer J, Moore PA, Rossie K, Myers D, Mongelluzzo MB, Block HM, Weyant R, Orchard T. Insulin-dependent diabetes mellitus and oral soft tissue pathologies: II. Prevalence and characteristics of Candida and Candidal lesions. Oral Surg Oral Med Oral Pathol Oral Radiol Endod. 2000;89:570–6.
103. Lalla E, Papapanou PN. Diabetes mellitus and periodontitis: a tale of two common interrelated diseases. Nat Rev Endocrinol. 2011;7:738–48.
104. Hameedaldeen A, Liu J, Batres A, Graves GS, Graves DT. FOXO1, TGF-beta regulation and wound healing. Int J Mol Sci. 2014;15:16257–69.
105. Desta T, Li J, Chino T, Graves DT. Altered fibroblast proliferation and apoptosis in diabetic gingival wounds. J Dent Res. 2010;89:609–14.
106. Ponugoti B, Xu F, Zhang C, Tian C, Pacios S, Graves DT. FOXO1 promotes wound healing through the up-regulation of TGF-beta1 and prevention of oxidative stress. J Cell Biol. 2013;203:327–43.
107. Xu F, Othman B, Lim J, Batres A, Ponugoti B, Zhang C, Yi L, Liu J, Tian C, Hameedaldeen A, Alsadun S, Tarapore R, Graves DT. Foxo1 inhibits diabetic mucosal wound healing but enhances healing of normoglycemic wounds. Diabetes. 2015;64:243–56.
108. Zhang C, Ponugoti B, Tian C, Xu F, Tarapore R, Batres A, Alsadun S, Lim J, Dong G, Graves DT. FOXO1 differentially regulates both normal and diabetic wound healing. J Cell Biol. 2015;209:289–303.

In brief, different salivary components can be used for the diagnosis of oral conditions and monitoring the course of treatment, such as enzymes and immunoglobulins, hormones of host origin, bacteria and bacterial products, ions, and volatile compounds. However, to date no one chemical agent is more important than any other. A significant number of the defense factors show additive or synergistic interactions against oral pathogens [8, 13]. A lack of saliva or saliva deficiency results in negative implications for both oral health and general body health [14].

7.3 Saliva Collection Methods

Salivary secretion occurs at two rates: resting (unstimulated) and stimulated. The resting rate is about 0.5 mL/min and is dependant on higher cortical brain activity; thus the flow rate is lower when asleep or zero when unconscious. The stimulated rate is increased by taste, chewing, and smell. Thus it can be collected and measured as unstimulated whole saliva or stimulated saliva. Unstimulated whole saliva covers, moisturizes and lubricates oral tissues and provides the normal baseline present in oral cavity during a 24-h period. It often correlates with systemic clinical conditions more accurately than stimulated saliva does due to the fact that the materials used to stimulate flow may change the composition of saliva [15].

Whole-mouth resting saliva is collected using one of the following methods: draining, drooling, spitting, swabbing, or suction [16]. More recently, the alternative technique of using filter paper placed in the sublingual pocket for unstimulated saliva collection has shown numerous potential advantages over other known procedures [17].

Mechanical, olfactory, gustatory, or pharmacological stimuli are used to stimulate saliva, which is collected either by having the patient chew on a piece of paraffin and/or by applying (approximately one drop of) citric acid to the tongue [16]. Saliva can also be probed from specific glands via cannulation of the glandular ducts, or through the use of certain collecting devices at the site where the glandular ducts emerge [18].

7.4 Salivary Changes with Periodontal Diseases

Periodontal disease is the most noteworthy condition that can be diagnosed via saliva analysis since various enzymes, cytokines, and biomarkers of bone turnover are present in saliva in cases of periodontitis in comparison to saliva pertaining to healthy periodontal status [19, 20]. In addition, while both locally and systemically derived biomarkers of host origin are contained in saliva, further microbial markers for periodontal disease are also present [11, 21]. Nevertheless, it must be noted that the detection of the active disease site is difficult in saliva analysis and, furthermore, such an analysis is affected by factors such as salivary flow rate, medications, and smoking [22].

Several studies have been conducted to investigate certain biomarkers in saliva for the early detection and management of periodontitis; the earliest biomarker investigated was interleukin-1 beta. Many studies concluded significantly higher levels of interleukin-1 beta in the periodontitis group comparing to the healthy ones [23–25], and others found same level of interleukin-1 beta with and without periodontitis [20, 26]. Alternative studies showed same levels of salivary TNF-α in patients with periodontitis and healthy subjects [24, 27]. In addition to the cytokines, significantly higher levels of adipokines such as visfatin and chemerin have been observed in saliva of patients with periodontitis compared to periodontal health subjects [28, 29]. Salivary RANK and OPG were measured in patients with periodontitis, some studies showed same levels of OPG in diseased and healthy groups [30, 31], and others mentioned that OPG levels declined following periodontal treatment [19, 32]. Another study mentioned that the salivary levels of RANKL were the same in periodontitis and healthy groups [33]; however other studies reported that salivary RANKL was significantly higher in periodontitis group in comparison to the healthy one [34, 35]. Further researches demonstrated significantly higher levels of growth factors in saliva with periodontitis [36–38].

7.5 Salivary Changes with Orthodontic Treatment

Orthodontic tooth movement takes place through bone remodeling that is used primarily to enable bones to respond and adapt to mechanical stress during orthodontic treatment [39, 40]. Due to such treatment, lactate dehydrogenase levels increase in saliva during bone remodeling. For the time being, clinical and radiographic follow-up examinations comprise the main method for patient evaluation. However, saliva analysis may provide a foundation for phase-specific screening in orthodontic tooth movement [41].

During orthodontic treatment, salivary flow rates have been shown to increase 1–3 months after the delivery of fixed appliances in comparison to the baseline measured both before starting treatment and at the control levels [42]. It has also been shown that 90–180 days after the completion of orthodontic treatment, the salivary flow rate moved closer to the baseline and control levels. Research into the cariogenic bacterial counts in whole saliva has found that 3 months after the delivery of the fixed appliance, bacterial counts increased significantly before returning to baseline levels [43]. Other studies illustrate associations between specific genotypes and susceptibility to root resorption [44], speed of orthodontic tooth movement [45], and primary eruption failure [46]. More recently, saliva, collected in sufficient volume, has been employed to identify genotyping instead of blood and buccal swabs [47].

The biocompatibility of orthodontic appliances was investigated by analyzing the metal ion released from the materials of the appliances into saliva; the results showed that most of the ions were released during the initial stage of the treatment [48]. Salivary alpha-amylase activity was measured as a possible indicator for pain during orthodontic treatment, and the results showed no correlation between pain intensity during orthodontic treatment and salivary levels of alpha-amylase [49]. Alternative study examined the salivary concentration of bone remodeling biomarkers such as deoxypyridinoline and bone-specific alkaline phosphatase during orthodontic treatment; the outcomes of this prospective follow-up study indicated that although both of these biomarkers may indicate increased bone remodeling, it appears that only deoxypyridinoline increased in the earlier phases of orthodontic tooth movement, whereas bone-specific alkaline phosphatase might act as an indicator of bone formation at the end of tooth movement [41]. A different research paper linked the salivary levels of soluble RANKL, OPG, and RANKL/OPG ratio to the phases of orthodontic treatment and reported that their levels might assist clinically in the monitoring of orthodontic treatment [50]. Interleukine-1 beta, TNF-α, malondialdehyde, nitric oxide, and 8-hydroxydeoxyguanosine were investigated in saliva of patients with fixed orthodontic appliances before treatment, and at first month and at sixth month of treatment; the findings showed that there were no significant changes in the levels of these analytes at any time point indicating that changes with orthodontic tooth movement do not exceed the physiological limits of these analytes in saliva [51]. The level of inflammation during orthodontic treatment was assessed by measuring MPO activity in saliva and GCF at baseline, 2 h, 7 days, and 14 days after the activation of orthodontic appliances. MPO activity increased until day 7 with the highest activity at 2 h and values were reduced to baseline level at day 14 in both GCF and saliva samples [52].

7.6 Salivary Changes with Obesity

It has been suggested that the determination of salivary adipokines may help to contribute to the elucidation of the physiology and role of adipokines not only in the development of obesity and insulin resistance, but also in inflammation, lack of energy balance, or the stress response. For instance, a study mentioned a positive correlation between the salivary adiponectin concentrations with its circulating concentration of healthy individuals. However, the levels of salivary visfatin did not correlate with the concentration in serum [53]. Some reports indicated a significant increase in the concentration of salivary CRP (a sensitive marker of systemic inflammation) in obese children than in children of normal weight [54, 55].

Similarly, correlation was also observed between serum and salivary uric acid, systolic and diastolic blood pressure, waist circumference, BMI, blood glucose, triglycerides, high-density lipoprotein, and number of cardiometabolic risk factors [56]. A reduced level of antioxidants was also observed in obese individuals not only in serum but also in saliva [57].

Three times higher amount of bacterial cells were observed in the subgingival biofilm with six more bacterial species in the saliva of obese adolescents in comparison to normal weight [58]. In the same field another study evaluated the salivary conditions of morbidly obese patients prior to and 6 months after bariatric surgery, and mentioned that obese patients undergoing bariatric surgery present higher microbiological level of mutans streptococci in saliva after 6 months of surgery [59].

Some salivary parameters have been examined in stimulated saliva in relation to BMI such as pH, flow rate, buffer capacity, protein concentration, phosphate, calcium, sialic acid, and peroxidase activity. The authors suggested that overweight and obesity lead to increase in concentrations of sialic acid and protein, and reduced phosphorus as well as peroxidase activity which may promote dental caries [60].

In more elaborated investigations salivary flow rate has also been measured in relation to obesity. A low salivary flow rate was observed with childhood obesity [61], with BMI >25 in adults less than 50 years old [62] and in morbidly obese (BMI >40) [63] which further strengthens the negative influence of obesity on oral health. Conversely, another study reported comparable salivary pattern in obese and normal-weight individuals [64].

7.7 Gingival Crevicular Fluid

Gingival crevicular fluid (GCF) is an exudate found in the gingival sulcus, between the tooth and marginal gingiva. In the healthy sulcus it is a transudate of interstitial fluid and is present in minute amounts. GCF is released into the crevicular sulcus at a flow rate close to 3 μL/h [65], and under stimulated or inflamed conditions reflects the concentration of metabolites in the serum, increasing in volume with the severity of the inflammation up to 44 μL/h.

The result of increased capillary fluid filtration in comparison to lymphatic uptake leads to fluid accumulation as edema and/or leaves the area as GCF. It largely comprises serum components, inflammatory cells, connective tissue, epithelium, and microbial flora established in the gingival sulcus and exhibits defense activities by flushing particles and bacteria from the sulcus, while its antimicrobial properties and antibodies improve inflammation resistance [66, 67].

7.8 Composition of GCF

The main origin of the aqueous portion of GCF is serum, yet its composition can be modified to a great extent by the products of microorganisms present in its pathway from the gingival tissue through to the sulcus. In general, it comprises cellular components that include bacteria, desquamated epithelial cells and leukocytes that pass through sulcular epithelium, such as PMNs, lymphocytes, monocytes, and erythrocytes. These cells originate from blood vessels in the gingival connective tissues as a result of the stimulation of cells and metabolic products in plaque close to the gingival sulcus [68].

Granulocytes form about 70–80% of GCF cells, while monocytes and macrophages account for around 10–20% and 5% each for mast cells and T-lymphocytes. Together with organic compounds such as carbohydrates, lipids, and protein, electrolytes like calcium, sodium, potassium, fluoride, and magnesium are also present. Certain levels may be higher in GCF, with glucose concentration 3–4 times that of serum, for example, while protein levels are much lower than in serum. Metabolic and bacterial products such as lactic acid, hydroxyl proline, urea, endotoxins, cytotoxic substances, prostaglandins, antibacterial factor, and hydrogen sulfide have also been shown to be present [69].

Other components of GCF are enzyme and enzyme inhibitors such as acid phosphatase, pyrophosphatase, alkaline phosphatase, lysozyme, B-glucuronidase, hyaluronidase, and proteolytic enzymes like mammalian proteinases that include cathepsin D, cathepsin G, elastase, plasminogen activators, bacterial proteinases (endo- and exopeptidases, collagenase, and lactic dehydrogenase serum proteinase inhibitors like alpha 2-macroglobulin, alpha 1-antichymotrypsin, alpha 1-antitrypsin [69]).

7.9 Rationale for the Study of GCF

The study of GCF was first introduced and pursued in the hope of improving clinical diagnosis and for prevention and treatment for gingivitis and periodontal disease. Early studies focused on the amount of GCF and its correlation with clinical and histological measures of inflammation. To summarize, cross-sectional and longitudinal studies find that the amount of GCF is positively correlated to clinical indices of gingival inflammation, while correlation between the amount of GCF and histological evidence of inflammation proves weaker and at times discordant [70].

Early efforts evaluated the differences among GCF constituents linked to healthy versus periodontally compromised teeth or sites, and various studies indicate that levels of GCF constituents reflect both local events and a donor's overall systemic response [71]. More recently, improved knowledge of humoral immune responses, genomics, and proteomics has broadened the prospective applications of GCF analysis. It has been suggested that if an individual's immune and inflammatory responses to stimuli can be measured and assessed to calculate periodontitis, then via careful analysis an assessment of the relative risks for an individual of developing other conditions could also be made, including diabetes mellitus, cardiovascular disease, HIV, and hepatitis [72, 73].

7.10 Collection of GCF

GCF can be collected using various techniques, such as gingival washing, immune-magnetic beads, micropipettes or capillary tubing, preweighed twisted threads, and absorbent strips of filter paper. The last of these methods is considered less traumatic when correctly performed and can be applied to the specific site both easily and quickly [74]. Intra-crevicular application can be employed, where the strip is inserted into the base of the pocket (Fig. 7.2), or extra-crevicular application, where the strip is placed over the crevice region or at its entrance to prevent traumatic irritation and the subsequent simulation of GCF secretion [75]. No significant differences were found in GCF collected from either site [67].

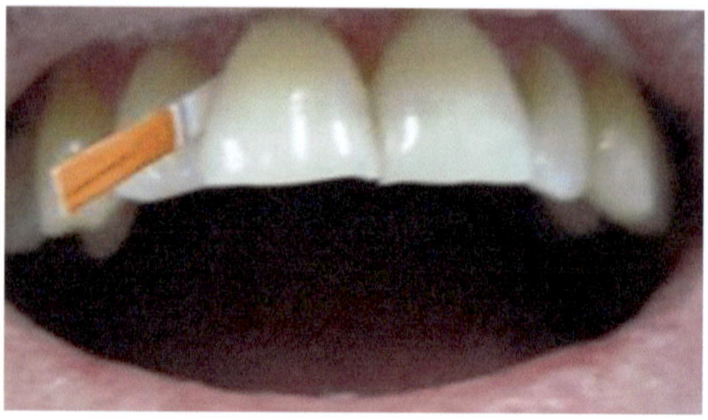

Fig. 7.2 Collection of gingival crevicular fluid collection

7.11 GCF Collection Difficulties

Major sources of GCF contamination constitute blood, plaque, and saliva, hence precise steps should be taken prior to GCF collection. These include isolation of the teeth with cotton rolls, followed by the gentle removal of the supragingival plaque and, finally, complete drying of the tooth's surface with an airstream [76].

In the case of GCF contamination with blood the sample should be discarded, while plaque contamination may affect the determination of the volume. The avoidance of saliva contamination requires sure isolation and alpha-amylase could be used to confirm or disprove contamination by saliva in a sample [77].

7.12 Volume of GCF

GCF volume is measured according to the surface area of the strip dampened by the fluid. For a more accurate measurement a staining of the

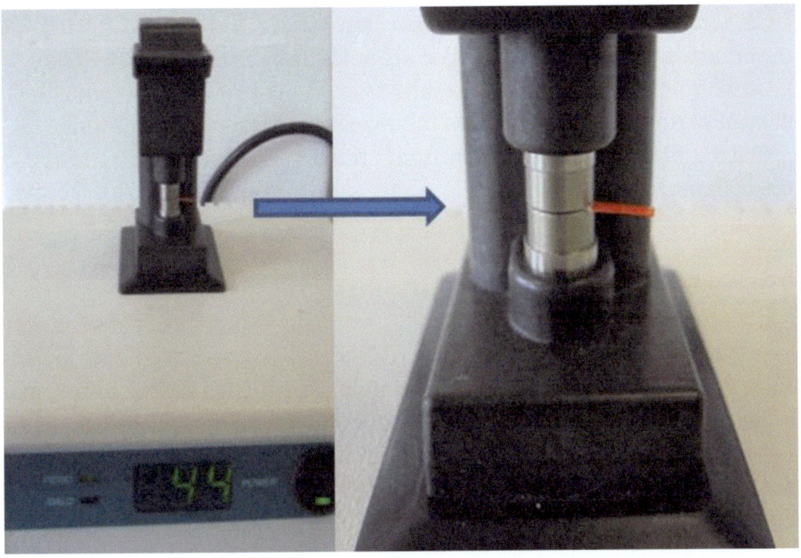

Fig. 7.3 Measurement of gingival crevicular fluid volume using Periotron

strip can be done, either with ninhydrin, which gives a purple color as an indicator, or by the systemic administration of fluorescein 2 h before GCF collection, followed by a UV light examination of the stained strip; the latter of these methods is more sensitive to protein staining. The volume of collected GCF can be measured by comparing the strip weight before and after sample collection. However, this is quite a sensitive process due to the small amount of fluid secreted from a healthy crevice. Periotron provides a more reliable method and involves an instrument that measures the volume and composition of the collected sample using an electrical current that is passed through the dampened strip [78]. This ultimate technique is fast and leaves no marked effects on the sample (Fig. 7.3).

Sample evaporation presents the main problem in GCF volume measurement, especially for volumes of less than 1 μL. Since such samples require further investigation, care should be taken to store each strip in a sealed container to avoid any fluid evaporation that may affect volume measurement. Recovery of the collected GCF is essential, with centrifugal elution demonstrating approximately 100% protein recovery [70]. One microliter is the maximum strip capacity and a longer collecting time may lead to the collection of more fluid than allowed for by the maximum capacity limit [66].

Previous evidence shows that GCF volume might prove a better indicator of gingival inflammation than standard clinical assessments [76]. Given that tissue remodeling incident to orthodontic tooth movement is elicited by an inflammatory process [79], it is hypothesized that the volume of GCF production will reflect these tissue changes [67]. Additionally, the clinical recording of periodontal conditions may exclude the idea that, due to plaque accumulation, gingival inflammation is responsible for an increase in GCF production rather than tooth movement. However, conflicting results are reported in the literature, with studies showing both increased and unchanged GCF volumes incident to orthodontic tooth movement [65].

Across numerous studies, very little or no statistically significant changes in GCF volume due to orthodontic tooth movement have been recorded. The changes that are seen are generally ascribed to clinical or subclinical inflammation following the placement of a fixed orthodontic appliance. Some such studies have compared the volume of GCF during distalization of the maxillary canine in split-mouth designs in adult and juvenile groups across four time points. They found no significant difference between moved and non-moved teeth at each time point or over time within the test and control teeth for both the adult and juvenile groups [80, 81]. Comparable results are cited by further studies that measure

GCF volume at different time points using varying orthodontic force [82]. A final study notes that GCF volume increase mainly results from subclinical periodontal inflammation following fixed appliance placement, rather than from orthodontic tooth movement [83].

7.13 Flow Rate and Constituents of GCF

Flow rate describes the movement of fluid from or into the gingival pocket and occurs at approximately a few microliters per hour. It can be measured by monitoring the volume of fluid that crosses a defined boundary at a given time. Its rate and composition can be altered through inflammation induced by orthodontic tooth movement or periodontitis. Both a flushing action and an isolation effect are the result of GCF flow, with the substances present in the gingival sulcus being easily rinsed out by the flushing action. Furthermore, the outward flow of GCF may prevent saliva entering the gingival sulcus, resulting in the isolation of the gingival sulcus from the oral cavity [68].

Under healthy conditions GCF flow rate remains stable over time, while it has been shown to increase with inflammation and decrease with therapy [66]. GCF flow rate was also reported to be increased [84] or remained with no significant changes [85, 86] during orthodontic treatment. A greater quantity of GCF can be collected from adolescents in comparison to adults, regardless of the presence of any stimuli [87]. Smokers provide a smaller amount of GCF than nonsmokers [88].

GCF sampling sequences may affect volume and for healthy sites the fourth and fifth samples have been shown to yield higher volumes of GCF than previous samples due to increased trauma [89]. By comparison, protein concentrations remain stable at healthy sites yet increase at inflamed sites with repeated sampling [90].

At this point, it is important to emphasize that GCF constituents are affected by both local and systemic conditions [72]. Periodontitis is an obvious example of a heterogeneous disease in which both the environment and genes affect severity [91].

7.14 GCF and Orthodontic Therapy

Studies concerning GCF and orthodontic therapy have both followed and reflected periodontal research, with the mechanical stimuli applied to teeth during orthodontic treatment being expected to affect the amount and constituents of GCF. Histological studies conducted in a guinea pig model demonstrate that orthodontic forces cause localized increased vascular permeability. This increased vascular permeability is similar to that seen in the inflammatory process [92] and appears to support the aforementioned expectation that the mechanical stimuli applied to teeth during orthodontic treatment affect the amount and constituents of GCF.

Numerous studies have examined the relationship between tooth movement and physical, cellular, or molecular changes in paradental tissues and modifications in GCF constituents during orthodontic tooth movement [93]. The results indicate noticeable effects on the quantity and constituents of GCF, together with alterations in the hemostatic condition of the periodontal space—an effect that led to biochemical and cellular changes that reshape the contour of the alveolar bone [94]. Less significant changes in GCF were observed in children with good oral hygiene undergoing orthodontic treatment when compared to the control group [95]. Meanwhile, GCF does not constitute an indication for tissue remodeling in orthodontic treatment [65].

In the late 1980s, in vitro and in vivo research began to report studies of the measurement of GCF constituents in relation to orthodontic force, with many of these studies being concerned with the identification of GCF markers linked to orthodontic stimuli. Some such studies compared markers at both the experimental and control sites using a split-mouth design study across various time points. Here GCF samples are collected before the application of mechanical stimulus to provide a baseline and further samples are gathered at a variety of time points after fixed appliance application. The results reveal that, on average, peak levels occur in the markers 1 or 2 days after the application of force, before returning to the baseline level after 1 week [81, 96, 97]. Similar results have been observed for inflammatory biomarkers during the retraction of maxillary canines [98], while other studies that have compared GCF biomarker levels before, during and after orthodontic treatment revealed no differences between untreated teeth, treated teeth, and teeth with retention [99, 100]. Others have employed GCF analysis in the testing of the effectiveness of preventive measures against plaque accumulation during orthodontic tooth movement [101, 102] and a selection of alternative studies have measured the levels of extracellular matrix proteins in GCF to analyze the presence and levels of root resorption (no resorption, mild, and marked) [103, 104]. For instance, a study reported that RANKL/OPG has been significantly increased with root resorption more than 2 mm in comparison to the control samples [105]. Additionally, age and growth status influence cytokine levels in GCF, which is shown to have an effect on the rate and amount of tooth movement. In one particular study, different mediators were found to increase in different age groups, with IL-6 and GM-CSF increasing in juveniles alone and PGE2 increasing in both juveniles and adults [87].

Previous studies mentioned that the forces used in orthodontic tooth movement lead to an initial increase in levels of bone-resorptive mediators and the associated receptors, namely IL-1β, IL-8, RANKL, and TNF-α; this occurs as early as 1 min [106] or 1 h into the procedure [97], and peaks after 24 h [82, 98, 107–109]. These mediators slowly decrease to the baseline at the following observation points: 48 h, 168 h, 14 days, and 21 days [97, 109]. Conversely, bone-forming mediators, such as OPG, exhibit an immediate decline on the application of orthodontic force on the site of retraction after 1 h [110, 111], and at 24 h [110, 112]. Furthermore, during the acute inflammatory phase, pain can be noted due to an increased concentration of inflammatory mediators. In particular, these include cytokines and prostaglandins in the gingival crevicular fluid of the moving teeth [113].

Various studies compare the compression and tension sites showing a decrease in OPG by 24 h at the compression side [112], as well as an increase in RANK and TGF-β1 after 7 days [112, 114]. Various other mediators show temporal variations on the compression side, with IL-1β increasing as soon as 1 min [106] or after 4 h [115]; RANKL increasing after 24 h in both juveniles and adults [81] or after 42 days [115]; and IL-8 increasing after 4 h [115] or after 10 days [116]. In contrast, the tension site has been shown in various studies to demonstrate an appreciable increase in TNF-α [115] and other bone-resorbing mediators, such as IL-1β, PGE2, and IL-8. However, this rise is shown to have occurred earlier than for compression and, across all of the observation points, with levels higher than those seen at the compression side [115, 116].

Conclusion

Saliva and GCF are adapted to their environments to enable to perform the functions to maintain health. Both contain proteins which bind, neutralize, and aggregate microorganisms and their products. In addition, both continuously flow and this has a major benefit to oral health—usually by flushing irritants and microorganisms away from the surfaces. A number of factors will affect the composition and flow rate of either GCF or saliva. Smoking has well-characterized effects on GCF flow rate but almost no effect on salivary flow rate. In contrast prescribed drugs have major effects on salivary flow but few effects on gingival flow. However, infection affects both. The gingival margin represents the interface of saliva and GCF and is probably the greatest site of bacterial colonization. Whether colonization leads to disease and damage is dependant on the tissue responses to quantity and constituents of the biofilm that forms.

References

1. Loo J, Yan W, Ramachandran P, Wong D. Comparative human salivary and plasma proteomes. J Dent Res. 2010;89:1016–23.
2. Humphrey SP, Williamson RT. A review of saliva: normal composition, flow, and function. J Prosthet Dent. 2001;85:162–9.

3. Welton H. The anatomy and physiology of salivary glands. In: Edgar WM, Dawes C, O'mullane DM, editors. Saliva and oral health. UK: Stephen Hancocks Limited; 2012.
4. Proctor GB, Carpenter GH. Regulation of salivary gland function by autonomic nerves. Auton Neurosci. 2007;133:3–18.
5. Pfaffe T, Cooper-White J, Beyerlein P, Kostner K, Punyadeera C. Diagnostic potential of saliva: current state and future applications. Clin Chem. 2011;57:675–87.
6. Aps JK, Martens LC. Review: the physiology of saliva and transfer of drugs into saliva. Forensic Sci Int. 2005;150:119–31.
7. Chiappin S, Antonelli G, Gatti R, De Palo EF. Saliva specimen: a new laboratory tool for diagnostic and basic investigation. Clin Chim Acta. 2007;383:30–40.
8. Mandel ID. The diagnostic uses of saliva. J Oral Pathol Med. 1990;19:119–25.
9. Streckfus C, Bigler L. Saliva as a diagnostic fluid. Oral Dis. 2002;8:69–76.
10. Nagler R, Salameh F, Reznick A, Livshits V, Nahir A. Salivary gland involvement in rheumatoid arthritis and its relationship to induced oxidative stress. Rheumatology. 2003;42:1234–41.
11. Kaufman E, Lamster IB. Analysis of saliva for periodontal diagnosis. J Clin Periodontol. 2000;27:453–65.
12. Numabe Y, Hisano A, Kamoi K, Yoshie H, Ito K, Kurihara H. Analysis of saliva for periodontal diagnosis and monitoring. Periodontology. 2004;40:115–9.
13. Kinney JS, Ramseier CA, Giannobile WV. Oral fluid–based biomarkers of alveolar bone loss in periodontitis. Ann N Y Acad Sci. 2007;1098:230–51.
14. Pinka R, Simeka J, Vondrakovab J, Faberb E, Michla P, Pazderaa J, Indrakb K. Saliva as a diagnostic medium. Biomed Pap Med Fac Univ Palacky Olomouc Czech Repub. 2009;153:103–10.
15. Munro CL, Grap MJ, Jablonski R, Boyle A. Oral health measurement in nursing research: state of the science. Biol Res Nurs. 2006;8:35–42.
16. Navazesh M. Methods for collecting saliva. Ann N Y Acad Sci. 1993;694:72–7.
17. Neu M, Goldstein M, Gao D, Laudenslager ML. Salivary cortisol in preterm infants: validation of a simple method for collecting saliva for cortisol determination. Early Hum Dev. 2007;83:47–54.
18. Edgar W, O'mullane D, Dawes C. Saliva and oral health. London: British Dental Association; 2004.
19. Kinney J, Morelli T, Braun T, Ramseier C, Herr A, Sugai J, Shelburne C, Rayburn L, Singh A, Giannobile W. Saliva/pathogen biomarker signatures and periodontal disease progression. J Dent Res. 2011;90:752–8.
20. Teles R, Likhari V, Socransky S, Haffajee A. Salivary cytokine levels in subjects with chronic periodontitis and in periodontally healthy individuals: a cross-sectional study. J Periodontal Res. 2009;44:411–7.
21. Koh DS-Q, Koh GC-H. The use of salivary biomarkers in occupational and environmental medicine. Occup Environ Med. 2007;64:202–10.
22. Gursoy UK, Könönen E, Pradhan-Palikhe P, Tervahartiala T, Pussinen PJ, Suominen-Taipale L, Sorsa T. Salivary MMP-8, TIMP-1, and ICTP as markers of advanced periodontitis. J Clin Periodontol. 2010;37:487–93.
23. Ebersole JL, Schuster JL, Stevens J, Dawson Iii D, Kryscio RJ, Lin Y, Thomas MV, Miller CS. Patterns of salivary analytes provide diagnostic capacity for distinguishing chronic adult periodontitis from health. J Clin Immunol. 2013;33:271–9.
24. Gursoy UK, Könönen E, Uitto VJ, Pussinen PJ, Hyvärinen K, Suominen-Taipale L, Knuuttila M. Salivary interleukin-1β concentration and the presence of multiple pathogens in periodontitis. J Clin Periodontol. 2009;36:922–7.
25. Tobón-Arroyave S, Jaramillo-González P, Isaza-Guzmán D. Correlation between salivary IL-1β levels and periodontal clinical status. Arch Oral Biol. 2008;53:346–52.
26. Christodoulides N, Floriano PN, Miller CS, Ebersole JL, Mohanty S, Dharshan P, Griffin M, Lennart A, Ballard KLM, King CP. Lab-on-a-chip methods for point-of-care measurements of salivary biomarkers of periodontitis. Ann N Y Acad Sci. 2007; 1098:411–28.
27. Mirrielees J, Crofford LJ, Lin Y, Kryscio RJ, Dawson Iii DR, Ebersole JL, Miller CS. Rheumatoid arthritis and salivary biomarkers of periodontal disease. J Clin Periodontol. 2010;37:1068–74.
28. Özcan E, Saygun NI, Serdar MA, Kurt N. Evaluation of the salivary levels of visfatin, chemerin, and progranulin in periodontal inflammation. Clin Oral Investig. 2015;19:921–8.
29. Tabari ZA, Azadmehr A, Nohekhan A, Naddafpour N, Ghaedi FB. Salivary visfatin concentrations in patients with chronic periodontitis. J Periodontol. 2014;85:1081–5.
30. Costa PP, Trevisan GL, Macedo GO, Palioto DB, Souza SL, Grisi MF, Novaes AB Jr, Taba M Jr. Salivary interleukin-6, matrix metalloproteinase-8, and osteoprotegerin in patients with periodontitis and diabetes. J Periodontol. 2010;81:384–91.
31. Miller CS, King CP, Langub MC, Kryscio RJ, Thomas MV. Salivary biomarkers of existing periodontal disease: a cross-sectional study. J Am Dent Assoc. 2006;137:322–9.
32. Sexton WM, Lin Y, Kryscio RJ, Dawson DR, Ebersole JL, Miller CS. Salivary biomarkers of periodontal disease in response to treatment. J Clin Periodontol. 2011;38:434–41.
33. Frodge BD, Ebersole JL, Kryscio RJ, Thomas MV, Miller CS. Bone remodeling biomarkers of periodontal disease in saliva. J Periodontol. 2008;79:1913–9.
34. Tabari ZA, Azadmehr A, Tabrizi MAA, Hamissi J, Ghaedi FB. Salivary soluble receptor activator of nuclear factor kappa B ligand/osteoprotegerin ratio

in periodontal disease and health. J Periodontal Implant Sci. 2013;43:227–32.
35. Tobón-Arroyave SI, Isaza-Guzmán DM, Restrepo-Cadavid EM, Zapata-Molina SM, Martínez-Pabón MC. Association of salivary levels of the bone remodelling regulators sRANKL and OPG with periodontal clinical status. J Clin Periodontol. 2012;39:1132–40.
36. Rudrakshi C, Srinivas N, Mehta D. A comparative evaluation of hepatocyte growth factor levels in gingival crevicular fluid and saliva and its correlation with clinical parameters in patients with and without chronic periodontitis: a clinico-biochemical study. J Indian Soc Periodontol. 2011;15:147.
37. Wilczyńska-Borawska M, Borawski J, Bagińska J, Małyszko J, Myśliwiec M. Hepatocyte growth factor in saliva of patients with renal failure and periodontal disease. Ren Fail. 2012;34:942–51.
38. Wilczynska-Borawska M, Borawski J, Kovalchuk O, Chyczewski L, Stokowska W. Hepatocyte growth factor in saliva is a potential marker of symptomatic periodontal disease. J Oral Sci. 2006;48:47–50.
39. Ariffin SHZ, Ellias MF, Wahab RMA, Bakar Y, Senafi S. Profiles of lactate dehydrogenase, tartrate resistant acid phosphatase and alkaline photophatase in saliva during orthodontic treatment. Sains Malaysiana. 2010;39:405–12.
40. Van Nieuw Amerongen A, Bolscher J, Veerman E. Salivary proteins: protective and diagnostic value in cariology? Caries Res. 2004;38:247–53.
41. Flórez-Moreno GA, Marín-Restrepo LM, Isaza-Guzmán DM, Tobón-Arroyave SI. Screening for salivary levels of deoxypyridinoline and bone-specific alkaline phosphatase during orthodontic tooth movement: a pilot study. Eur J Orthod. 2013b;35:361–8.
42. Li Y, Hu B, Liu Y, Ding G, Zhang C, Wang S. The effects of fixed orthodontic appliances on saliva flow rate and saliva electrolyte concentrations. J Oral Rehabil. 2009;36:781–5.
43. Peros K, Mestrovic S, Anic-Milosevic S, Slaj M. Salivary microbial and nonmicrobial parameters in children with fixed orthodontic appliances. Angle Orthod. 2011;81:901–6.
44. Al-Qawasmi RA, Hartsfield JK, Everett ET, Flury L, Liu L, Foroud TM, Macri JV, Roberts WE. Genetic predisposition to external apical root resorption. Am J Orthod Dentofac Orthop. 2003;123:242–52.
45. Iwasaki L, Chandler J, Marx D, Pandey J, Nickel J. IL-1 gene polymorphisms, secretion in gingival crevicular fluid, and speed of human orthodontic tooth movement. Orthod Craniofac Res. 2009;12:129–40.
46. Frazier-Bowers SA, Simmons D, Wright JT, Proffit WR, Ackerman JL. Primary failure of eruption and PTH1R: the importance of a genetic diagnosis for orthodontic treatment planning. Am J Orthod Dentofac Orthop. 2010;137(160):E1–160.E7.
47. Pulford DJ, Mosteller M, Briley JD, Johansson KW, Nelsen AJ. Saliva sampling in global clinical studies: the impact of low sampling volume on performance of dna in downstream genotyping experiments. BMC Med Genet. 2013;6:20.
48. Mikulewicz M, Chojnacka K. Trace metal release from orthodontic appliances by in vivo studies: a systematic literature review. Biol Trace Elem Res. 2010;137:127–38.
49. Campos MJDS, Raposo NRB, Ferreira AP, Vitral RWF. Salivary alpha-amylase activity: a possible indicator of pain-induced stress in orthodontic patients. Pain Med. 2011;12:1162–6.
50. Flórez-Moreno GA, Isaza-Guzmán DM, Tobón-Arroyave SI. Time-related changes in salivary levels of the osteotropic factors srankl and opg through orthodontic tooth movement. Am J Orthod Dentofac Orthop. 2013a;143:92–100.
51. Atuğ Özcan SS, Ceylan İ, Özcan E, Kurt N, Dağsuyu İM, Çanakçi CF. Evaluation of oxidative stress biomarkers in patients with fixed orthodontic appliances. Dis Markers. 2014;2014:7.
52. Navarro-Palacios A, García-López E, Meza-Rios A, Armendariz-Borunda J, Sandoval-Rodríguez A. Myeloperoxidase enzymatic activity is increased in patients with different levels of dental crowding after initial orthodontic activation. Am J Orthod Dentofac Orthop. 2014;146:92–7.
53. Mamali I, Roupas ND, Armeni AK, Theodoropoulou A, Markou KB, Georgopoulos NA. Measurement of salivary resistin, visfatin and adiponectin levels. Peptides. 2012;33:120–4.
54. Goodson JM, Kantarci A, Hartman M-L, Denis GV, Stephens D, Hasturk H, Yaskell T, Vargas J, Wang X, Cugini M. Metabolic disease risk in children by salivary biomarker analysis. PLoS One. 2014;9:E98799.
55. Naidoo T, Konkol K, Biccard B, Dubose K, Mckune A. Elevated salivary C-reactive protein predicted by low cardio-respiratory fitness and being overweight in African children. Cardiovasc J Afr. 2012;23:501.
56. Soukup M, Biesiada I, Henderson A, Idowu B, Rodeback D, Ridpath L, Bridges EG, Nazar AM, Bridges KG. Salivary uric acid as a noninvasive biomarker of metabolic syndrome. Diabetol Metab Syndr. 2012;4:1.
57. Bhardwaj S, Misra A, Khurana L, Gulati S, Shah P, Vikram NK. Childhood obesity in Asian Indians: a burgeoning cause of insulin resistance, diabetes and sub-clinical inflammation. Asia Pac J Clin Nutr. 2008;17:172–5.
58. Zeigler CC, Persson GR, Wondimu B, Marcus C, Sobko T, Modéer T. Microbiota in the oral subgingival biofilm is associated with obesity in adolescence. Obesity. 2012;20:157–64.
59. Hashizume LN, Bastos LF, Cardozo DD, Hilgert JB, Hugo FN, Stein AT, Souto KE, Meinhardt NG. Impact of bariatric surgery on the saliva of patients with morbid obesity. Obes Surg. 2015;25:1550–5.
60. Pannunzio E, Amancio OMS, Vitalle MSDS, Souza DND, Mendes FM, Nicolau J. Analysis

of the stimulated whole saliva in overweight and obese school children. Rev Assoc Med Bras. 2010;56:32–6.
61. Modéer T, Blomberg CC, Wondimu B, Julihn A, Marcus C. Association between obesity, flow rate of whole saliva, and dental caries in adolescents. Obesity. 2010;18:2367–73.
62. Flink H, Bergdahl M, Tegelberg Å, Rosenblad A, Lagerlöf F. Prevalence of hyposalivation in relation to general health, body mass index and remaining teeth in different age groups of adults. Community Dent Oral Epidemiol. 2008;36:523–31.
63. Knaś M, Maciejczyk M, Sawicka K, Hady HR, Niczyporuk M, Ładny JR, Matczuk J, Waszkiel D, Żendzian-Piotrowska M, Zalewska A. Impact of morbid obesity and bariatric surgery on antioxidant/oxidant balance of the unstimulated and stimulated human saliva. J Oral Pathol Med. 2016;45(6):455–64.
64. Epstein LH, Paluch R, Coleman KJ. Differences in salivation to repeated food cues in obese and non-obese women. Psychosom Med. 1996;58:160–4.
65. Perinetti G, Primožič J, Castaldo A, Di Lenarda R, Contardo L. Is gingival crevicular fluid volume sensitive to orthodontic tooth movement? A systematic review of split-mouth longitudinal studies. Orthod Craniofac Res. 2013;16:1–19.
66. Goodson JM. Gingival crevice fluid flow. Periodontol. 2003;31:43–54.
67. Griffiths GS. Formation, collection and significance of gingival crevice fluid. Periodontology. 2003;2000(31):32–42.
68. Kavadia-Tsatala S, Kaklamanos E, Tsalikis L. Effects of orthodontic treatment on gingival crevicular fluid flow rate and composition: clinical implications and applications. Int J Adult Orthodon Orthognath Surg. 2001;17:191–205.
69. Zia A, Khan S, Bey A, Gupta N, Mukhtar-Un-Nisar S. Oral biomarkers in the diagnosis and progression of periodontal diseases. Biol Med. 2011;3:45–52.
70. Cimasoni G. Crevicular Fluid Updated. Monogr Oral Sci. 1982;12(III-VII):1–152.
71. Champagne CM, Buchanan W, Reddy MS, Preisser JS, Beck JD, Offenbacher S. Potential for gingival crevice fluid measures as predictors of risk for periodontal diseases. Periodontology. 2003;2000(31):167–80.
72. Ebersole JL. Humoral immune responses in gingival crevice fluid: local and systemic implications. Periodontology. 2003;2000(31):135–66.
73. Lamster IB, Ahlo JK. Analysis of gingival crevicular fluid as applied to the diagnosis of oral and systemic diseases. Ann N Y Acad Sci. 2007;1098:216–29.
74. Tzannetou S, Efstratiadis S, Nicolay O, Grbic J, Lamster I. Interleukin-1β and B-glucuronidase in gingival crevicular fluid from molars during rapid palatal expansion. Am J Orthod Dentofac Orthop. 1999;115:686–96.
75. Newman MG, Takei HH, Carranza FA, Saunders WB. Clinical periodontology. Philadelphia: WB Saunders; 2006.
76. Griffiths G, Sterne J, Wilton J, Eaton K, Johnson N. Associations between volume and flow rate of gingival crevicular fluid and clinical assessments of gingival inflammation in a population of British male adolescents. J Clin Periodontol. 1992;19:464–70.
77. Griffiths G, Wilton J, Curtis M. Permeability of the gingival tissues to IGM during an experimental gingivitis study in man. Arch Oral Biol. 1997;42:129–36.
78. Golub L, Kleinberg I. Gingival crevicular fluid: a new diagnostic aid in managing the periodontal patient. Oral Sci Rev. 1976;8:49–61.
79. Krishnan V, Davidovitch ZE. Cellular, molecular, and tissue-level reactions to orthodontic force. Am J Orthod Dentofac Orthop. 2006;129(469):469. E1–E32.
80. Dilsiz A, Kiliç N, Aydin T, Ates FN, Zihni M, Bulut C. Leptin levels in gingival crevicular fluid during orthodontic tooth movement. Angle Orthod. 2010;80:504–8.
81. Kawasaki K, Takahashi T, Yamaguchi M, Kasai K. Effects of aging on RANKL and OPG levels in gingival crevicular fluid during orthodontic tooth movement. Orthod Craniofac Res. 2006;9:137–42.
82. Luppanapornlarp S, Kajii TS, Surarit R, Iida J. Interleukin-1β levels, pain intensity, and tooth movement using two different magnitudes of continuous orthodontic force. Eur J Orthod. 2010;32:596–601.
83. Drummond S, Canavarro C, Perinetti G, Teles R, Capelli J. The monitoring of gingival crevicular fluid volume during orthodontic treatment: a longitudinal randomized split-mouth study. Eur J Orthod. 2012;34(1):109–13.
84. Baldwin P, Pender N, Last K. Effects on tooth movement of force delivery from nickel-titanium archwires. Eur J Orthod. 1999;21:481–9.
85. Miyajima K, Ohno Y, Iwata T, Tanida K, Iizuka T. The lactic acid and citric acid content in the gingival fluid of orthodontic patients. Aichi Gakuin Dent Sci. 1990;4:75–82.
86. Uematsu S, Mogi M, Deguchi T. Increase of transforming growth factor-B 1 in gingival crevicular fluid during human orthodontic tooth movement. Arch Oral Biol. 1996;41:1091–5.
87. Ren Y, Maltha JC, Von Den Hoff JW, Kuijpers-Jagtman AM, Zhang D. Cytokine levels in crevicular fluid are less responsive to orthodontic force in adults than in juveniles. J Clin Periodontol. 2002;29:757–62.
88. Hedin C, Ronquist G, Forsberg O. Cyclic nucleotide content in gingival tissue of smokers and non-smokers. J Periodontal Res. 1981;16:337–43.
89. Chapple IL, Socransky SS, Dibart S, Glenwright DH, Matthews JB. Chemiluminescent assay of alkaline phosphatase in human gingival crevicular fluid: investigations with an experimental gingivitis model and studies on the source of the enzyme within crevicular fluid. J Clin Periodontol. 1996;23:587–94.

90. Curtis M, Griffiths G, Price S, Coulthurst S, Johnson N. The total protein concentration of gingival crevicular fluid. J Clin Periodontol. 1988;15:628–32.
91. Kinane DF, Shiba H, Hart TC. The genetic basis of periodontitis. Periodontology. 2005;2000(39):91–117.
92. Storey E. The nature of tooth movement. Am J Orthod. 1973;63:292–314.
93. Krishnan V, Davidovitch Z. On a path to unfolding the biological mechanisms of orthodontic tooth movement. J Dent Res. 2009;88:597–608.
94. Toms SR, Lemons JE, Bartolucci AA, Eberhardt AW. Nonlinear stress-strain behavior of periodontal ligament under orthodontic loading. Am J Orthod Dentofac Orthop. 2002;122:174–9.
95. Tersin J. Studies of gingival conditions in relation to orthodontic treatment. IV. The effect of oral hygiene measures on gingival exudation during the course of orthodontic treatment. Swed Dent J. 1978;2:131.
96. Giannopoulou C, Dudic A, Kiliaridis S. Pain discomfort and crevicular fluid changes induced by orthodontic elastic separators in children. J Pain. 2006;7:367–76.
97. Karacay S, Saygun I, Bengi AO, Serdar M. Tumor necrosis factor-α levels during two different canine distalization techniques. Angle Orthod. 2007;77:142–7.
98. Alikhani M, Raptis M, Zoldan B, Sangsuwon C, Lee YB, Alyami B, Corpodian C, Barrera LM, Alansari S, Khoo E. Effect of micro-osteoperforations on the rate of tooth movement. Am J Orthod Dentofac Orthop. 2013;144:639–48.
99. Griffiths G, Moulson A, Petrie A, James I. Evaluation of osteocalcin and pyridinium crosslinks of bone collagen as markers of bone turnover in gingival crevicular fluid during different stages of orthodontic treatment. J Clin Periodontol. 1998;25:492–8.
100. Pender N, Samuels R, Last K. The monitoring of orthodontic tooth movement over a 2-year period by analysis of gingival crevicular fluid. Eur J Orthod. 1994;16:511–20.
101. Paschos E, Limbach M, Teichmann M, Huth KC, Folwaczny M, Hickel R, Rudzki-Janson I. Orthodontic attachments and chlorhexidine-containing varnish effects on gingival health. Angle Orthod. 2008;78:908–16.
102. Sköld-Larsson K, Yucel-Lindberg T, Twetman S, Modéer T. Effect of a triclosan-containing dental gel on the levels of prostaglandin I2 and interleukin-1 β in gingival crevicular fluid from adolescents with fixed orthodontic appliances. Acta Odontol. 2003;61:193–6.
103. Balducci L, Ramachandran A, Hao J, Narayanan K, Evans C, George A. Biological markers for evaluation of root resorption. Arch Oral Biol. 2007;52:203–8.
104. Mah J, Prasad N. Dentine phosphoproteins in gingival crevicular fluid during root resorption. Eur J Orthod. 2004;26:25–30.
105. George A, Evans C. Detection of root resorption using dentin and bone markers. Orthod Craniofac Res. 2009;12:229–35.
106. Dudic A, Kiliaridis S, Mombelli A, Giannopoulou C. Composition changes in gingival crevicular fluid during orthodontic tooth movement: comparisons between tension and compression sides. Eur J Oral Sci. 2006;114:416–22.
107. Kaya FA, Hamamcı N, Uysal E, Yokuş B. Identification of tumor necrosis factor-α levels around miniscrews during canine distalization. Korean J Orthod. 2011;41:36–41.
108. Kuroki H, Miyagawa Y, Shimomura-Kuroki J, Endo T, Shimomura H. Identification of marker proteins by orthodontic treatment: relationship of RANKL in the gingival crevicular fluid and of amylase in whole saliva with orthodontic treatment. Odontology. 2014;102:303–9.
109. Sari E, Uçar C. Interleukin 1β levels around microscrew implants during orthodontic tooth movement. Angle Orthod. 2007;77:1073–8.
110. Toygar HU, Kircelli BH, Bulut S, Sezgin N, Tasdelen B. Osteoprotegerin in gingival crevicular fluid under long-term continuous orthodontic force application. Angle Orthod. 2008;78:988–93.
111. Tuncer BB, Özdemir BÇ, Boynueğri D, Karakaya İB, Ergüder İ, Yücel AA, Aral LA, Özmeriç N. OPG-RANKL levels after continuous orthodontic force. Gazi Med J. 2013;24:144.
112. Barbieri G, Solano P, Alarcón JA, Vernal R, Rios-Lugo J, Sanz M, Martín C. Biochemical markers of bone metabolism in gingival crevicular fluid during early orthodontic tooth movement. Angle Orthod. 2012;83:63–9.
113. Krishnan V, Davidovitch ZE. Biologic orthodontics: methods to accelerate or decelerate orthodontic tooth movement. In: Krishnan V, editor. Biological mechanisms of tooth movement. New Jersey: John Wiley & Sons, Ltd; 2015. p. 191.
114. Nishijima Y, Yamaguchi M, Kojima T, Aihara N, Nakajima R, Kasai K. Levels of RANKL and OPG in gingival crevicular fluid during orthodontic tooth movement and effect of compression force on releases from periodontal ligament cells in vitro. Orthod Craniofac Res. 2006;9:63–70.
115. Grant M, Wilson J, Rock P, Chapple I. Induction of cytokines, MMP9, TIMPS, RANKL and OPG during orthodontic tooth movement. Eur J Orthod. 2013;35(5):644–51.
116. Tuncer BB, Özmeriç N, Tuncer C, Teoman İ, Çakilci B, Yücel A, Alpar R, Balos K. Levels of interleukin-8 during tooth movement. Angle Orthod. 2005;75:631–6.

8 Oral Cancer: Recent Developments and Future Challenges

E. Kenneth Parkinson

8.1 Introduction

The most common form of oral cancer is squamous cell carcinoma (OSCC) and is the sixth most frequent cancer worldwide [1]. The major aetiological factors for oral cavity tumours are tobacco and areca nut use, alcohol and poor oral hygiene [2, 3], although at other head and neck sites such as the tonsil there is an increasingly frequent role for human papilloma virus (HPV) subtypes [4]. OSCC is an important disease because it has a poor prognosis, especially when diagnosed late. In addition, treatment is very expensive and is associated with severe morbidity post-therapy and the disease is increasing in young people for reasons that are presently unclear [2].

OSCC has been the subject of intensive investigation over the last two decades but advances in treatments have been limited and largely due to improvements in surgery rather than targeted therapies. However, very recently the development of prophylactic HPV vaccines [5] and the disclosure that new immunotherapies appear to show promise, http://www.biospace.com/News/bristol-myers-squibb-release-checkmate-141-a/407013?type=email&source=CS_012816, have given cause for renewed optimism that the disease may at last decrease in frequency and the requirement for expensive and debilitating surgery be reduced. However, clonal evolution [6, 7] and field cancerisation [8] remain significant problems in the treatment of many patients and the recent evidence that this may be exacerbated by chemotherapy and radiotherapy [7] highlights the requirement for a thoughtful application of these new discoveries as well as the development of early non-invasive diagnostic markers.

In this chapter I review the latest research in the OSCC field and highlight what important issues and challenges still remain.

8.2 Mechanisms of OSCC Development and Progression

The classical view of OSCC development is that an initiated cell progresses to hyperplasia and through various grades of dysplasia to carcinoma in situ and finally invasive and metastatic carcinoma [9, 10]. This model of progression (Fig. 8.1) has historically been challenged on the grounds that OSCC often arises without any noticeable sign of the premalignant stages [11]. However, the barriers that many cancers must

E. Kenneth Parkinson (✉)
Centre for Oral Immunobiology and Regenerative Medicine, Institute of Dentistry Barts and The London School of Medicine and Dentistry, Queen Mary University of London, London, UK

Centre for Clinical and Diagnostic Oral Sciences, Blizard Institute, London, UK
e-mail: e.k.parkinson@qmul.ac.uk

L.A. Bergmeier (ed.), *Oral Mucosa in Health and Disease*,
https://doi.org/10.1007/978-3-319-56065-6_8

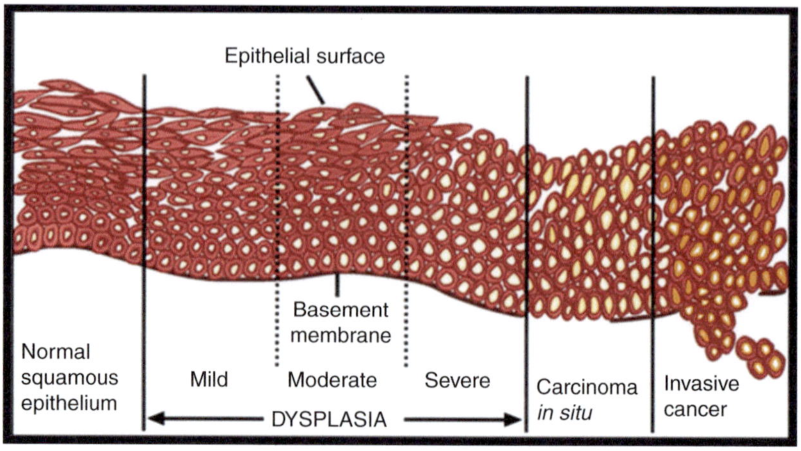

Fig. 8.1 A schematic representation of the histological changes that take place as a normal squamous epithelium gives rise to increasing grades of dysplasia, then carcinoma in situ and finally invasive squamous cell carcinoma

bypass in order to progress are senescence, an irreversible cell cycle arrest that eventually leads to cell cycle arrest, mitochondrial dysfunction, increased cell size and secretion of numerous proteins and metabolites [12–14]. Senescence is associated with shortening of telomeres, the TTAGGG base repeats in DNA that protect the ends of chromosomes with each round of replication. When telomeres reach a critical length, DNA double-strand breaks occur, DNA damage foci assemble and the phosphorylation of multiple proteins known as the DNA damage response (DDR) [15, 16]. The senescence programme is intact in many premalignant lesions [17–21] along with the DDR [15, 16] and the senescence effectors p16^{INK4A} and p53 [21]. It has been hypothesised that the DDR can result from either telomere shortening [15] or alternatively via stalled replication forks as a result of excessive growth factor signalling following oncogene or tumour-suppressor gene mutation [15, 16]. Malignant progression frequently results in an attenuation of the DDR and reduced senescence [15, 17–21]. One hypothesis (Fig. 8.2) is that when the level of oncogenic signalling (e.g. via the epidermal growth factor receptor) increases to the level required for invasion and malignancy, senescence is triggered [22]. Senescence in turn results in the production of an array of cytokines and other secreted proteins termed the senescence-associated secretory phenotype (SASP) [23] and this triggers senescent cell clearance via the innate and adaptive immune systems (Fig. 8.3a) [24, 25]; moreover, senescence bypass or blockade of the innate immune system results in progression to malignancy (Fig. 8.3b) [24]. So why are premalignant oral lesions so rarely seen? The answer may be that progression to malignancy may often take place microscopically from a field of genetically altered keratinocytes and support for this hypothesis has emerged from the study of a transgenic mouse expressing a fluorescently tagged p16^{INK4A} gene promoter sequence. When this mouse was crossed with 14 models of cancer the fluorescent cells were visible well before the premalignant lesions were visible or even detectable by magnetic resonance imaging [26].

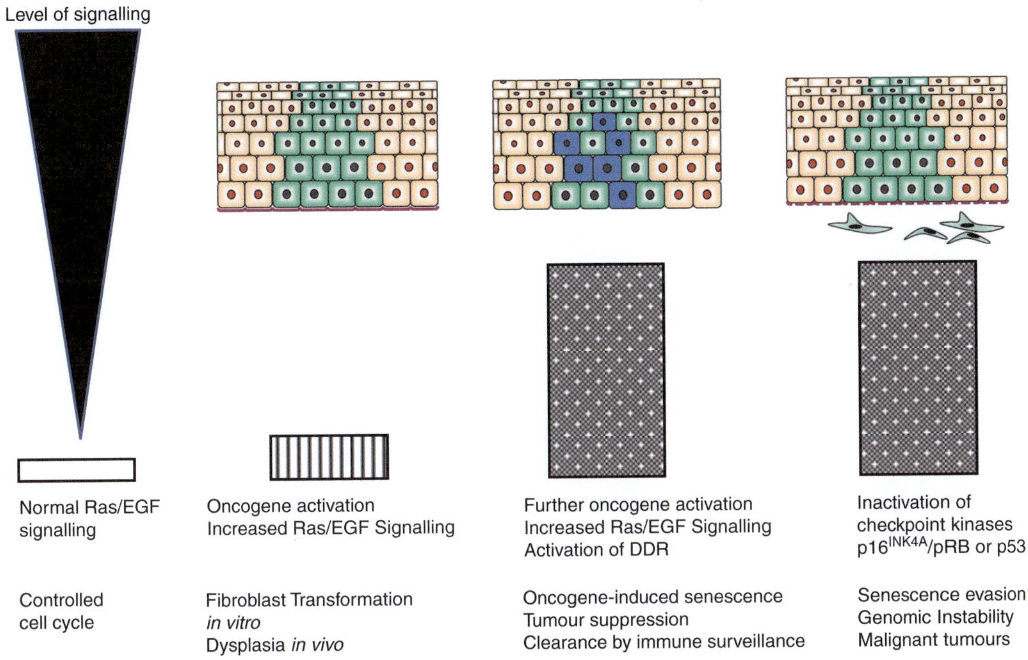

Fig. 8.2 A schematic representation showing how increasing oncogenic signalling through the epidermal growth factor receptor pathway might transform cells in vitro and lead to the growth of dysplasia in vivo but eventually on increasing to the level required for malignancy and tumorigenicity in vivo senescence is triggered leading to clearance of the dysplastic cells by the immune system. Inactivation of the senescence programme leads to increased proliferation and evasion of the immune system leading to telomere erosion, crisis, genetic instability and malignant conversion. The normal cells are depicted with red cytoplasm and the neoplastic cells with green cytoplasm (growing) and blue cytoplasm (senescent). See [15, 22]

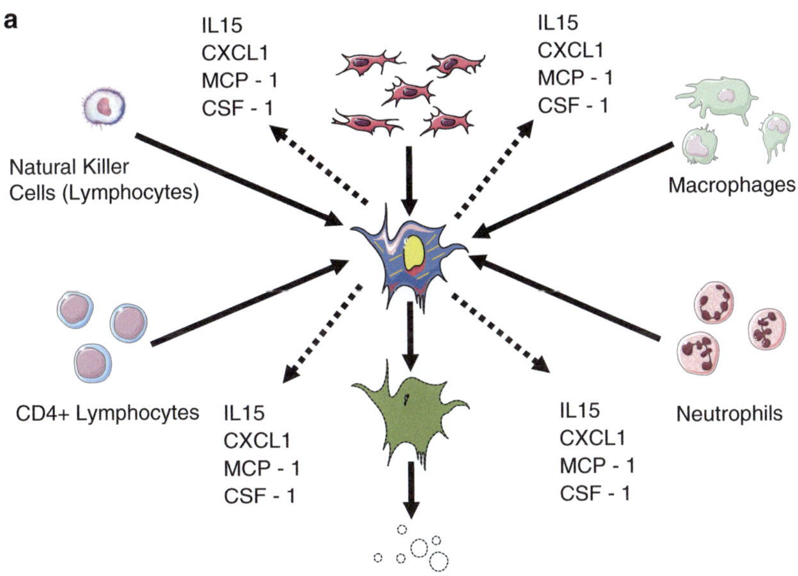

Fig. 8.3 (**a**) Depicts the secretion of an array of inflammatory cytokines, dashed arrows, that lead to the targeting and clearance of the senescent cells. (**b**) Shows that when the immune system (black crosses) or senescence (red cross) is inhibited, senescent cell clearance is blocked (blue crosses) and the senescent cells proliferate and progress to malignancy (see [24, 25])

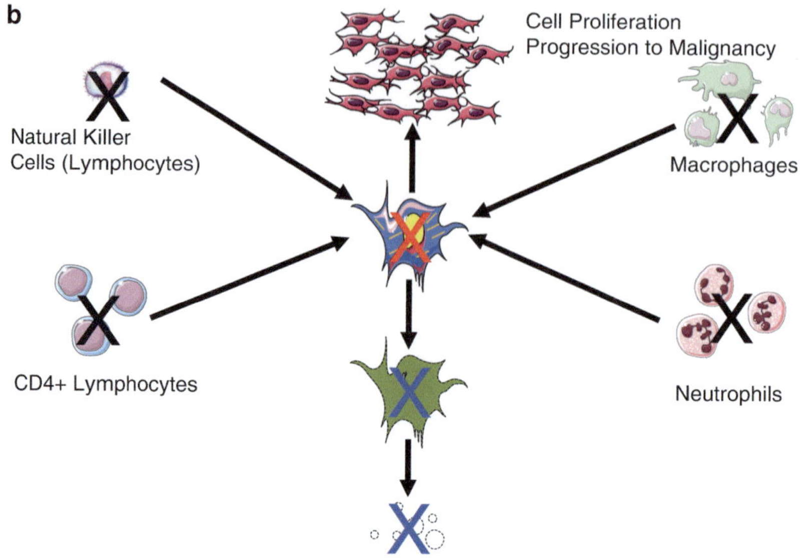

Fig. 8.3 (continued)

8.3 Senescence Bypass Results in Crisis and Chromosomal Instability

Senescence is bypassed in around 90% of oral and head and neck tumours by the concerted loss of p16^{INK4A} and p53 or the expression of HPV E6 and E7 proteins (The Cancer Genome Atlas Network 2015). However, the expression of HPV is largely confined to the tonsil, especially in tumours of the tonsil and oropharynx [27]. Following senescence bypass the cells erode their telomeres further and eventually reach a point where no telomere sequence remains; this leads to extensive chromosome fusions, dicentric chromosomes and formation of anaphase bridges [28, 29]. The anaphase bridges either cause irreversible cell cycle arrest or break resulting in loss and gain of chromosomal material. The consequence of this is mainly in cell death and the phenotype historically known as crisis [30]. At crisis the cells continue to cycle but the population barely expands due to extensive cell death until one or more cells inherit the genetic alterations sufficient to deregulate telomerase and immortalise the cells. This hypothesis is supported by the coexistence of p16^{INK4A}, p53 and telomerase genetic alterations in most tumours and immortal OPML/OSCC lines [14, 31–36]. These include promoter mutations in the gene encoding the catalytic component of telomerase [32], *TERT*, and amplifications of both *TERT* [37] and the RNA component, *TERC* [38], loci. Furthermore, the introduction of telomerase activity coupled with p53 and p16^{INK4A} deficiency into normal keratinocytes is necessary and sufficient to immortalise normal keratinocytes [39]. In mouse models the combination of p53 deficiency and telomeric attrition actually provokes the generation of carcinomas [28] that have similar chromosomal alterations to those of human carcinomas [40] supporting the hypothesis that crisis drives the early chromosomal instability that is associated with tumour progression and resistance to therapies.

8.4 HPV Integration Also Compromises the p53 and p16^{INK4A} Pathways and Results in an Upregulation of Telomerase

HPV infections are common in the human oral cavity but usually the viral DNA remains in the circular form and in this state the expression of the E6 and E7 genes is repressed by the E2 protein [41]. However, upon random integration into the host genome E2 is often lost, thus allowing derepression of the E6 and E7 genes. E6 targets p53 for degradation by ubiquitin ligase [4] and E7 binds pRb, the downstream target of p16^{INK4A}, and targets it for degradation via the cullin 2 ubiquitin ligase [4]. In addition, E6 activates telomerase by upregulating *TERT* transcription; it does this by upregulating *CMYC*, a known positive regulator of *TERT*, in conjunction with NFX1-123 [42] and by targeting a regulator of chromatin known as NFX1-91 for degradation [42]. NFX1-91 regulates chromatin by interacting with the corepressor complex mSin3A/histone deacetylase at the hTERT promoter [43].

8.5 Other Gene Mutations in OSCC Reveal Key Pathways Involved in Their Development

Several other pathways are disrupted in OSCC in addition to those involved in the bypass of senescence. These include the *NOTCH* genes, which although only mutated in around 20% of OSCC [44–47] may also be compromised by p53 dysfunction [48] and hence HPV integration. NOTCH signalling has been estimated to be compromised in around two-thirds of human OSCC [46]. Another gene is *FAT1* which is mutated [44–47] or deleted [49] in a high fraction of OSCCs. *FAT1* is structurally similar to E-cadherin and thought to regulate keratinocyte differentiation and migration by regulation of the β-catenin signalling pathway [50]. The gene encoding caspase 8, *CASP8*, is also mutated in a subset of OSCCs; they tend to have *HRAS* mutations and have a good prognosis [44, 46] whereas PI3 kinase alpha subunit (*PIKCA*) mutations tend to be associated with HPV integration [44, 46]. Interestingly, the frequency of apoptosis increases during OSCC progression [51] but the presence of *CASP8* mutations [44–47] in around 8% of SCC-HN suggests that resistance to the extrinsic apoptosis pathway involving FAS, TRAIL and tumour necrosis factor may be disrupted in some tumours. However, gingival OSCC from Indian patients have a higher frequency of *CASP8* and *FAT1* mutations perhaps indicative of a different aetiology or pathology in tumours from this geographical area [52]. In addition, many gene copy number alterations (gains and losses) are found in OSCC [44, 46] and SCC-HN [44] and these include high copy number gains of the *ERBB1* and *PIKCA* genes [44, 46] that are associated with overexpression of the related proteins, EGFR and PI3K, respectively. As a result, both EGFR and PI3K have been assessed as therapeutic targets for SCC-HN (see below).

8.6 Transcriptional Profiling and Genetic Analysis Support the Hypothesis That OSCCs Arise from Premalignant Lesions and the Bypass of Senescence Occurs Early

Several studies have shown that dysplasias and OSCC share many transcriptional alterations whether isolated from the same patient or not, although OSCCs possess many additional alterations when compared to their premalignant counterparts [9], and similar data has been obtained from the genetic analysis of cultured keratinocytes from these types of lesion [53]. Thus these data support the hypothesis that dysplastic lesions can give rise to OSCC but malignant conversion may in most cases occur at the microscopic level [26].

One of the issues that has compromised the study of OSCC and aerodigestive cancer is the problem that biopsies may not necessarily reflect the properties of the whole lesion. This was originally noted some years ago at the histological level for oral dysplasia [54] and recently confirmed by extensive genetic analyses of lung cancer [6]. Another problem is that OSCC is widely held to emerge from a preneoplastic field of clones that harbour critical cancer-driver mutations [8] and if like other SCCs they may contain more than one cancer-driver mutation per clone [6]. Indeed, it is very rare to find OPML lines that contain only p16^{INK4A} loss without p53 and/or telomerase dysfunction, although two examples have been reported [34, 55]. Although other explanations are possible, this may mean that dysplasias that bypass senescence have done so at the microscopic level and may arise from pre-existing $p16^{INK4A}$ and p53 dysfunctional clones. There is evidence that around 50% of OPMLs have p53 mutations [56], express telomerase [57] and fail to express $p16^{INK4A}$ in vivo [58], strongly suggesting that senescence and crisis are bypassed [14, 36] prior to malignancy. However, immortal OMPLs are very genetically unstable and have almost as many DNA copy number variations as immortal OSCCs (Nalin Thakker—unpublished data) despite being non-tumorigenic in immunosuppressed mice (S. Prime—personal communication). These data indicate that the clonal variations that are thought to underpin drug resistance and confound attempts at targeted therapy occur very early in OSCC development.

8.7 Cancer Stem Cells and OSCC

The classical hierarchical model of specialist stem cells is shown in Fig. 8.4 and consists of a slow-cycling stem cell that then gives rise to fast-cycling progenitors that ultimately produce one or more postmitotic differentiated cell types [59, 60]. The stem cells are also reported to be very sensitive to apoptosis and to segregate their DNA so that they always retain the original template strand, thus avoiding replication errors [61]. The logic behind this hypothesis is that the properties of the stem cell reduce its probability of surviving with potentially damaging mutations and that the progenitors and their offspring would eventually differentiate before they could accumulate sufficient mutations to form tumours.

The original cancer stem cell (CSC) hypothesis extended this model to suggest that in cancers this proliferative hierarchy is preserved with the stem cells being the cells responsible for the regeneration of the tumours following chemo- and radiotherapy (Fig. 8.4) [62].

The first evidence in support of the CSC hypothesis came from the haematopoietic system where only the cells with the characteristics of the CD34++/CD38− stem cell were shown to be capable of the engraftment of acute myeloid leukaemia in non-obese diabetic mice with severe combined immunodeficiency disease (NOD/SCID) mice [63] and several similar studies followed in other human cancer systems, including SCC of the head and neck (SCC-HN [64]) that includes OSCC. Numerous cell surface markers of CSCs have been characterised and used to demonstrate several of their properties, including several normal stem cell characteristics [64] and radio- and chemo-resistance [65].

However, the application of lineage tracing technology has suggested that the original stem cell hypothesis may need modification and that epithelial tissues may be more plastic than originally realised. In particular, the discovery of fast-cycling cells that are positive for leucine-rich repeat-containing G protein-coupled receptor 5 or 6 (LGR5 and LGR6) [66–68] and the inability to confirm the existence of fast-cycling progenitor cells [69] have led to the hypothesis that there are two types of stem cell in epithelial tissues and the LGR5/6-positive type are the permanent residents [66, 67]. The LGR5/6-positive cells are fast cycling, not sensitive to apoptosis, do not segregate their DNA to retain an 'immortal' DNA strand and are telomerase positive, at least in the mouse [66, 67, 70]. This new data has resulted in experiments showing that the LGR5-positive cells are required for tumour maintenance at least in the mouse small intestine, making the cells good candidates for CSCs [67, 71].

More recently, several modifications have been made to the CSC hypothesis in an attempt to explain how CSCs metastasise and the generation of epithelial-mesenchymal transition (EMT) has been reported to drastically increase the number of CSCs. EMT is known to precipitate epithelial cell discohesion, dedifferentiation and invasion in a variety of cell and animal model systems and when the tumour cells spread to new sites the cells are proposed to undergo mesenchymal-epithelial transition (MET) and revert back to the differentiated tissue that is actually observed in most human metastatic deposits [72]. However, more recently the essential role for at least certain types of EMT in the generation of invasive and metastatic CSCs has been questioned [73, 74], although a role for EMT in certain types of chemotherapeutic resistance and resultant metastasis was confirmed and this was associated with a

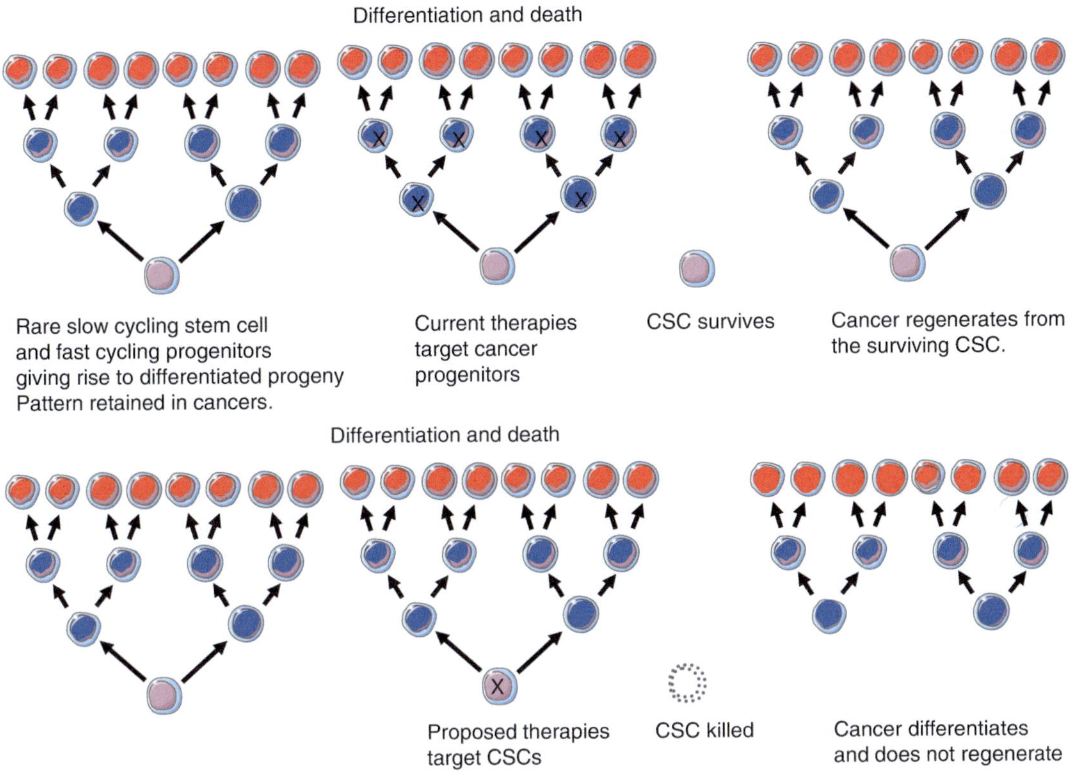

Fig. 8.4 Depicts the classical cancer stem cell (CSC) hypothesis showing the selective survival of CSCs following chemo- or radiotherapy (top panel) and the proposal that targeting the CSC will lead to a more sustained therapeutic response as the non-CSC cells will undergo terminal differentiation and die (bottom panel)

slower proliferation rate and the expression of certain drug transporters [73, 74]. Additionally, although the role of EMT in the CSC phenotype was not addressed in vivo, the ablation of EMT by the deletion of twist or snail did not affect the short-term potential of mouse pancreatic cancer cells to form tumour spheres which is associated with the putative CSC phenotype [74].

The CSC hypothesis relies largely on injecting cancer cells into various sites in immunosuppressed mice such as SCID/NOD and in vitro assays such as the spheroid assay and as such many authors prefer the term tumour-initiating cell (TIC) as this is a better description of what the CSC experiments actually demonstrate and does not commit to any association of the CSC with its normal counterpart.

Recent research in the last 10 years has resulted in fierce debates that have questioned whether CSCs/TICs are actually that rare and whether the nature of the SCID/NOD mouse assay is actually too severe to be physiologically relevant because rendering the mouse more immunodeficient and injecting the cancer cells with basement membrane components in the form of Matrigel do increase the number of assayable CSCs/TICs, especially in cell lines [75], although this has been disputed. Related to this point, many carcinoma cells undergo either suspension-induced apoptosis (anoikis) or terminal differentiation [76] and the cells that survive this assay better have long been known to be more tumorigenic [76]. Therefore, as the same pathways in drug resistance and anoikis overlap this could also explain why CSCs/TICs from carcinomas are resistant to chemo- radiotherapy [62]. In addition the original experiments performed on acute myeloid leukaemia have been reassessed and CSCs/TICs have been shown to be mainly present in CD38+ and CD34+ compartments, although the CD34−/CD38+ cells were shown to have superior serial tumour transplantation capacity [77]. Finally, several recent papers have highlighted the instability of the CSC/TIC populations, especially in melanoma.

Despite the intensive debate on the potential importance of CSCs/TICs in the spread and recurrence of human cancer, understanding their nature remains an important issue. It has recently been questioned on the grounds of probability whether CSCs/TICs are derived from mutated normal stem cells [78] and the recently demonstrated plasticity of both normal [67] and CSC/TICs [79] supports this hypothesis and also suggests that the targeting of CSC/TICs alone would not result in as drastic an improvement in therapy as originally anticipated.

Additional problems with the CSC hypothesis are that it does not explain why multiple clones of genetically distinct cancer cells exist and that this clonal heterogeneity is stimulated by chemotherapy [7]. In addition, the telomeres of most human cancers are short yet normal stem cells do have some telomerase activity or are slow cycling (Fig. 8.5a, b) which would be expected to slow the rate of telomere attrition [80] and both of these stem cell types have been reported to initiate tumorigenesis [67, 71] (Fig. 8.5a, b). In many cancers telomere shortening precedes overt malignancy [81, 82] and these data collectively are more consistent with the hypothesis that cancers, including OSCC, progress from telomerase-deficient non-stem cells [78, 83] that ultimately

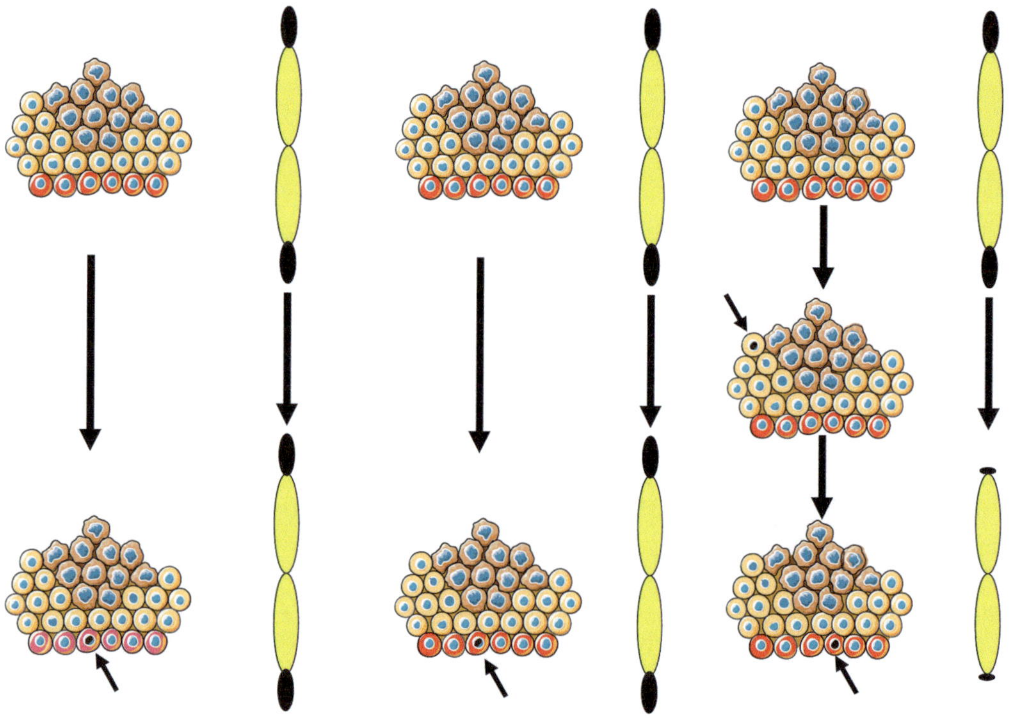

Fig. 8.5 Depicts a revised model of how different stem cells might contribute to cancer. The left-hand panel shows quiescent Bmi1-postive stem cells giving rise to tumours with long telomeres as they rarely divide normally and express telomerase when they do divide. The middle panel shows the fast-cycling LGR5/6-positive stem cells which are telomerase positive also giving rise to tumours with long telomeres. The right-hand panel shows the mutated stem cell arising from the telomerase-negative compartment of the tissue, undergoing telomere attrition, and then under conditions of NFκB-mediated inflammation or regeneration finding new CSC clones that may be genetically distinct. This model explains both telomere attrition in most human cancers and genetic diversity within CSC populations. See [67]

bypass senescence and crisis to become genetically unstable and generate multiple genetically distinct clones (Fig. 8.5c). Admittedly, these cells may found new CSCs that could well be genetically distinct from one another and contribute to therapeutic resistance [7, 78] and so the original CSC hypothesis needs modification to take account of these observations.

8.8 The Failure of Targeted Therapies and Prospects for the Future Management of OSCC

8.8.1 Cetuximab and EGF Kinase Inhibitors

Cetuximab is a monoclonal antibody directed against the epidermal growth factor receptor (EGFR) and would be predicted to block all forms of downstream signalling from the EGFR including the phosphatidylinositol 3-kinase (PI3K) pathway and mitogen-activated kinase cascades. However, only one clinical trial reported any benefit of cetuximab in recurrent metastatic disease and even then it had to be combined with cisplatin and 5-fluorouracil but resulted in an increased survival of 2.7 months [84]. Other monoclonal antibodies directed against EGFR such as panitumumab gave no survival advantage in clinical trials and also resulted in more toxicity and treatment-related deaths. The EGFR tyrosine kinase inhibitors gefitinib and erlotinib have not yielded positive results but some of the broader spectrum ERB2 receptor family inhibitors such as afatinib and dacomitinib have yielded slightly better results than cetuximab in some trials and in the latter case better effects are observed in tumours without PI3K mutations [84].

8.8.2 PI3K and TORC1 Inhibitors

The gene encoding the alpha subunit of PI3K, *PIKCA*, is mutated in a significant fraction of SCC-HN and the pathway is estimated to be altered in about 30–50% overall but especially in HPV-positive tumours. However, so far PI3K inhibitors have not shown any clear benefit to SCC-HN patients in clinical trials. Drugs directed against one of the downstream targets of PI3K, TORC1, have been tested in the clinic and although they are poorly tolerated in combination with EGFR inhibitors one of them, temsirolimus, gave a good response in patients refractory to cetuximab therapy [84].

8.9 Immunotherapy

The evidence for cancer cells evading the immune system was presented 5 years ago by Hanahan and Weinberg [72] and the most compelling evidence was the occurrence of a cancer in an immunosuppressed patient that was of donor origin [85]. Since then it has been shown that cancer cells of various types attenuate the cytotoxic T cell response by expressing the ligand for the PD-1 receptor on the T cells, PD ligand 1 (PD-L1) on their surface [86]. This suppresses the proliferation of the T cells and their cytokine production allowing the cancer cells to evade destruction. Several preclinical studies subsequently showed that inhibiting the interaction between PD-1 and PD-L1 enhanced the immune response and mediated an antitumour effect [86–88]. Additionally, PD-L1 is expressed on the surface of several human cancers, including SCC-HN as well as tumour-infiltrating cells in the tumour environment [86]. Dual therapy using antibodies against PD-1 and PD-L1 have been shown to be highly successful in clinical trials, including most recently SCC-HN [86] and http://www.biospace.com/News/bristol-myers-squibb-release-checkmate-141-a/407013?type=email&source=CS_012816. The dual therapy works by targeting the immune checkpoint in the cognitive phase where the PD-1 is expressed on the antigen-presenting cells in the lymph node and also in the effector phase in the cancer microenvironment where the PD-1 is expressed on the cancer cells and tumour-infiltrating lymphocytes [86]. Furthermore, the dual therapy is likely to be most effective in tumour types with the heaviest mutation load as the more mutations in the cancer the greater the number of cancer-associated antigens [86] and it has recently been shown that tumours that have a durable response to PD-1 blockade are enriched for tumour-specific clonal neoantigens [7].

8.10 Clonal Heterogeneity and Cancer Therapy

One of the problems with most adult cancers is that by the time they are diagnosed they have amassed a large number of cells approximately and these cell populations contain tremendous genetic heterogeneity, thus providing the platform for resistance to virtually all current forms of cancer therapy. A T1-stage oral cancer for example could contain more than 10^{12} cells. This problem is exacerbated by recent data showing that conventional chemo- and radiotherapy actually increases clonal heterogeneity and mutational load and this was associated with resistance to dual immunotherapy possibly by creating subclonal neoantigens [7].

8.11 Early Diagnosis

In the light of recent reports that conventional chemo- and radiotherapies exacerbate clonal heterogeneity [7] and hence likely therapeutic resistance the ability to detect cancer early now assumes an even greater importance than previously and preferably before genetically unstable cancers expand to a critical mass where clonal heterogeneity compromises therapeutic strategies. However, invasive technologies such as biopsy and even oral brushings may not represent a genetically heterogenous premalignant cell population accurately. This said, detecting small clones of genetically unstable cancer cells is technically difficult as the development of genetic instability may well occur microscopically [26] and this coupled with biopsy inconsistency [6] makes any invasive strategy difficult to interpret. Attempts have been made to identify secreted proteins and metabolites [89, 90] from serum [91] and other body fluids such as saliva [89, 92] but as yet these studies have largely been confined to established cancers where confounding factors such as hypoxia and other indirect consequences of having cancer remain a problem; inconsistency in the results from different techniques and a failure to understand their cell and molecular basis are also outstanding issues.

More recently, it has been reported that the driver mutations reported above can be detected in the saliva and serum of all head and neck SCC patients [93] and clearly this has great potential in the early detection of OSCC provided that these methods are proven to be sufficiently cheap and specific to cancer cells. However, certain driver mutations may not be as useful as others because they can be found at a high frequency in normal epidermal keratinocytes at sun-exposed sites [94] and presumably this will also apply to the oral mucosa of heavy smokers. Further work is required to resolve these issues.

8.12 HPV Vaccination

Recently a 9-valent vaccine (Gardasil-9 from Merck) has been developed and subjected to clinical trials in nearly 6000 young individuals. It has activity against HPV subtypes 6, 11, 16, 18, 31, 33, 45, 52 and 58 and reduces the frequency of cervical intraepithelial neoplasia grade 2 (CIN2) premalignant lesions and persistent viral infections by 96% (reviewed by [95]. However, it is currently difficult to assess whether this vaccine will have a similar effect on SCCs of the oropharynx as it is not easy to identify premalignant lesions of this site [95]. Although premalignant lesions of other sites are observed and HPV presence has been reported [96] it was not clear from these studies whether HPV was integrated. Most SCCs in the head and neck region arise without any visible premalignant lesion; furthermore, most dysplastic lesions that are visible at other sites do not progress [11] and those that do are generally HPV negative [97].

8.13 Summary and Conclusions

In summary, whilst OSCC remains a very difficult type of cancer to treat two novel therapy strategies hold considerable promise. Firstly, the development of prophylactic HPV vaccines

[5, 95] could be predicted to reduce the frequency of the type of SCC-HN (mainly tonsil and oropharynx) where HPV is integrated in parallel with a reduction in genital SCCs in the next 20 years. Furthermore, as the role of episomal HPV in OSCC and SCC-HN is not really known, the effect of these vaccines on sites other than tonsil may be more dramatic than anticipated [4]. Secondly, the novel dual-targeting immunotherapy techniques [86] that have recently been licensed for melanoma treatment have also been reported to work for 33% of HPV-negative and 25% of SCC-HN-positive SCC-HN, http://www.biospace.com/News/bristol-myers-squibb-release-checkmate-141-a/407013?type=email&source=CS_012816, and this represents the first promising advance in the treatment for OSCC and OSCC-HN although the manner of its use may need to be assessed to achieve optimum results especially in the light of recent results showing that conventional therapy may exacerbate clonal heterogeneity and hence drug resistance to all therapies, including dual targeting of PD-1 and PD-L1 [7].

References

1. Parkin DM, Bray F, Ferlay J, Pisani P. Global cancer statistics, 2002. CA Cancer J Clin. 2005;55(2):74–108.
2. Majchrzak E, Szybiak B, Wegner A, Pienkowski P, Pazdrowski J, Luczewski L, et al. Oral cavity and oropharyngeal squamous cell carcinoma in young adults: a review of the literature. Radiol Oncol. 2014;48(1):1–10.
3. Paterson IC, Eveson JW, Prime SS. Molecular changes in oral cancer may reflect aetiology and ethnic origin. Eur J Cancer B Oral Oncol. 1996;32B(3):150–3.
4. Chung CH, Gillison ML. Human papillomavirus in head and neck cancer: its role in pathogenesis and clinical implications. Clin Cancer Res. 2009;15(22):6758–62.
5. Yang DY, Bracken K. Update on the new 9-valent vaccine for human papillomavirus prevention. Can Fam Physician. 2016;62(5):399–402.
6. de Bruin EC, McGranahan N, Mitter R, Salm M, Wedge DC, Yates L, et al. Spatial and temporal diversity in genomic instability processes defines lung cancer evolution. Science. 2014;346(6206):251–6.
7. McGranahan N, Furness AJ, Rosenthal R, Ramskov S, Lyngaa R, Saini SK, et al. Clonal neoantigens elicit T cell immunoreactivity and sensitivity to immune checkpoint blockade. Science. 2016;351(6280):1463–9.
8. Braakhuis BJ, Tabor MP, Kummer JA, Leemans CR, Brakenhoff RH. A genetic explanation of Slaughter's concept of field cancerization: evidence and clinical implications. Cancer Res. 2003;63(8):1727–30.
9. Ha PK, Benoit NE, Yochem R, Sciubba J, Zahurak M, Sidransky D, et al. A transcriptional progression model for head and neck cancer. Clin Cancer Res. 2003;9(8):3058–64.
10. Mithani SK, Mydlarz WK, Grumbine FL, Smith IM, Califano JA. Molecular genetics of premalignant oral lesions. Oral Dis. 2007;13(2):126–33.
11. Hunter KD, Parkinson EK, Harrison PR. Profiling early head and neck cancer. Nat Rev Cancer. 2005;5(2):127–35.
12. Campisi J, d'Adda di Fagagna F. Cellular senescence: when bad things happen to good cells. Nat Rev Mol Cell Biol. 2007;8(9):729–40.
13. James EL, Michalek RD, Pitiyage GN, de Castro AM, Vignola KS, Jones J, et al. Senescent human fibroblasts show increased glycolysis and redox homeostasis with extracellular metabolomes that overlap with those of irreparable DNA damage, aging, and disease. J Proteome Res. 2015;14(4):1854–71.
14. Parkinson EK, James EL, Prime SS. Senescence-derived extracellular molecules as modulators of oral cancer development: a mini-review. Gerontology. 2016;62(4):417–24.
15. Bartkova J, Horejsi Z, Koed K, Kramer A, Tort F, Zieger K, et al. DNA damage response as a candidate anti-cancer barrier in early human tumorigenesis. Nature. 2005;434(7035):864–70.
16. Di Micco R, Fumagalli M, Cicalese A, Piccinin S, Gasparini P, Luise C, et al. Oncogene-induced senescence is a DNA damage response triggered by DNA hyper-replication. Nature. 2006;444(7119):638–42.
17. Bartkova J, Rezaei N, Liontos M, Karakaidos P, Kletsas D, Issaeva N, et al. Oncogene-induced senescence is part of the tumorigenesis barrier imposed by DNA damage checkpoints. Nature. 2006;444(7119):633–7.
18. Braig M, Lee S, Loddenkemper C, Rudolph C, Peters AH, Schlegelberger B, et al. Oncogene-induced senescence as an initial barrier in lymphoma development. Nature. 2005;436(7051):660–5.
19. Chen Z, Trotman LC, Shaffer D, Lin HK, Dotan ZA, Niki M, et al. Crucial role of p53-dependent cellular senescence in suppression of Pten-deficient tumorigenesis. Nature. 2005;436(7051):725–30.
20. Collado M, Gil J, Efeyan A, Guerra C, Schuhmacher AJ, Barradas M, et al. Tumour biology: senescence in premalignant tumours. Nature. 2005;436(7051):642.
21. Michaloglou C, Vredeveld LC, Soengas MS, Denoyelle C, Kuilman T, van der Horst CM, et al. BRAFE600-associated senescence-like cell cycle arrest of human naevi. Nature. 2005;436(7051):720–4.
22. Sarkisian CJ, Keister BA, Stairs DB, Boxer RB, Moody SE, Chodosh LA. Dose-dependent oncogene-induced senescence in vivo and its evasion during mammary tumorigenesis. Nat Cell Biol. 2007;9(5):493–505.
23. Coppe JP, Patil CK, Rodier F, Sun Y, Munoz DP, Goldstein J, et al. Senescence-associated secretory phenotypes reveal cell-nonautonomous functions of oncogenic RAS and the p53 tumor suppressor. PLoS Biol. 2008;6(12):2853–68.

24. Kang TW, Yevsa T, Woller N, Hoenicke L, Wuestefeld T, Dauch D, et al. Senescence surveillance of pre-malignant hepatocytes limits liver cancer development. Nature. 2011;479(7374):547–51.
25. Xue W, Zender L, Miething C, Dickins RA, Hernando E, Krizhanovsky V, et al. Senescence and tumour clearance is triggered by p53 restoration in murine liver carcinomas. Nature. 2007;445(7128):656–60.
26. Burd CE, Sorrentino JA, Clark KS, Darr DB, Krishnamurthy J, Deal AM, et al. Monitoring tumorigenesis and senescence in vivo with a p16(INK4a)-luciferase model. Cell. 2013;152(1–2):340–51.
27. Braakhuis BJ, Snijders PJ, Keune WJ, Meijer CJ, Ruijter-Schippers HJ, Leemans CR, et al. Genetic patterns in head and neck cancers that contain or lack transcriptionally active human papillomavirus. J Natl Cancer Inst. 2004;96(13):998–1006.
28. Artandi SE, Chang S, Lee SL, Alson S, Gottlieb GJ, Chin L, et al. Telomere dysfunction promotes non-reciprocal translocations and epithelial cancers in mice. Nature. 2000;406(6796):641–5.
29. Capper R, Britt-Compton B, Tankimanova M, Rowson J, Letsolo B, Man S, et al. The nature of telomere fusion and a definition of the critical telomere length in human cells. Genes Dev. 2007;21(19):2495–508.
30. Wright WE, Pereira-Smith OM, Shay JW. Reversible cellular senescence: implications for immortalization of normal human diploid fibroblasts. Mol Cell Biol. 1989;9(7):3088–92.
31. Edington KG, Loughran OP, Berry IJ, Parkinson EK. Cellular immortality: a late event in the progression of human squamous cell carcinoma of the head and neck associated with p53 alteration and a high frequency of allele loss. Mol Carcinog. 1995;13(4):254–65.
32. Killela PJ, Reitman ZJ, Jiao Y, Bettegowda C, Agrawal N, Diaz LA Jr, et al. TERT promoter mutations occur frequently in gliomas and a subset of tumors derived from cells with low rates of self-renewal. Proc Natl Acad Sci U S A. 2013;110(15):6021–6.
33. Loughran O, Clark LJ, Bond J, Baker A, Berry IJ, Edington KG, et al. Evidence for the inactivation of multiple replicative lifespan genes in immortal human squamous cell carcinoma keratinocytes. Oncogene. 1997;14(16):1955–64.
34. McGregor F, Muntoni A, Fleming J, Brown J, Felix DH, MacDonald DG, et al. Molecular changes associated with oral dysplasia progression and acquisition of immortality: potential for its reversal by 5-azacytidine. Cancer Res. 2002;62(16):4757–66.
35. Munro J, Stott FJ, Vousden KH, Peters G, Parkinson EK. Role of the alternative INK4A proteins in human keratinocyte senescence: evidence for the specific inactivation of p16INK4A upon immortalization. Cancer Res. 1999;59(11):2516–21.
36. Parkinson EK. Senescence as a modulator of oral squamous cell carcinoma development. Oral Oncol. 2010;46(12):840–53.
37. Zhang A, Zheng C, Lindvall C, Hou M, Ekedahl J, Lewensohn R, et al. Frequent amplification of the telomerase reverse transcriptase gene in human tumors. Cancer Res. 2000;60(22):6230–5.
38. Soder AI, Hoare SF, Muir S, Going JJ, Parkinson EK, Keith WN. Amplification, increased dosage and in situ expression of the telomerase RNA gene in human cancer. Oncogene. 1997;14(9):1013–21.
39. Rheinwald JG, Hahn WC, Ramsey MR, Wu JY, Guo Z, Tsao H, et al. A two-stage, p16(INK4A)- and p53-dependent keratinocyte senescence mechanism that limits replicative potential independent of telomere status. Mol Cell Biol. 2002;22(14):5157–72.
40. O'Hagan RC, Chang S, Maser RS, Mohan R, Artandi SE, Chin L, et al. Telomere dysfunction provokes regional amplification and deletion in cancer genomes. Cancer Cell. 2002;2(2):149–55.
41. Ordonez RM, Espinosa AM, Sanchez-Gonzalez DJ, Armendariz-Borunda J, Berumen J. Enhanced oncogenicity of Asian-American human papillomavirus 16 is associated with impaired E2 repression of E6/E7 oncogene transcription. J Gen Virol. 2004;85(Pt 6):1433–44.
42. Gewin L, Myers H, Kiyono T, Galloway DA. Identification of a novel telomerase repressor that interacts with the human papillomavirus type-16 E6/E6-AP complex. Genes Dev. 2004;18(18):2269–82.
43. Xu M, Luo W, Elzi DJ, Grandori C, Galloway DA. NFX1 interacts with mSin3A/histone deacetylase to repress hTERT transcription in keratinocytes. Mol Cell Biol. 2008;28(15):4819–28.
44. The Cancer Genome Atlas Network. Comprehensive genomic characterization of head and neck squamous cell carcinomas. Nature. 2015;517(7536):576–82.
45. Agrawal N, Frederick MJ, Pickering CR, Bettegowda C, Chang K, Li RJ, et al. Exome sequencing of head and neck squamous cell carcinoma reveals inactivating mutations in NOTCH1. Science. 2011;333(6046):1154–7.
46. Pickering CR, Zhang J, Yoo SY, Bengtsson L, Moorthy S, Neskey DM, et al. Integrative genomic characterization of oral squamous cell carcinoma identifies frequent somatic drivers. Cancer Discov. 2013;3(7):770–81.
47. Stransky N, Egloff AM, Tward AD, Kostic AD, Cibulskis K, Sivachenko A, et al. The mutational landscape of head and neck squamous cell carcinoma. Science. 2011;333(6046):1157–60.
48. Lefort K, Mandinova A, Ostano P, Kolev V, Calpini V, Kolfschoten I, et al. Notch1 is a p53 target gene involved in human keratinocyte tumor suppression through negative regulation of ROCK1/2 and MRCKalpha kinases. Genes Dev. 2007;21(5):562–77.
49. Nakaya K, Yamagata HD, Arita N, Nakashiro KI, Nose M, Miki T, et al. Identification of homozygous deletions of tumor suppressor gene FAT in oral cancer using CGH-array. Oncogene. 2007;26(36):5300–8.
50. Nishikawa Y, Miyazaki T, Nakashiro K, Yamagata H, Isokane M, Goda H, et al. Human FAT1 cadherin controls cell migration and invasion of oral squamous cell carcinoma through the localization of beta-catenin. Oncol Rep. 2011;26(3):587–92.
51. Hague A, Eveson JW, MacFarlane M, Huntley S, Janghra N, Thavaraj S. Caspase-3 expression is reduced, in the absence of cleavage, in terminally differentiated normal oral epithelium but is increased in

oral squamous cell carcinomas and correlates with tumour stage. J Pathol. 2004;204(2):175–82.
52. India, Project, Team, International, Cancer, Genome, et al. Mutational landscape of gingivo-buccal oral squamous cell carcinoma reveals new recurrently-mutated genes and molecular subgroups. Nat Commun. 2013;4:2873.
53. Hunter KD, Thurlow JK, Fleming J, Drake PJ, Vass JK, Kalna G, et al. Divergent routes to oral cancer. Cancer Res. 2006;66(15):7405–13.
54. Jeong WJ, Paik JH, Cho SW, Sung MW, Kim KH, Ahn SH. Excisional biopsy for management of lateral tongue leukoplakia. J Oral Pathol Med. 2012;41(5):384–8.
55. Dickson MA, Hahn WC, Ino Y, Ronfard V, Wu JY, Weinberg RA, et al. Human keratinocytes that express hTERT and also bypass a p16(INK4a)-enforced mechanism that limits life span become immortal yet retain normal growth and differentiation characteristics. Mol Cell Biol. 2000;20(4):1436–47.
56. Qin GZ, Park JY, Chen SY, Lazarus P. A high prevalence of p53 mutations in pre-malignant oral erythroplakia. Int J Cancer. 1999;80(3):345–8.
57. Mao L, El-Naggar AK, Fan YH, Lee JS, Lippman SM, Kayser S, et al. Telomerase activity in head and neck squamous cell carcinoma and adjacent tissues. Cancer Res. 1996;56(24):5600–4.
58. Papadimitrakopoulou V, Izzo J, Lippman SM, Lee JS, Fan YH, Clayman G, et al. Frequent inactivation of p16INK4a in oral premalignant lesions. Oncogene. 1997;14(15):1799–803.
59. Allen TD, Potten CS. Fine-structural identification and organization of the epidermal proliferative unit. J Cell Sci. 1974;15(2):291–319.
60. Mackenzie IC. Relationship between mitosis and the ordered structure of the stratum corneum in mouse epidermis. Nature. 1970;226(5246):653–5.
61. Potten CS. Keratinocyte stem cells, label-retaining cells and possible genome protection mechanisms. J Investig Dermatol Symp Proc. 2004;9(3):183–95.
62. Costea DE, Tsinkalovsky O, Vintermyr OK, Johannessen AC, Mackenzie IC. Cancer stem cells—new and potentially important targets for the therapy of oral squamous cell carcinoma. Oral Dis. 2006;12(5):443–54.
63. Bonnet D, Dick JE. Human acute myeloid leukemia is organized as a hierarchy that originates from a primitive hematopoietic cell. Nat Med. 1997;3(7):730–7.
64. Birkeland AC, Owen JH, Prince ME. Targeting head and neck cancer stem cells: current advances and future challenges. J Dent Res. 2015;94(11):1516–23.
65. Gemenetzidis E, Gammon L, Biddle A, Emich H, Mackenzie IC. Invasive oral cancer stem cells display resistance to ionising radiation. Oncotarget. 2015;6(41):43964–77.
66. Barker N, van Es JH, Kuipers J, Kujala P, van den Born M, Cozijnsen M, et al. Identification of stem cells in small intestine and colon by marker gene Lgr5. Nature. 2007;449(7165):1003–7.
67. Clevers H. Stem cells. What is an adult stem cell? Science. 2015;350(6266):1319–20.
68. Snippert HJ, Haegebarth A, Kasper M, Jaks V, van Es JH, Barker N, et al. Lgr6 marks stem cells in the hair follicle that generate all cell lineages of the skin. Science. 2010;327(5971):1385–9.
69. Clayton E, Doupe DP, Klein AM, Winton DJ, Simons BD, Jones PH. A single type of progenitor cell maintains normal epidermis. Nature. 2007;446(7132):185–9.
70. Schepers AG, Vries R, van den Born M, van de Wetering M, Clevers H. Lgr5 intestinal stem cells have high telomerase activity and randomly segregate their chromosomes. EMBO J. 2011;30(6):1104–9.
71. Barker N, Ridgway RA, van Es JH, van de Wetering M, Begthel H, van den Born M, et al. Crypt stem cells as the cells-of-origin of intestinal cancer. Nature. 2009;457(7229):608–11.
72. Hanahan D, Weinberg RA. Hallmarks of cancer: the next generation. Cell. 2011;144(5):646–74.
73. Fischer KR, Durrans A, Lee S, Sheng J, Li F, Wong ST, et al. Epithelial-to-mesenchymal transition is not required for lung metastasis but contributes to chemoresistance. Nature. 2015;527(7579):472–6.
74. Zheng X, Carstens JL, Kim J, Scheible M, Kaye J, Sugimoto H, et al. Epithelial-to-mesenchymal transition is dispensable for metastasis but induces chemoresistance in pancreatic cancer. Nature. 2015;527(7579):525–30.
75. Quintana E, Shackleton M, Sabel MS, Fullen DR, Johnson TM, Morrison SJ. Efficient tumour formation by single human melanoma cells. Nature. 2008;456(7222):593–8.
76. Rheinwald JG, Beckett MA. Defective terminal differentiation in culture as a consistent and selectable character of malignant human keratinocytes. Cell. 1980;22(2 Pt 2):629–32.
77. Taussig DC, Miraki-Moud F, Anjos-Afonso F, Pearce DJ, Allen K, Ridler C, et al. Anti-CD38 antibody-mediated clearance of human repopulating cells masks the heterogeneity of leukemia-initiating cells. Blood. 2008;112(3):568–75.
78. Chaffer CL, Weinberg RA. How does multistep tumorigenesis really proceed? Cancer Discov. 2015;5(1):22–4.
79. Ye X, Weinberg RA. Epithelial-mesenchymal plasticity: a central regulator of cancer progression. Trends Cell Biol. 2015;25(11):675–86.
80. Allen ND, Baird DM. Telomere length maintenance in stem cell populations. Biochim Biophys Acta. 2009;1792(4):324–8.
81. Hastie ND, Dempster M, Dunlop MG, Thompson AM, Green DK, Allshire RC. Telomere reduction in human colorectal carcinoma and with ageing. Nature. 1990;346(6287):866–8.
82. Roger L, Jones RE, Heppel NH, Williams GT, Sampson JR, Baird DM. Extensive telomere erosion in the initiation of colorectal adenomas and its association with chromosomal instability. J Natl Cancer Inst. 2013;105(16):1202–11.

83. Shay JW, Wright WE. The reactivation of telomerase activity in cancer progression. Trends Genet. 1996;12(4):129–31.
84. Sacco AG, Cohen EE. Current treatment options for recurrent or metastatic head and neck squamous cell carcinoma. J Clin Oncol. 2015;33(29):3305–13.
85. Strauss DC, Thomas JM. Transmission of donor melanoma by organ transplantation. Lancet Oncol. 2010;11(8):790–6.
86. Hamanishi J, Mandai M, Matsumura N, Abiko K, Baba T, Konishi I. PD-1/PD-L1 blockade in cancer treatment: perspectives and issues. Int J Clin Oncol. 2016;21(3):462–73.
87. Blank C, Brown I, Peterson AC, Spiotto M, Iwai Y, Honjo T, et al. PD-L1/B7H-1 inhibits the effector phase of tumor rejection by T cell receptor (TCR) transgenic CD8+ T cells. Cancer Res. 2004;64(3):1140–5.
88. Iwai Y, Ishida M, Tanaka Y, Okazaki T, Honjo T, Minato N. Involvement of PD-L1 on tumor cells in the escape from host immune system and tumor immunotherapy by PD-L1 blockade. Proc Natl Acad Sci U S A. 2002;99(19):12293–7.
89. James EL, Parkinson EK. Serum metabolomics in animal models and human disease. Curr Opin Clin Nutr Metab Care. 2015;18(5):478–83.
90. Yakob M, Fuentes L, Wang MB, Abemayor E, Wong DT. Salivary biomarkers for detection of oral squamous cell carcinoma—current state and recent advances. Curr Oral Health Rep. 2014;1(2):133–41.
91. Tiziani S, Lopes V, Gunther UL. Early stage diagnosis of oral cancer using 1H NMR-based metabolomics. Neoplasia. 2009;11(3):269–76.
92. Sugimoto M, Wong DT, Hirayama A, Soga T, Tomita M. Capillary electrophoresis mass spectrometry-based saliva metabolomics identified oral, breast and pancreatic cancer-specific profiles. Metabolomics. 2010;6(1):78–95.
93. Wang Y, Springer S, Mulvey CL, Silliman N, Schaefer J, Sausen M, et al. Detection of somatic mutations and HPV in the saliva and plasma of patients with head and neck squamous cell carcinomas. Sci Transl Med. 2015;7(293):293ra104.
94. Martincorena I, Roshan A, Gerstung M, Ellis P, Van Loo P, McLaren S, et al. Tumor evolution. High burden and pervasive positive selection of somatic mutations in normal human skin. Science. 2015;348(6237):880–6.
95. Guo T, Eisele DW, Fakhry C. The potential impact of prophylactic human papillomavirus vaccination on oropharyngeal cancer. Cancer. 2016;122(15):2313–23.
96. Jayaprakash V, Reid M, Hatton E, Merzianu M, Rigual N, Marshall J, et al. Human papillomavirus types 16 and 18 in epithelial dysplasia of oral cavity and oropharynx: a meta-analysis, 1985–2010. Oral Oncol. 2011;47(11):1048–54.
97. Natarajan E, Saeb M, Crum CP, Woo SB, McKee PH, Rheinwald JG. Co-expression of p16(INK4A) and laminin 5 gamma2 by microinvasive and superficial squamous cell carcinomas in vivo and by migrating wound and senescent keratinocytes in culture. Am J Pathol. 2003;163(2):477–91.

9 Epigenetics in the Oral Cavity

Lena Larsson

Epigenetics consider the interface between genetics and environmental factors, resulting in the phenotype. The term epigenetics was coined by Conrad H. Waddington already in the 1940s [1] as a term to describe the "causal mechanisms" by which the genes in the genotype give rise to the phenotype. The modern definition of epigenetics is changes in gene expression that are not encoded in the DNA sequence [2]. Epigenetic modifications include chemical alterations of DNA and associated proteins, leading to remodeling of the chromatin and activation or inactivation of a gene. Changes in the epigenome contribute to the development and maintenance of cancer and autoimmune or inflammatory diseases. Interestingly, epigenetic modifications are considered reversible and therefore present a potential treatment model. Epigenetics in the field of dental research is only at an early stage. However, several reports in particular related to inflammation and inflammatory markers have emerged, as well as reports on the influence of environmental factors affecting oral health [3, 4] (Fig. 9.1). In this chapter the two major epigenetic modifications, DNA methylation and histone acetylation and methylation, are described. In addition, the influence of environmental factors on the epigenome and the relation to disease susceptibility as well as their potential use as treatment models are also discussed.

9.1 Epigenetics

Our genetic material in the form of the DNA helix is packaged in the nucleus as chromatin. The chromatin can be packed in the nucleus either in the form of euchromatin which is loosely packed and therefore available for gene expression or as heterochromatin that is very densely packed leading to silencing of gene expression [5]. The structure of the chromatin is highly regulated, in which the epigenetic mechanisms play a vital role by modifying the accessibility for the transcriptional machinery, thereby regulating gene expression. The building blocks of chromatin are the nucleosomes, consisting of 146 bp of DNA and a core histone complex. This complex includes two copies each of histones H2A, H2B, H3, and H4 and a linker histone H1 that connects the nucleosomes forming the primary chromatin structure, often referred to as "beads-on-string." Histones can be acetylated or methylated at amino acid tails that protrude from the nucleosome [5]. Histone acetylation is regulated by histone acetyltransferases (HATs) and histone deacetylases (HDACs). Removal of acetyl groups by histone

L. Larsson
Department of Periodontology, The Sahlgrenska Academy at University of Gothenburg, Gothenburg, Sweden
e-mail: lena.larsson@odontologi.gu.se

L.A. Bergmeier (ed.), *Oral Mucosa in Health and Disease*,
https://doi.org/10.1007/978-3-319-56065-6_9

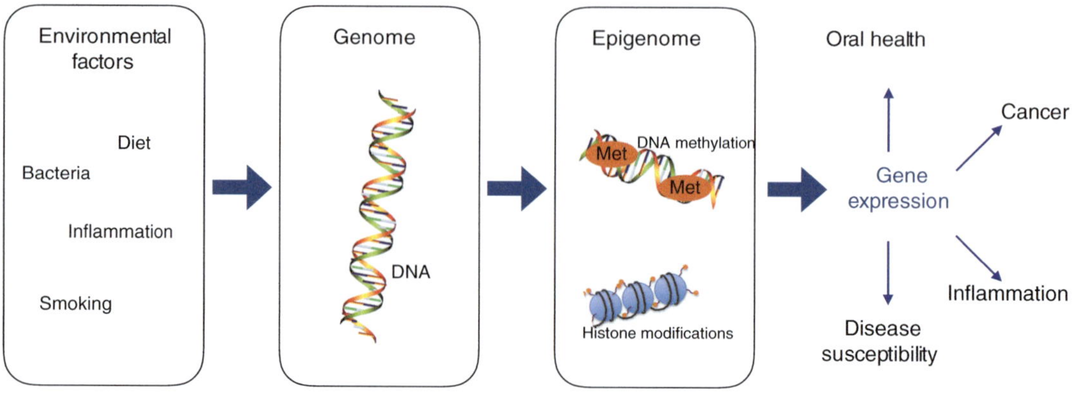

Fig. 9.1 Schematic drawing of the interaction between the environment and genetics

deacetylases (HDACs) leads to a dense packing of DNA around histones and inactivation of genes. In contrast, hyperacetylation by histone acetyltransferases (HATs) is associated with transcriptionally active chromatin. Histones H3 and H4 can also be methylated and occur at certain lysine molecules. Importantly, a lysine can be mono-, di-, or trimethylated, further adding to the various functions of methylated histones [6]. In the classical DNA methylation model, the addition of methyl groups to cytosine bases (5mC) occurs at specific sites in the DNA sequence, so-called CpG islands or CpG sites, by the de novo DNA methyltransferases (DNMTs) DNMT3a and DNMT3b [2, 7]. This modification alters the configuration of the DNA and the binding of transcription factors, resulting in changes in gene expression. Transcriptionally active genes are associated with low levels of DNA methylation. To add another level to the concept of DNA methylation in 2009 it was discovered that 5mC could be further oxidized into 5-hydroxymethylcytosine (5hmC) by the ten-eleven translocation (TET) family of enzymes [8]. The TET enzymes can then convert 5hmC into unmethylated cytosine, resulting in DNA hypomethylation [9]. The biological function of 5hmC is not clear, but it has been suggested to be an intermediate leading to demethylation of 5mC and thereby re-expression of genes silenced by DNA methylation [10]. Histone modifications and DNA methylation are not separate events but are linked resulting in a unique tissue- and cell-specific transcription pattern and response to intrinsic and extrinsic stimuli (Fig. 9.2).

In contrast to our genome, which is the same in all cells and throughout our life, the epigenome is dynamic and changes during life in response to diseases and environmental factors. This is especially visible in twin studies, since homozygote twins have the same genome but different epigenome. At early years twins are epigenetically identical but during the years the epigenetic patterns are influenced by environmental factors. These differences increase as the twins grow older and by differences in disease, and the epigenetic pattern differs more in twins that grow up in different environments. Epigenetic changes increase during life, and age itself is therefore considered a "risk factor" for epigenetic changes [11, 12].

9.2 Diet, Smoking, and Environmental Factors

The epithelium covering the oral cavity is continually being renewed, the differentiated cells are exfoliated, and the epithelial stem cells residing in the basal cell layers forming new cells in turn undergo differentiation when leaving the basal layer. This transition has been suggested to be associated with epigenetic changes involving both DNA methylation and histone modifications [13]. The oral epithelium is constantly being exposed to environmental factors, such as smoking, diet, and bacteria, factors known to influence

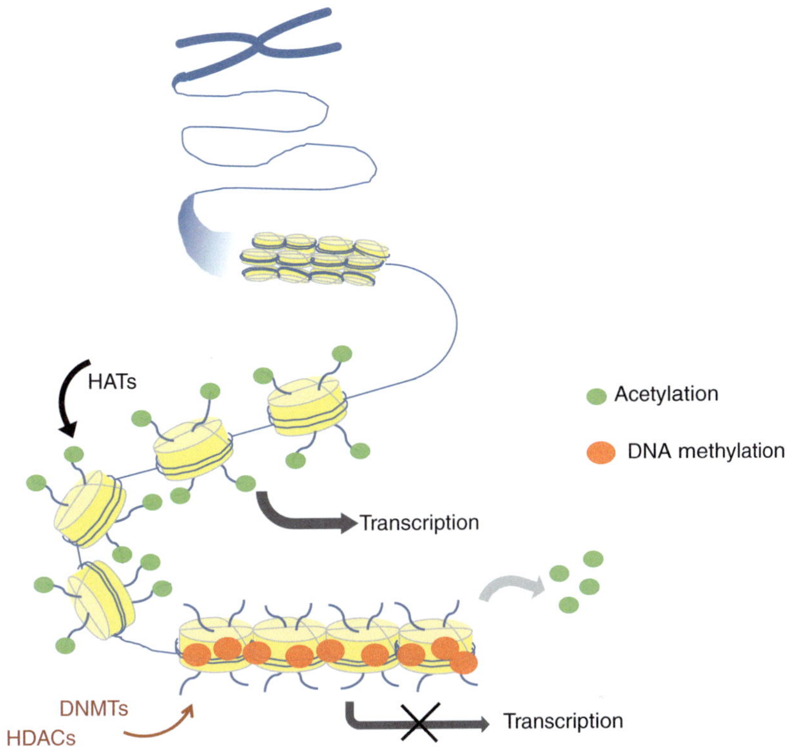

Fig. 9.2 Schematic illustration of the influence of epigenetics on chromatin configuration and gene transcription. *HATs* histone acetyltransferases, *HDACs* histone deacetylases, *DNMT* DNA methyltransferas

epigenetic mechanisms. Studies indicate that these factors have the potential to affect the oral health by altering the epigenome [3].

Diet has been found to be one of the most influential environmental factors due to its influence on gene expression. The impact of nutrients on epigenetic mechanism has so far mostly been studied in the field of cancer and it has been shown to influence cellular longevity and cancer incidence and prognosis [14, 15]. Not only during embryonic development but also in adult life the diet influences the epigenome, and studies are emerging on the influence of bioactive dietary compounds on health and disease. In addition, the epigenetic changes caused by a certain diet do not only affect the individual at the time of intake but can have long-term effects [3]. A number of nutrients are considered to influence epigenetic mechanisms, thereby becoming a great interest to use for both treatment and prevention of disease. The term epigenetic diet refers to the consumption of food such as cruciferous vegetables, green tea, grapes, and soy that have been shown to induce epigenetic changes protecting against cancer and aging [14, 16]. At present, there is a lack of research on these effects in the oral mucosa, but there are reports on the increased risk for oral squamous cell carcinoma (OSCC) with a decrease in folate, found in green leafy vegetables, and on the effect of EGCG from green tea on inhibition of oral cancer [14, 17].

Another level of the influence of our dietary compounds is that they can act as vehicles for toxic substances, i.e., arsenic, cadmium, nickel, chromium, and mercury, that we are exposed to through ingestion, inhalation, or exposure (i.e., nickel in jewelry). Cadmium and mercury, that can be found in fish, shellfish, cereals, and vegetables, influence our DNA methylation pattern, while nickel alters histone acetylation pattern [18, 19].

Smoking together with diet is known to affect the oral health. Smokers are known to have more severe periodontitis associated with increase in attachment loss. A comparison of monozygotic twins showed that the twin that smoked had a higher degree of disease [20]. Smoking causes long-term hypo- and hypermethylation changes in the DNA that can be found also in former smokers, indicating that the epigenetic alterations of the chromatin in smokers have a long-term effect.

9.3 Inflammation

It has been suggested that induction of epigenetic changes in the oral mucosa can occur as a result of a bacterial challenge and that the following inflammatory processes in turn further modulate these changes [21]. Pathogen recognition receptors such as Toll-like receptors (TLRs) mediate the process of recognition of the bacteria and initiate a remodeling of the chromatin to promote an induction of gene expression of inflammatory mediators. It has been hypothesized that pathobiont-induced periodontal disease mediates acetylation of the chromatin of oral epithelial cells. It was found that pathogens, such as *Porphyromonas gingivalis* and *Fusobacterium nucleatum*, induced epigenetic modifications such as acetylation of histones and downregulation of DNMT1. These changes alter the chromatin configuration leading to enhanced transcription of inflammatory genes, including the p300/CBP histone acetyltransferase target gene NFkB and several proinflammatory cytokines, commonly upregulated in chronic inflammatory diseases. In addition, pathogen recognition receptors, PRRs, such as NOD1 and TLR1/2 and 4 induce histone modifications [22]. Importantly, it has been suggested that epigenetics is a vital factor in determining the outcome of an inflammatory response, e.g., chronic inflammation vs. inflammatory resolution [21]. Hypoacetylated histones and hypermethylated CpGs in chronic inflammatory diseases may result in persistent inflammation [6].

Gingivitis is a periodontal disease characterized by inflammation in the gingival connective tissue. Periodontitis in turn is characterized by chronic inflammation in the gingival tissues but with associated loss of connective tissue attachment and loss of supportive bone. Periodontitis is a common disorder and severe forms of the disease occur in about 10% of an adult population [23]. Periodontitis has been characterized as a result of bacteria colonizing the tooth surface and the host response, which in turn is influenced by both genetic and epigenetic components [24]. Several studies have investigated the methylation status of several cytokines in gingival tissue samples and oral epithelial cells, respectively [25, 26]. The findings in these studies indicate that changes in the methylation pattern in promoter region of genes involved in inflammation may be caused by either the presence of periodontal pathogens or the inflammatory process. Several cytokines influence the epigenetic pattern and their epigenetic pattern can in turn be influenced by bacteria. Several studies have shown that periodontitis patients have an altered epigenetic pattern in inflammatory cytokines such as IL-8, IL-10, IL-4, IL-6, and Toll-like receptors (reviewed in [3, 27]). As mentioned previously, age is considered a risk factor for disease through alterations in epigenetic pattern and it has been suggested that the epigenome of older periodontitis patients may differ from that of younger patients even though they may have the same clinical phenotype. Not only is there an age-related epigenetic change in the collagen in periodontal ligament but also inflammatory cells such as lymphocytes have an altered phenotype [3].

Reactive oxygen species (ROS) produced in response to bacteria and inflammatory molecules induce breaks in the DNA strand. In order for the DNA repair system to get access to the site of damage, epigenetic mechanisms are activated to loosen the chromatin structure and to recruit the DNA repair machinery to the site of damage. In individuals with chronic inflammatory disease or in individuals that are exposed to environmental stimuli, such as smoking, these repeated changes of the chromatin structure may result in permanent epigenetic changes influencing disease progression and severity. It has been suggested that the production of reactive oxygen species (ROS) can influence the epigenetic pattern in epithelial cells [13]. Not only does production of ROS induce DNA breaks and tissue damage, but it also induces phenotype changes of other inflammatory cells as well as activate the production of pro-inflammatory cells leading to and maintaining a chronic inflammation [28].

Chronic inflammatory diseases such as periodontitis have specific target tissue in which the inflammation is persistent and tissue destruction occurs. Periodontitis is also site specific with only some teeth affected indicating a different response to the same bacteria at different places in the oral mucosa, hence suggesting a local change in the regulation of genes associated with inflammation.

9.4 Tumorigenesis

Epigenetic changes in the oral mucosa are not only associated with chronic inflammatory conditions such as periodontitis. Evidence is emerging that oral squamous cell carcinomas (OSCC) are linked to epigenetic alterations, in particular DNA methylation, even though few studies are emerging regarding histone modifications. OSCC is a neoplastic form of cancer with a multistep process influenced by endogenous and environmental factors, of which tobacco and alcohol are considered risk factors [29]. Interestingly, epigenetics was suggested as a potential link between inflammation and cancer [30]. Chronic inflammation induced by IL-6 may lead to hypermethylation of tumor-suppressor genes, thus being suggested to be a factor contributing to the development of OSCC [31]. Aberrant methylation might also be triggered by inflammation caused by a specific population of oral pathogens, linking microflora, inflammation, and tumorigenesis [32].

The importance of epigenetics in the oral mucosa is not only the potential use for new treatment models, but also their use as prediction markers for disease progression, as changes in DNA methylation pattern occur during all stages of tumorigenesis from premalignant to oral cancer. One potential marker is p16, which has been found to have an increase in hypermethylation in dysplastic lesions compared to non-dysplastic lesions. In addition, a higher methylation was found in those dysplastic lesions that subsequently turned into cancer [33]. However, it must be remembered that this is based on a few studies and it is important that further studies include also lifestyle, diet, and sociodemographic data.

Two types of premalignant lesions in the oral cavity are oral leukoplakia and oral lichen planus (OLP). Oral leukoplakia does in some cases develop into OSCC, but so far the mechanisms for this transformation are still unknown and finding early markers is important for an early identification of patients at risk for developing OSCC [34]. An aberrant methylation pattern similar to OSCC for certain genes has been reported in oral leukoplakia, indicating that this epigenetic pattern may be linked to malignant transformation of oral leukoplakia [34, 35]. Oral lichen planus is characterized by a chronic inflammatory disease in the oral mucosa [36]. As for leukoplakia, the exact mechanisms behind oral lichen planus are unknown and there is not much research on the role of epigenetics in the disease. However, a correlation between increase in acetylation of histone H3 and poor response to therapy of clinically more severe lesions was reported [37].

9.5 Clinical Application

Knowledge of epigenetics contributes to a better understanding of the interactions between genes and the environment and may provide explanations to why patients with the same clinical phenotype respond differently to treatment [3]. The fact that epigenetic mechanisms are reversible makes them attractive targets for new treatment models in both cancer and inflammatory diseases. The term epidrugs was coined by Ivanov and coworkers as "drugs that inhibit or activate disease-associated epigenetic proteins ameliorating, curing or preventing the disease" [38]. In the field of cancer there are numerous studies on the use of epidrugs and more recently nutrients as treatment models; however at present there is a lack of research on this in relation to oral health. Reports are emerging on the use of epidrugs in inflammatory diseases. It has been found that HDAC inhibitors suppress bone loss in rheumatoid arthritis (RA) as well as in periodontitis and they have been suggested as potential treatment models for these diseases [39, 40]. In addition to be potential treatment models epigenetic factors may be used as risk indicators for disease susceptibility and disease progression. A recent review indicated that aberrant methylation could be an early indicator of disease development and progression in OSCC [33]. For periodontitis, which is a site-specific disease and also differs in patient susceptibility to disease, identifying epigenetic markers may present biomarkers for identifying patients at risk to develop periodontitis and also present a means to individualize the treatment plan as a part of moving towards the concept of personalized medicine. Another clinical application for epigenetics is in tissue engineering, in which epidrugs have been suggested as a tool for modulating cell differentiation, thereby improving regeneration of tissue [41]. In addition, the possible use for buccal swabs or scraping of the oral mucosa for epigenetic analysis makes it clinically feasible as a diagnostic tool [34].

Conclusions

Epigenetic modifications are tissue specific and therefore analysis of epigenetic modifications in the oral mucosa is of great importance in order to gain knowledge of the effect of these mechanisms on oral health and the influence of environmental factors on oral health. There is currently a void in the research on epigenetics in relation to oral health. Many studies have shown the importance of epigenetics on disease susceptibility and progression. Several key factors, such as diet, smoking, environment, bacteria, and inflammation, known to induce epigenetic alterations are in contact with the oral mucosa and may affect the oral health through changes in genes involved in the immune response in the oral mucosa.

References

1. Waddington CH. The epigenotype. Endeavor. 1942;1:18–20. Reprinted in Int J Epidemiol 2012;41: 10-13.
2. Bird A. DNA methylation patterns and epigenetic memory. Genes Dev. 2002;16:6–21.
3. Lod S, et al. The influence of epigenetics in relation to oral health. Int J Dent Hyg. 2014;12:48–54.
4. Seo J-Y, et al. Epigenetics: general characteristics and implications for oral health. Restor Dent Endod. 2015;40:14–22.
5. Jenuwein T, Allis CD. Translating the histone code. Science. 2001;293:1074–80.
6. Bäckdahl L, et al. Inflammatory signaling as mediator of epigenetic modulation in tissue-specific chronic inflammation. Int J Biochem Cell Biol. 2009;41:176–84.
7. Robertson KD, Wolffe AP. DNA methylation in health and disease. Nat Rev Genet. 2000;1:11–9.
8. Tahiliani M, et al. Conversion of 5-methylcytosine to 5-hydroxymethylcytosine in mammalian DNA by MLL partner TET1. Science. 2009;324:930–5.
9. Ito S, et al. Tet proteins can convert 5-metylcytosine to 5-formylcytosine and 5-carboxylcytosine. Science. 2011;333:1300–3.
10. Kraus TF, et al. Low values of 5-hydroxymethylcytosine (5hmC), the 'sixth base', are associated with anaplasia in human brain tumors. Int J Cancer. 2012;131:1577–90.
11. Barros SP, Offenbacher S. Epigenetics: connecting environment and genotype to phenotype and disease. J Dent Res. 2009;88:400–8.
12. Fraga MF, et al. Epigenetic differences arise during the lifetime of monozygotic twins. PNAS. 2005;102:10604–9.
13. Iglesias-Bartolome R, et al. Control of the epithelial stem cell epigeneome: the shaping of epithelial stem cell identity. Curr Opin Cell Biol. 2013;25:162–9.
14. Daniel M, Tollefsbol TO. Epigenetic linkage of aging, cancer and nutrition. J Exp Biol. 2015;218:59–70.
15. Su LJ, et al. Epigenetic contributions to the relationship between cancer and dietary intake of nutrients, bioactive food components, and environmental toxicants. Front Genet. 2012;2:91.
16. Hardy TM, Tollefsbol TO. Epigenetic diet: impact on the epigenome and cancer. Epigenomics. 2011;3:503–18.
17. Bishop KS, Ferguson LR. The interaction between epigenetics, nutrition and the development of cancer. Forum Nutr. 2015;7:922–47.
18. Colcancino JA, et al. Pretreatment dietary intake is associated with tumor suppressor DNA methylation in head and neck squamous cell carcinomas. Epigenetics. 2012;7:883–91.
19. Torano EG, et al. The impact of external factors on the epigenome: *In utero* and over lifetime. Biomed Res Int. 2016;2016. article ID 2568635:17.
20. Seddon JM, et al. Smoking, dietary betaine, methionine, and vitanmin D in monozygotic twins with discordant macular degeneration: epigenetic implications. Ophtalmology. 2011;118:1386–94.
21. Barros SP, Offenbacher S. Modifiable risk factors in periodontal disease. Epigenetic regulation of gene expression in the inflammatory response. Periodont 2000. 2014;64:95–110.

22. Martins MD, et al. Epigenetic modifications of histones in periodontal disease. J Dent Res. 2016;95:215–22.
23. Hugoson A, et al. Trends over 30 years, 1973-2003, in the prevalence and severity of periodontal disease. J Clin Periodontol. 2008;35:405–14.
24. Kornman KS. Mapping the pathogenesis of periodontitis: a new look. J Periodontol. 2008;79:1560–8.
25. Oliveira NFP, et al. DNA methylation status of the IL8 gene promoter in oral cells of smokers and non-smokers with chronic periodontitis. J Clin Periodontol. 2009;36:719–25.
26. Zhang S, et al. Alteration of PTGS2 promoter methylation in chronic periodontitis. J Dent Res. 2010; 89:133–7.
27. Larsson L, et al. Epigenetics and its role in periodontal diseases: a state-of-the-art review. J Periodontol. 2015;86:556–68.
28. Rahman I, et al. Redox modulation of chromatin remodeling: impact on histone acetylation and deacetylation, NFkB and pro-inflammatory gene expression. Biochem Pharmacol. 2004;68:1255–67.
29. González-Ramírez I, et al. Histones and long non-coding RNAs: the new insights of epigenetic deregulation involved in oral cancer. Oral Oncol. 2014;50:691–5.
30. Kundu JK, Surh Y. Inflammation: gearing the journey to cancer. Mutat Res. 2008;659:15–30.
31. Gasche JA, et al. Interleukin-6 promotes tumorigenesis by altering DNA methylation in oral cancer cells. Int J Cancer. 2011;129:1053–63.
32. Bebek G, et al. Microbiomic subprofiles and MDR1 promoter methylation in head and neck squamous cell carcinoma. Hum Mol Genet. 2012;21:1557–65.
33. Shridhar K, et al. DNA methylation markers for oral pre-cancer progression: a critical review. Oral Oncol. 2016;53:1–19.
34. Abe M, et al. High-risk oral leukoplakia is associated with aberrant promoter methylation of multiple genes. BMC Cancer. 2016;16:350.
35. Asokan GS, et al. Promoter hypermethylation profile of tumor suppressor genes in oral leukoplakia and oral squamous cell carcinoma. J Clin Diagn Res. 2014;8:ZC09–12.
36. Alrashdan MS, et al. Oral lichen planus: a literature review and update. Arch Dermatol Res. 2016;308(8):539–51. https://doi.org/10.1007/s00403-016-1667-2.
37. Dillenburg CS, et al. Epigenetic modifications and accumulation of DNA double-strand breaks in oral lichen planus lesions presenting poor response to therapy. Medicine. 2015;30:e997.
38. Ivanov M, et al. Epigenetic mechanisms of importance for drug treatment. Trends Pharmacol Sci. 2014;35:384–96.
39. Cantley MD, et al. Histone deacetylase inhibitors and periodontal bone loss. J Periodontal Res. 2011a;46:697–703.
40. Cantley M, et al. Compounds that inhibit histone deacetylases in class I and class II effectively suppress human osteoclasts in vitro. J Cell Physiol. 2011b; 22:22684.
41. Kim TI, et al. Analysis of histone deacetylase inhibitor-induced responses in human periodontal ligament fibroblasts. Biotechnol Lett. 2013;35:129–33.

10 Infections of the Oral Mucosa and Immune Responses

Lesley Ann Bergmeier

10.1 Introduction

The British Dental Health Foundation, a charity-promoting oral health, has reported that only one in six people realises that gum disease may increase their risk of stroke and diabetes (http://www.dentalhealth.org/). An extensive report by the Surgeon General of the United States in 2000 clearly enunciated the link between oral and general health and set out a framework for action, https://www.nidcr.nih.gov/DataStatistics/SurgeonGeneral/Report/ExecutiveSummary.htm. In a large cross-sectional study of over 17,000 men and women a strong association was seen between the presence of oral disease and systemic inflammation and a history of cardiovascular diseases [1]. Worldwide there are considerable inequalities in oral healthcare provision that has a deleterious effect on general health and well-being [2, 3].

The media, both print and electronic, emphasise the importance of health and fitness and oral healthcare is part of that message—and yet in 2012 at the London Olympics 18% of the ELITE athletes said that oral discomfort has seriously affected their performance and almost half of the athletes had not seen a dentist in the last year [4]. The consequences of a high-carbohydrate diet, often part of an athletes training programme, may lead to dental decay and gum disease and the inflammation that accompanies these changes in the homeostasis of the oral cavity has serious and now well-documented effects to general health and well-being.

A recent report of the Joint European Federation of Periodontology and American Academy of Periodontology (EFP/AAP) [5] demonstrated the link between chronic inflammatory states and the onset of diabetes. Other systemic health risks from chronic periodontitis include atherosclerosis, adverse pregnancy outcomes, rheumatoid arthritis, aspirational pneumonia and cancer [6–12]. Head and neck cancers are said to be the fastest growing group of tumours, representing 3% of all cancers in the USA [13], https://www.cancer.gov/research/progress/snapshots/head-and-neck.

The oral cavity is a complex biological environment protected by one of the most sophisticated and important fluids in the body, saliva, which is actively secreted into the mouth. This is a unique body cavity due to the presence of hard tissue (teeth) which is abutted with a mucosal epithelium. This junction, at the gingival margin, allows for access of substances and cells from the systemic circulation into the gingival crevicular fluid and has a considerable effect on the defence of the oral mucosa and development of disease (Chap. 7).

L.A. Bergmeier
Centre for Oral Immunology and Regenerative Medicine, Institute of Dentistry, Barts and The London School of Medicine and Dentistry, London, UK
e-mail: l.a.bergmeier@qmul.ac.uk

L.A. Bergmeier (ed.), *Oral Mucosa in Health and Disease*,
https://doi.org/10.1007/978-3-319-56065-6_10

The *mucosal immune system* contributes to the protection of the oral cavity (see Chaps. 4 and 5) through both innate and adaptive immune functions. Despite an enormous antigenic challenge, the oral mucosa in healthy individuals rarely shows signs of inflammatory responses and is generally regarded as a tolerogenic environment.

Oral pathologies can be driven by infections and/or by dysregulation of homeostasis and many pathologies reflect a contribution of the immune system in either resolution of disease or exaggerated host responses that drive pathology. Failures of the immune system can also contribute to disease through immunodeficiency, where the immune system either fails to recognise a pathogen or is defective in its response. Hypersensitivity, an overreaction to innocuous materials and autoimmunity and the development of immune responses to self-antigen have all been shown to contribute to oral mucosal diseases.

The oral cavity is subject to infections by bacteria, viruses and fungi (Fig. 10.1). Some of the organisms are part of the normal oral flora but become pathogenic due to loss of homeostatic regulation in the tissues. Some parasitic infections also induce oral manifestation but since these are rare they will not be discussed [14].

Two of the most common diseases, caries and periodontal disease, result in damage and loss of the hard tissues of the oral cavity (teeth and bone). An immune response to the antigens of cariogenic organisms can protect against dental caries and the potential for the development of a caries vaccine has been investigated over many decades. In contrast, the host response to periodontal pathogens contributes to pathology. The clinical management of these two diseases will not be discussed.

10.1.1 Bacterial Infections

Dental caries is one of the most common diseases of humans [15] and recent papers have presented evidence of the disease in prehistory [16–18]. Increased consumption of refined sugars has resulted in an increased rate of caries, and in developing countries the lesions often go untreated due to poverty and lack of provision of oral healthcare.

The major causative organism for dental caries in humans is *Streptococcus mutans*. This highly acidogenic organism is usually found in low numbers where caries is absent. However, in the presence of caries there is an abundance of *Strep. mutans*. Acid production, through fermentation of dietary carbohydrates, erodes the enamel and dentine resulting in carious lesions.

Although *caries* is a disease of the hard tissues, the capacity for its resolution and indeed prevention is dependent on the gingival mucosa and the passage of specific antibodies from the serum into the gingival crevicular fluid. Many vaccine studies have been carried out in animal models that demonstrate that protection from caries can be achieved when IgG antibodies specific for *S. mutans* antigens reach the gingival margin by serum transudation. A selection of early studies is outlined below.

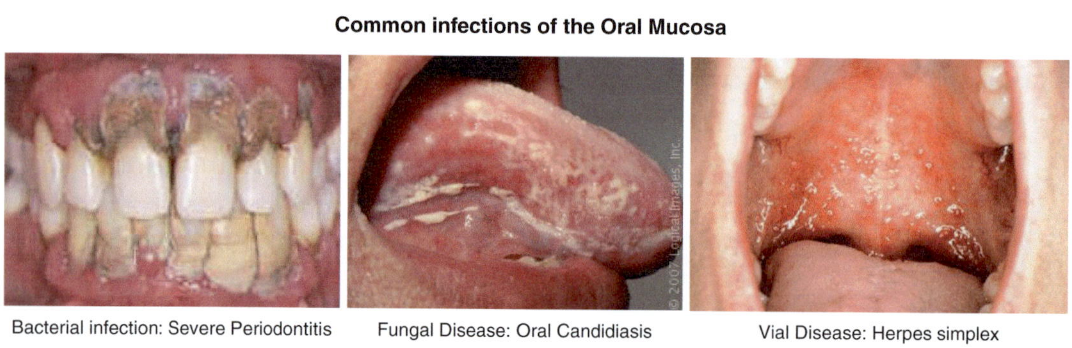

Fig. 10.1 Common infections of the oral cavity

S. mutans has two major antigenic proteins which have been shown to be highly immunogenic: glucosyltransferase (GTF) and the surface adhesion streptococcal antigen I/II (SAI/II). The characterisation of SAI/II [19] led to studies in experimental vaccine design based on SAI/II [20, 21] where protection from caries was established to be due to the induction of IgG antibodies which gained access to the tooth surface through the gingival margin [22, 23]. Passive immunisation studies in rhesus macaques had shown that antibodies applied to the gingival margin could protect against colonisation with *S. mutans* [24]. Observations in human subjects established that the presence of IgG anti SAI/II in serum correlated with low caries incidence [25–27]. Monoclonal antibodies to *Strep. mutans* antigens were later developed and used in passive immunisation studies in both non-human primates and in human clinical trials [28–30]. Passive immunisation using monoclonal antibodies showed that recolonisation was prevented over a period of about 18 months [29, 31–33]. Glucosyltransferase (Gtf) vaccine candidates have been investigated in a phase I clinical trial [34] and induced salivary IgA antibodies which also reduced recolonisation.

In more recent animal model studies nasal adjuvants have been used to enhance the immunogenicity of anti-caries DNA vaccine candidates [35]. Sublingual immunisation with a recombinant phosphate-binding protein (PstP) derived from *Strep. mutans* and administered with a mucosal adjuvant resulted in decreased colonisation and increased IgA antibody-secreting cells in draining lymph nodes in a rodent model [36].

These tools have enormous potential for reenergising the dental caries vaccine field.

Poor oral hygiene can lead to the accumulation of plaque and the development of gingivitis and while this is reversible in some patients it can become a chronic inflammation. Chronic marginal gingivitis (CMG) is the most common form of the inflammatory response to plaque constituents, which include bacterial toxins and antigens.

Periodontal disease (PD) is a destructive chronic inflammatory disease of the connective tissues and bone supporting the teeth and is a major source of tooth loss in adults. It occurs when there is irreversible damage to the periodontium that includes loss of collagen fibres in the periodontal ligament, deepening of the periodontal pocket and bone resorption. There are many different types of periodontal disease, some of which are caused by or exacerbated by systemic diseases. The most common form of adult periodontal disease is known as chronic inflammatory periodontal disease (CIPD).

Periodontitis is the result of a dramatic change from a symbiotic community of mostly facultative organisms to a *dysbiotic* community consisting of anaerobic organisms that have evolved to thrive in an inflammatory environment including the acquisition of numerous virulence factors [37, 38]. Microbial *dysbiosis* is defined as a reduction in the number of symbiotic organisms and/or an increase of pathogenic species that have an effect on both the innate and adaptive immune responses [39].

Recent research has built on the knowledge that certain "keystone" organisms have a disproportionate effect on their environment relative to their abundance [40]. *P. gingivalis* is an example of a *pathobiont*, an organism that co-operates with others to remodel its microenvironment into a dysbiotic and disease-provoking microbiota. These new communities can subvert the normally tolerogenic immune system to maintain an inflammatory environment that promotes their survival in the face of a robust immune response [40–42]. However, dysbiosis alone does not always lead to periodontitis. Other risk factors such as genetic susceptibility [43], stress and behaviours such as smoking and diet are important in precipitating disease [44, 45].

The innate immune response to the periodontal organisms includes an increased influx of neutrophils into the periodontal pockets to maintain homeostasis [46, 47]. Individuals with defects in neutrophil function are highly susceptible to periodontal disease [48–50]. However, the effect of the *pathobiont* is to hyperactivate these cells inducing the overproduction of toxic substances and proteolytic enzymes that can initiate tissue destruction [47, 51–53].

In response to infection, macrophages produce pro-inflammatory cytokines, such as IL-1β and TNFα. If these cytokines are produced in high concentrations or for prolonged periods they can induce the production of other inflammatory mediators including IL-6 and prostaglandin E2(PGE2) and metalloproteinases (MMPs). Both IL-1β and TNFα induce bone resorption and inhibit bone formation.

In generalised aggressive periodontitis, a higher IL-1β/IL-10 ratio was observed compared to healthy subjects, suggesting an imbalance between pro- and anti-inflammatory cytokines [54]. In some studies, periodontal therapy was shown to decrease IL-1β levels in GCF while increasing IL-10 [55]. IL-10 is an important anti-inflammatory cytokine that has been shown to suppress MMPs and stimulate an inhibitor of bone resorption (osteoprotegerin). Studies of IL-10 knockout mice demonstrated an increased susceptibility to bone destruction induced by *P. gingivalis* infection [56, 57].

The *complement cascade* is an important mechanism for controlling bacterial infection. Subversion of the cascade can be affected by the gingipains, kaylysins and interpain-A, produced in concert by *P. gingivalis, T. forsythia, T. dentolytica* and *P. intermedia.* The mechanism of action is through interruption of the C5aR-TLR2 signalling cascade and the degradation of C5b, thereby preventing the formation of the membrane attack complex which should destroy invading bacteria [58–60].

Human β defensins are also inhibited in periodontal disease [61], when the gingival epithelium response to *Fusobacterium nucleatum* is subverted by *T. dentolytica* which blocks defensin production. *P. gingivalis* is also capable of degrading antimicrobial peptides (reviewed in [42]).

The adaptive immune response is also important in periodontitis, and antibody responses are readily demonstrated to many of the virulence factors of the keystone pathogens including gingipains (Table 10.1). In experiments, like those carried out for caries vaccines, monoclonal antibodies to *P. gingivalis* were developed and used both in animal models and human clinical trials [62–64]. More recently a therapeutic experimental vaccine demonstrated that IgG1 antibodies protect against experimental periodontitis in mice [65]. Immune responses have also been demonstrated to microbial and human heat-shock proteins [66]. Responses to these chaperone molecules have also been implicated in the immunopathogenesis of ulcerative diseases such as recurrent aphthous stomatitis (RAS) and Behçet's disease (see below).

Table 10.1 Balance between some beneficial and harmful host defences in periodontal disease

Periodontal disease	
Complex interplay between bacterial species and host defences	
Beneficial effects	**Harmful effects**
• IgG ↑ phagocytosis of bacteria • IgA ↓ bacterial adherence	• Ag/Ab complexes: hypersensitivity—attract leucocytes—release proteases—tissue damage
• Complement: activated by endotoxin—initiates inflammatory process—recruits neutrophils	• Complement: could induce hypersensitivity
• Cytokines: IL-12 recruits immune cells • Phagocytes: destroy bacteria • T cells: help produce antibody	• Cytokines: too much pro-inflammatory cytokine secretion contributes to tissue damage, e.g. osteoclast-activating factor
• B cell: produce antibodies against perio-organisms	• Phagocytes: release proteases—tissue damage

T cells also contribute to PD mainly through the production of cytokines that support antibody production which are prevalent in chronic disease [67, 68]. Both regulatory T cells (T_{regs}) and Th17 cells have been shown to be present in periodontal lesions [69].

There is increasing evidence that the presence of IL-17 (produced by both Th17 cells and other cells) in human PD lesions is associated with disease severity. Several animal studies indicate the potential role of Th17 cells in gingival inflammation and bone destruction in PD [70, 71].

This association of *Th17 cells* with chronic inflammation and dysregulation of homeostasis is a common feature of other oral mucosal diseases described in chapter 11 and its role in oral immunity and shaping of the oral microbiome was recently reviewed [72].

Periodontitis can present without any associated disease but is frequently a major component of systemic disorders such as acquired immunodeficiency disease (AIDS), leukaemia, Crohn's disease, diabetes mellitus, Down syndrome, sarcoidosis and syndromes associated with polymorphonuclear leucocyte defects (Chédiak-Higashi syndrome, agranulocytosis and cyclic neutropenia). Periodontitis can also contribute to systemic disorders and has been found to be an etiological factor in several important diseases including infective endocarditis, pulmonary and brain abscesses and adverse pregnancy outcomes [73–75]. In a recent review, the overlap between the inflammatory processes in PD and in cancer was outlined with respect to the function of chemokines. The main function of these molecules is the recruitment of immune cells into sites of infection and inflammation, but they also have the capacity to prolong inflammatory processes which can exacerbate disease progression including the development of cancer [76].

In a recent study by the Ebersole group, the transcriptome of the B cell compartment was investigated in a non-human primate model of natural periodontal disease and in the healthy gingiva of aging humans. This study revealed complex changes in the expression of genes involved in antigen-dependent activation and proliferation, T cell interaction and maturation of B cells both in adult periodontitis and in aged non-human primates. In healthy aging, gingival homeostasis is maintained by adaptive B cell responses which modulate tissue-destructive gene expression. These functions are lost in periodontitis resulting in immune dysfunction and enhanced inflammation [77].

In an extensive review of the research into the immunobiology of periodontal disease, over the last 40 years, it has been suggested that the transition point between health and disease might be the new frontier for oral biology research [78].

10.1.2 Oral Manifestations of Other Bacterial Infections

The barrier function of the oral epithelium, the normal oral flora and the innate and adaptive immune defence mechanisms of the oral cavity make infection by pathogenic bacteria a rare event. The antimicrobial components of the saliva and the desquamation of the epithelium limit colonisation by pathogens.

Some organisms evade the immune system by becoming coated with host proteins that make them less susceptible to neutralising antibodies as is thought to be the case of *T. pallidum* (syphilis). While other organisms have evolved mechanisms that either evade or subvert immune responses such as the changes in antigen serotype expression seen in *Strep. pneumoniae*.

Intracellular pathogens such as *Mycobacterium leprae*, although a very rare condition, illustrate a classic example of immune subversion. Two clinical forms of disease are seen that result in very different patterns of immune activity. In *tuberculoid leprosy* the immune response shows strong Th1 responses and activated macrophages control but do not eliminate the infection. Tuberculoid lesions contain granulomas and inflammation is local, causing only local effects such as peripheral nerve damage. Normal immunoglobulin levels and T cell responses to *M. leprae* antigens are seen and the disease has low infectivity (Table 10.2).

In contrast, *lepromatous leprosy* shows a profound suppression of cell-mediated immunity, leading to anergy, and there is a significant shift from a Th1 cytokine profile (supporting macrophage activation and T cell cytotoxicity) to a Th2 profile. The Th2 cytokines include IL-4 which is thought to inhibit bactericidal activity in macrophages. The mechanism might be like that seen in *Mycobacterial tuberculosis* infection where the organism is taken up by macrophages but the fusion of the phagosome and lysosome is prevented, thus avoiding the bactericidal actions of the lysosomal contents.

In *lepromatous leprosy*, *M. leprae* shows high growth rates in macrophages (suggesting the failure of the phagolysosomal killing of the organ-

Table 10.2 Granulomatous infections

- **Actinomycosis**
  - Endogenous polymicrobial infections
  - Submandibular swelling
  - Chronic suppuration—multiple sinuses
  - "Sulphur" granules
- **Syphilis**
  - Primary: chancre
  - Secondary: snail-track ulcers, mucous patches
  - Tertiary: gumma, lingual leukoplakia
  - Congenital: dental anomalies; "dished face"
- **Tuberculosis**[a]
  - Oral usually secondary to pulmonary
  - Painless, chronic lingual ulcer
- **Leprosy**[a]
  - Oral lesions in lepromatous type
  - Secondary to nasal involvement
  - Nodular masses palate and anterior maxilla

Adapted from Soames and Southam: Oral Pathology 4th Edition
[a]Intracellular infections

ism) and the infection becomes disseminated with diffuse nerve damage. Hypergammaglobulinemia is observed, but there is a poor or absent T cell response to *M. leprae* antigens. This form of the disease is highly infectious. Oral lesions occur almost exclusively in the lepromatous form of the disease and are reported in about 50% of cases. Patients present with inflammatory masses that tend to ulcerate and resolve with fibrosis. The hard and soft palate, anterior maxilla and tongue are most frequently affected. The oral lesions are usually secondary to nasal involvement [79].

10.1.3 Viral Infections

There are several families of viruses which cause disease in the oral cavity or have oral manifestations following systemic infection (Table 10.3).

Picornaviruses, such as *coxsackie A* and *enterovirus 71*, have been associated with hand, foot and mouth disease which causes blistering in the mouth and on the hands and feet of children. While this disease is not usually severe, there have been reports of neurological complications and fatalities when enterovirus 71 is involved. Several serotypes of coxsackie virus (particularly serotypes 1–10, 16 and 22) have also been associated with herpangina, where an oropharyngitis results in oral vesicle which breaks down into small ulcers.

Table 10.3 Virus infections with oral mucosal presentations

Virus	Disease
Herpes simplex(HSV1 and 2)	Herpetic stomatitis (primary/recurrent)
Varicella zoster	Chickenpox/shingles
Coxsackie A	Herpangina; hand, foot and mouth disease
Epstein Barr	Infectious mononucleosis; hairy leukoplakia
Paramyxovirus	Measles; mumps
Human papilloma virus	Viral warts; epithelial hyperplasia
Cytomegalovirus	Associated with HIV infection
HIV	Necrotising periodontal disease; Kaposi sarcoma; candidiasis

10.1.3.1 Herpesviruses

There are eight known human *herpesviruses* with most well known being the herpes simplex viruses (HSV). The two types of *HSV* differ in their serological, biological and clinical presentation. Most commonly type 1 is associated with skin and oral epithelium and has a pattern of recurrence due to reactivation of latent virus from the trigeminal sensory ganglion. Primary infections with HSV-1 are sometimes asymptomatic, but can cause pharyngitis, tonsillitis or herpetic gingivostomatitis. In some patients, progression results in a widespread gingivitis which can be erythematous and oedematous. Vesicles form which can ulcerate and become secondarily infected.

These infections usually heal within 2–3 weeks but during this time the virus has been transported to nerve cells of the oral cavity where they remain latent, often in the trigeminal ganglion, until reactivated. In secondary infection, *herpes simplex labialis* ("cold sore") results from viral migration from the latently infected neurons to keratinocytes where it replicates. Visible lesions recur at the same site.

Inflammatory responses, and papule and vesicle formation, result from the death of infected keratinocytes while neutralising antibody production, and the release of IFNγ from $CD4^+$ and

CD8+ T cells as well as cytotoxic killing of infected cells play a role in limiting the spread of infection and in resolving lesions [80].

HSV-1 has been implicated in the immunopathogenesis of *Behçet's disease* (BD) as it has been isolated from saliva, peripheral blood leukocytes and genital ulcers in these patients. In a mouse model, immunisation with HSV gave rise to BD-like symptoms that were improved with antiviral treatments [81]. However, isolation of HSV is inconsistent and antiviral therapy failed to alleviate BD symptoms. It has been suggested that altered immune responses to HSV in BD compared to healthy individuals may allow persistence of the infection [82, 83].

HSV type 2 is less common and is a sexually transmitted infection; however, changes in sexual habits have resulted in increased oral infections. HSV-2 also has a pattern of latency.

Latency is an immune evasion strategy; the epithelial infection is controlled by the immune response; however, the virus that has migrated to the sensory neurons is quiescent and produces few viral peptides. This along with low expression of MHC I in the neurons makes recognition by cytotoxic cells difficult and results in survival of virus and latent infection.

Reactivation most commonly occurs due to a variety of stressors which impair host defences. In immunocompetent individuals, the lesions are self-limiting due to the induction of both specific neutralising antibodies and CD8+ cytotoxic T cell responses which can suppress reactivation [84, 85]. In immunocompromised patients, these recurrences can be intractable and severe.

Chickenpox and herpes zoster (shingles) are caused by *varicella zoster virus* (VSV/HHV-3). Chickenpox frequently presents first in the oral mucosa as small ulcers, followed by the characteristic skin lesions. Like HSV, there is a latent phase, and while reactivation is uncommon it is more severe and causes a painful neuralgia. The location is usually on the trunk but can also occur in the oral mucosa and the eye. Reactivation usually occurs in response to a variety of stressors which have induced suppression of host defence mechanisms.

Other herpesvirus infections with oral presentations include Epstein–Barr virus (EBV/HHV-4) and cytomegalovirus (CMV/HHV-5) and frequently involve enlargement of the lymph nodes and/or salivary glands.

CMV productively infects several types of lymphocytes and has a latent phase in macrophages. CMV has been isolated from periodontal pockets and from other oral mucosal lesions but there is little evidence of cause and effect, other than the presence of infected cells in inflammatory infiltrates. While this virus does not usually cause disease in immunocompetent individuals it is an important opportunistic pathogen in immunocompromised hosts such as AIDS patients and organ-transplant recipients.

CMV escapes immune control by producing four proteins that downregulate MHC class I expression on infected cells so that no viral peptides can be expressed on the cell surface for recognition by cytotoxic cells. The virus also produces a homologue of MHC-I molecules that inhibit the activation of NK cells which under normal circumstances would recognise the loss of the MHC-I and kill the infected cell.

Many people become infected with *Epstein–Barr virus (EBV)* in childhood and show no symptoms of primary infection. A primary infection with EBV later in life is the causative agent of infectious mononucleosis and is transmitted in saliva. Lymph node enlargement occurs and inflammation and ulceration of the oral mucosa at the junction of the hard and soft palates is a frequent presentation.

However, of greater significance is the association of EBV with Burkitt's lymphoma. EBV was the first virus to be shown to have a clear relationship to cancer. This B cell lymphoma is endemic in equatorial Africa and accounts for more than 50% of malignant diseases in children. While the disease tends to be multifocal, in Africa, a jaw tumour is the predominant feature in more than half of the cases.

EBV is also associated with hairy leukoplakia and has been detected in oral squamous cell carcinoma, and there is some evidence of involvement in some types of periodontal disease, but this might be coincidental with lymphocyte infiltration of infected B cells [86].

In other oral diseases EBV shedding has been demonstrated in the saliva of both RAS and Behçet's disease patients [87]. It was suggested that this might reflect a difference in the expression of TLRs and a dysregulation of the oral microbiota in BD and RAS [88, 89].

The remaining HHVs, namely HHV-6, -7 and -8, are the most recently described human herpesviruses. HHV-6 and -7 appear to be closely related to CMV and both cause childhood disease (HHV-6-exanthema subitum) and both are detectable in saliva. HHV-7 is thought to cause complications in transplant patients due to immune suppression. The target cell for HHV-7 is thought to be $CD4^{+}$T cells.

HHV-8 is considered the primary etiological agent of Kaposi's sarcoma (KS), primary effusion lymphoma (PEL) and multicentric Castleman's disease. KS is endemic in Uganda where 50% of the population are seropositive compared to less than 5% in the USA and Northern Europe. In non-endemic regions, infection with HIV is an important risk factor for the development of KS and HIV-1 infection was associated with an increased prevalence of infection with HHV-8 [90, 91].

10.1.3.2 Papilloma Virus

Human papilloma virus (HPV) has been implicated as a causative agent in several types of cancer and an association is thought to exist with oral cancer, particularly nasopharyngeal carcinomas [92–94] and salivary gland tumours [92]. Focal hyperplasia has been associated with viruses that seem to be restricted to the oral cavity, namely, HPV-13 and HPV-32.

HPV-16 (and 18) has a known association with invasive cervical carcinoma with HPV-16 detected in 50% of all cervical cancers in a study carried out in the USA [95]. The effectiveness of the HPV vaccine in reducing cervical cancer has been one of the great success stories of vaccine discovery and the potential for preventing oropharyngeal cancers is yet to be tested.

10.1.3.3 HIV

By far the most investigated virus in the last 30 years is the human immunodeficiency virus (HIV) and the catastrophic failure of the immune response, due to loss of $CD4^{+}$T cells, leads to a variety of oral lesions outlined in Table 10.4 which can be regarded as opportunistic infections in the absence of a robust immune response. In some respects, these might be regarded as a historical association as the success of highly active antiretroviral therapy has reduced the incidence of these oral manifestations in large areas of the world (but by no means all).

Hairy leukoplakia (EBV infection) and pseudomembranous candidiasis are the most frequent oral manifestations of HIV (Table 10.4 and Fig. 10.2). The presence of these conditions

Table 10.4 Oral lesions of HIV infection

Group 1: major associations	Group II: less common associations	Group III: possible associations
Candidiasis:	**Atypical oropharyngeal ulceration**	**Bacterial infections (not gingivitis/perio)**
Erythematous		
Hyperplastic		
Pseudomembranous		
Hairy leukoplakia (EBV)	**Idiopathic thrombocytopenic purpura**	**Fungal infections (not candidiasis)**
HIV-associated periodontal disease	**Salivary gland disorders**	**Melanotic hyperpigmentation**
HIV-gingivitis	Dry mouth, decreased flow rates	
Necrotising ulcerative gingivitis	Swelling of the major glands	
HIV-periodontitis		
Necrotising stomatitis		
Kaposi sarcoma (HHV8)	**Viral infections (Non-EBV)**	**Neurological disturbances**
	Cytomegalovirus	Facial palsy
	Herpes simplex virus	Trigeminal neuralgia
	Human papilloma virus	
	Varicella zoster virus	
Non-Hodgkin's lymphoma		

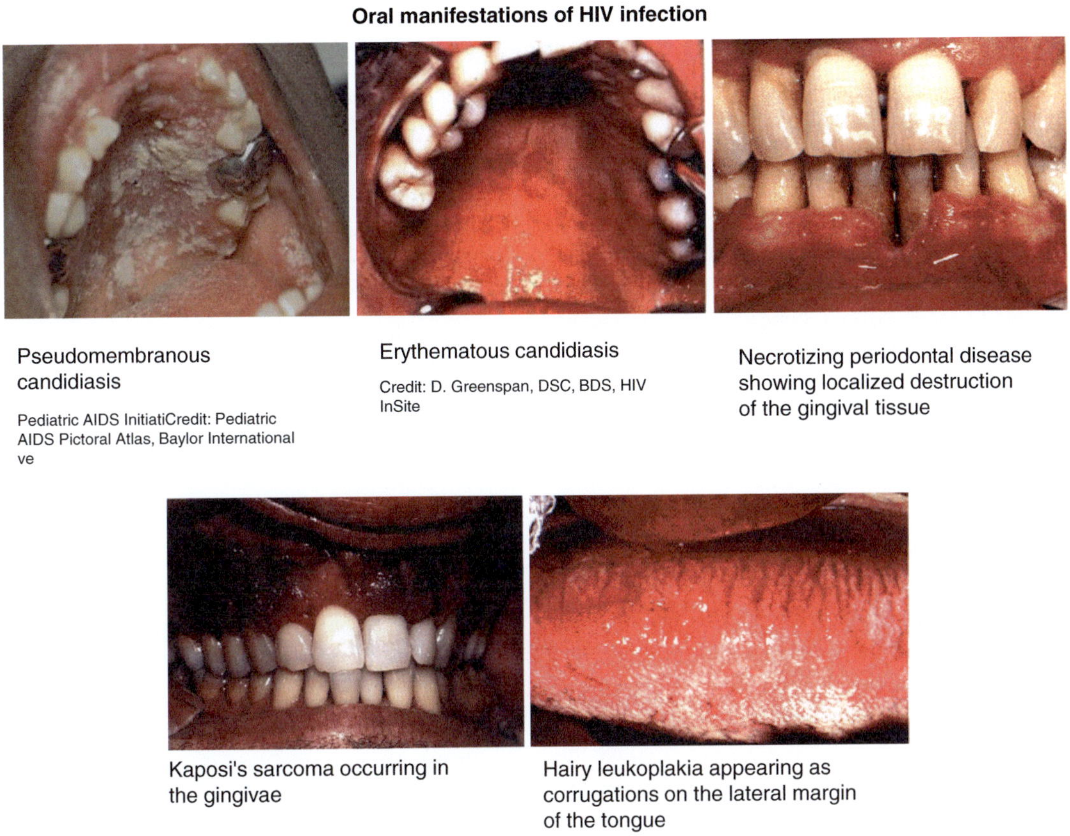

Fig. 10.2 Oral manifestations of HIV

increases with time following seroconversion and is an indication that the patient is progressing from the asymptomatic phase to AIDS [96, 97].

While the oral cavity succumbs to infections and disease induced by the dysregulated immune system in HIV infection, the HIV virus is rarely found in the oral tissues and no productive infection is seen in oral epithelial cells [98–101].

10.1.4 Fungal Infections

Candidiasis can present both in the oral cavity and elsewhere, such as the vagina. The causative organism, *Candida albicans,* is a normal component of the oral flora in about 40% of the population. Clinical infection is influenced by the immune status of the individual as well as the strain of *Candida sp*. that is present and the overall composition of the individuals' oral flora.

The three clinical forms of candidiasis that occur most frequently are pseudo-membranous (thrush), erythematous and hyperplastic. There are also several variations within each group [102, 103]. The observations that oral candidiasis is frequent in HIV (Fig. 10.2) and in other immunodeficiency states, where T cell immunity is absent or compromised, strongly suggest a role for cellular immunity in preventing overt infection or transition from the commensal yeast to pathogenic hyphal forms [103, 104].

Many of the factors which predispose towards Candida infection can be directly linked to immune dysregulation, for example, extremes of age (both young and old) where the immune system is either not fully developed or is waning. Smoking is also known to change the types and functions of antigen-presenting cells (especially dendritic cells) in the oral mucosa and affects susceptibility to infection [105–107]. Pseudomembranous candidiasis has also been associated with the use of asthma inhalers—which contain immune-modulatory pharmaceuticals [108].

IgA-deficient individuals are also highly susceptible to candidiasis suggesting a role for secretory IgA [109, 110].

Neutrophils are a key component in protection against candidiasis and IL-17 seems to be important in recruiting these cells into the oral epithelium [111, 112]. More recently the IL-17 receptor (IL-17R) signalling pathways in oral epithelial cells have been shown to be crucial in protection against candidiasis in a mouse model, through the production of the antimicrobial peptide, β-defensin 3 [113].

IL-17 has multiple roles in both immune protection and immunopathology. There is a well-recognised role in surveillance at mucosal and barrier surfaces [114]. However, IL-17 has also been implicated in driving autoimmunity and chronic inflammation [115, 116].

The sources of IL-17 are somewhat controversial as several different cell types can produce this cytokine including NKT, γδ T cells, macrophages and the relatively newly described *innate lymphoid* cells as well as Th17 helper cells [72, 117–119].

Conclusion

The most common oral mucosal infections are caused by bacteria, viruses and fungi. Some Helminth and protozoal organisms also have the potential to produce oral manifestations of disease but are much rarer and have not been explored. The immune system contributes to both protection from infection and resolution of the inflammation induced by infection but also the immunopathological consequences of dysregulation and dysbiosis. The importance of immune mechanisms will be explored in the next chapter where noninfectious oral mucosal disease will be explored.

References

1. Fedele S, Sabbah W, Donos N, Porter S, D'aiuto F. Common oral mucosal diseases, systemic inflammation, and cardiovascular diseases in a large cross-sectional US survey. Am Heart J. 2011;161:344–50.
2. Williams DM. Reducing inequalities in oral disease. Br Dent J. 2011;210:393. England.
3. Williams DM, Sheiham A, Honkala E. Addressing oral health inequalities in the Africa and middle east region. J Dent Res. 2015;94(7):875.
4. Needleman I, Ashley P, Petrie A, Fortune F, Turner W, Jones J, Niggli J, Engebretsen L, Budgett R, Donos N, Clough T, Porter S. Oral health and impact on performance of athletes participating in the London 2012 olympic games: a cross-sectional study. Br J Sports Med. 2013;47:1054–8.
5. Chapple ILC, Genco R, Workshop WGOTJEA. Diabetes and periodontal diseases: consensus report of the joint efp/aap workshop on periodontitis and systemic diseases. J Periodontol. 2013;84:S106–12.
6. Genco RJ, Van Dyke TE. Prevention: reducing the risk of CVD in patients with periodontitis. Nat Rev Cardiol. 2010;7:479–80.
7. Han YW, Wang X. Mobile microbiome: oral bacteria in extra-oral infections and inflammation. J Dent Res. 2013;92:485–91.
8. Kebschull M, Demmer RT, Papapanou PN. "Gum bug, leave my heart alone!"—epidemiologic and mechanistic evidence linking periodontal infections and atherosclerosis. J Dent Res. 2010;89:879–902.
9. Kholy KE, Genco RJ, Van Dyke TE. Oral infections and cardiovascular disease. Trends Endocrinol Metab. 2015;26:315–21.
10. Lundberg K, Wegner N, Yucel-Lindberg T, Venables PJ. Periodontitis in RA—the citrullinated enolase connection. Nat Rev Rheumatol. 2010;6:727–30.
11. Madianos PN, Bobetsis YA, Offenbacher S. Adverse pregnancy outcomes (APOs) and periodontal disease: pathogenic mechanisms. J Clin Periodontol. 2013;40(Suppl 14):S170–80.
12. Whitmore SE, Lamont RJ. Oral bacteria and cancer. PLoS Pathog. 2014;10:E1003933.
13. Young D, Xiao CC, Murphy B, Moore M, Fakhry C, Day TA. Increase in head and neck cancer in younger patients due to human papillomavirus (HPV). Oral Oncol. 2015;51:727–30.
14. Bergquist R. Parasitic infections affecting the oral cavity. Periodontol 2000. 2009;49:96–105.
15. Marcenes W, Kassebaum NJ, Bernabé E, Flaxman A, Naghavi M, Lopez A, Murray CJ. Global burden of oral conditions in 1990–2010: a systematic analysis. J Dent Res. 2013;92(7):592.
16. Topic B, Rascic-Konjhodzic H, Cizek Sajko M. Periodontal disease and dental caries from krapina neanderthal to contemporary man—skeletal studies. Acta Med Acad. 2012;41:119–30.
17. Weyrich LS, Dobney K, Cooper A. Ancient DNA analysis of dental calculus. J Hum Evol. 2015;79:119–24.
18. Weyrich LS, Duchene S, Soubrier J, Arriola L, Llamas B, Breen J, Morris AG, Alt KW, Caramelli D, Dresely V, Farrell M, Farrer AG, Francken M, Gully N, Haak W, Hardy K, Harvati K, Held P, Holmes EC, Kaidonis J, Lalueza-Fox C, De La Rasilla M, Rosas A, Semal P, Soltysiak A, Townsend G, Usai D, Wahl J, Huson DH, Dobney K, Cooper

A. Neanderthal behaviour, diet, and disease inferred from ancient DNA in dental calculus. Nature. 2017. Advance Online Publication.

19. Russell MW, Lehner T. Characterisation of antigens extracted from cells and culture fluids of streptococcus mutans serotype C. Arch Oral Biol. 1978;23:7–15.
20. Lehner T, Russell MW, Caldwell J, Smith R. Immunization with purified protein antigens from streptococcus mutans against dental caries in rhesus monkeys. Infect Immun. 1981;34:407–15.
21. Lehner T, Russell MW, Wilton JM, Challacombe SJ, Scully CM, Hawkes JE. Passive immunization with antisera to streptococcus mutans in the prevention of caries in rhesus monkeys. Adv Exp Med Biol. 1978;107:303–15.
22. Challacombe SJ, Russell MW, Hawkes J. Passage of intact igg from plasma to the oral cavity via crevicular fluid. Clin Exp Immunol. 1978;34:417–22.
23. Challacombe SJ, Russell MW, Hawkes JE, Bergmeier LA, Lehner T. Passage of immunoglobulins from plasma to the oral cavity in rhesus monkeys. Immunology. 1978;35:923–31.
24. Lehner T, Russell MW, Challacombe SJ, Scully CM, Hawkes JE. Passive immunisation with serum and immunoglobulins against dental caries in rhesus monkeys. Lancet. 1978;1:693–5.
25. Challacombe SJ, Bergmeier LA, Czerkinsky C, Rees AS. Natural antibodies in man to streptococcus mutans: specificity and quantification. Immunology. 1984;52:143–50.
26. Challacombe SJ, Bergmeier LA, Rees AS. Natural antibodies in man to a protein antigen from the bacterium streptococcus mutans related to dental caries experience. Arch Oral Biol. 1984;29:179–84.
27. Kirtaniya BC, Chawla HS, Tiwari A, Ganguly NK, Sachdev V. Natural prevalence of antibody titers to glucosyltransferase of streptococcus mutans in serum in high and low caries active children. J Indian Soc Pedod Prev Dent. 2010;28:91–4.
28. Ma JK, Hunjan M, Smith R, Lehner T. Specificity of monoclonal antibodies in local passive immunization against streptococcus mutans. Clin Exp Immunol. 1989;77:331–7.
29. Ma JK, Smith R, Lehner T. Use of monoclonal antibodies in local passive immunization to prevent colonization of human teeth by streptococcus mutans. Infect Immun. 1987;55:1274–8.
30. Smith R, Lehner T, Beverley PC. Characterization of monoclonal antibodies to streptococcus mutans antigenic determinants I/Ii, I, Ii, and Iii and their serotype specificities. Infect Immun. 1984;46:168–75.
31. Ma JK, Hiatt A, Hein M, Vine ND, Wang F, Stabila P, Van Dolleweerd C, Mostov K, Lehner T. Generation and assembly of secretory antibodies in plants. Science. 1995;268:716–9.
32. Ma JK, Hikmat BY, Wycoff K, Vine ND, Chargelegue D, Yu L, Hein MB, Lehner T. Characterization of a recombinant plant monoclonal secretory antibody and preventive immunotherapy in humans. Nat Med. 1998;4:601–6.
33. Ma JK, Lehner T. Prevention of colonization of streptococcus mutans by topical application of monoclonal antibodies in human subjects. Arch Oral Biol. 1990;35(Suppl):115s–22s.
34. Smith DJ, Taubman MA. Oral immunization of humans with streptococcus sobrinus glucosyltransferase. Infect Immun. 1987;55:2562–9.
35. Yan YH, Yu F, Zeng C, Cao LH, Zhang Z, Xu QA. Ccl17 combined with Ccl19 as a nasal adjuvant enhances the immunogenicity of an anti-caries DNA vaccine in rodents. Acta Pharmacol Sin. 2016;37:1229–36.
36. Ferreira EL, Batista MT, Cavalcante RC, Pegos VR, Passos HM, Silva DA, Balan A, Ferreira LC, Ferreira RC. Sublingual immunization with the phosphate-binding-protein (Psts) reduces oral colonization by streptococcus mutans. Mol Oral Microbiol. 2016;31:410–22.
37. Hajishengallis G, Liang S, Payne MA, Hashim A, Jotwani R, Eskan MA, Mcintosh ML, Alsam A, Kirkwood KL, Lambris JD, Darveau RP, Curtis MA. Low-abundance biofilm species orchestrates inflammatory periodontal disease through the commensal microbiota and complement. Cell Host Microbe. 2011;10:497–506.
38. Lamont RJ, Hajishengallis G. Polymicrobial synergy and dysbiosis in inflammatory disease. Trends Mol Med. 2015;21:172–83.
39. Petersen C, Round JL. Defining dysbiosis and its influence on host immunity and disease. Cell Microbiol. 2014;16:1024–33.
40. Hajishengallis G, Darveau RP, Curtis MA. The keystone-pathogen hypothesis. Nat Rev Microbiol. 2012;10:717–25.
41. Curtis MA, Zenobia C, Darveau RP. The relationship of the oral microbiotia to periodontal health and disease. Cell Host Microbe. 2011;10:302–6.
42. Hajishengallis G. Periodontitis: from microbial immune subversion to systemic inflammation. Nat Rev Immunol. 2015;15:30–44.
43. Stabholz A, Soskolne WA, Shapira L. Genetic and environmental risk factors for chronic periodontitis and aggressive periodontitis. Periodontol 2000. 2010;53:138–53.
44. Laine ML, Crielaard W. Functional foods/ingredients and periodontal diseases. Eur J Nutr. 2012;51(Suppl 2):S27–30.
45. Laine ML, Moustakis V, Koumakis L, Potamias G, Loos BG. Modeling susceptibility to periodontitis. J Dent Res. 2013;92:45–50.
46. Nussbaum G, Shapira L. How has neutrophil research improved our understanding of periodontal pathogenesis? J Clin Periodontol. 2011;38(Suppl 11):49–59.
47. Sima C, Glogauer M. Neutrophil dysfunction and host susceptibility to periodontal inflammation: current state of knowledge. Current Oral Health Reports. 2014;1:95–103.
48. Darveau RP. Periodontitis: a polymicrobial disruption of host homeostasis. Nat Rev Microbiol. 2010;8:481–90.
49. Hajishengallis E, Hajishengallis G. Neutrophil homeostasis and periodontal health in children and adults. J Dent Res. 2014;93:231–7.

50. Moutsopoulos NM, Konkel J, Sarmadi M, Eskan MA, Wild T, Dutzan N, Abusleme L, Zenobia C, Hosur KB, Abe T, Uzel G, Chen W, Chavakis T, Holland SM, Hajishengallis G. Defective neutrophil recruitment in leukocyte adhesion deficiency type I disease causes local Il-17-driven inflammatory bone loss. Sci Transl Med. 2014;6:229ra40.
51. Chapple IL, Matthews JB. The role of reactive oxygen and antioxidant species in periodontal tissue destruction. Periodontol 2000. 2007;43:160–232.
52. Eskan MA, Jotwani R, Abe T, Chmelar J, Lim JH, Liang S, Ciero PA, Krauss JL, Li F, Rauner M, Hofbauer LC, Choi EY, Chung KJ, Hashim A, Curtis MA, Chavakis T, Hajishengallis G. The leukocyte integrin antagonist Del-1 inhibits Il-17-mediated inflammatory bone loss. Nat Immunol. 2012;13:465–73.
53. Ryder MI. Comparison of neutrophil functions in aggressive and chronic periodontitis. Periodontol 2000. 2010;53:124–37.
54. Teles RP, Gursky LC, Faveri M, Rosa EA, Teles FR, Feres M, Socransky SS, Haffajee AD. Relationships between subgingival microbiota and GCF biomarkers in generalized aggressive periodontitis. J Clin Periodontol. 2010;37:313–23.
55. De Lima Oliveira AP, De Faveri M, Gursky LC, Mestnik MJ, Feres M, Haffajee AD, Socransky SS, Teles RP. Effects of periodontal therapy on GCF cytokines in generalized aggressive periodontitis subjects. J Clin Periodontol. 2012;39:295–302.
56. Sasaki H, Okamatsu Y, Kawai T, Kent R, Taubman M, Stashenko P. The interleukin-10 knockout mouse is highly susceptible to porphyromonas gingivalis-induced alveolar bone loss. J Periodontal Res. 2004;39:432–41.
57. Sasaki H, Suzuki N, Kent R, Kawashima N, Takeda J, Stashenko P. T cell response mediated by myeloid cell-derived Il-12 is responsible for porphyromonas gingivalis-induced periodontitis in Il-10-deficient mice. J Immunol. 2008;180:6193–8.
58. Jusko M, Potempa J, Karim AY, Ksiazek M, Riesbeck K, Garred P, Eick S, Blom AM. A metalloproteinase karilysin present in the majority of tannerella forsythia isolates inhibits all pathways of the complement system. J Immunol. 2012;188:2338–49.
59. Malm S, Jusko M, Eick S, Potempa J, Riesbeck K, Blom AM. Acquisition of complement inhibitor serine protease factor I and its cofactors C4b-binding protein and factor H by prevotella intermedia. PLoS One. 2012;7:E34852.
60. Popadiak K, Potempa J, Riesbeck K, Blom AM. Biphasic effect of gingipains from porphyromonas gingivalis on the human complement system. J Immunol. 2007;178:7242–50.
61. Shin JE, Baek KJ, Choi YS, Choi Y. A periodontal pathogen treponema denticola hijacks the fusobacterium nucleatum-driven host response. Immunol Cell Biol. 2013;91:503–10.
62. Booth V, Ashley FP, Lehner T. Passive immunization with monoclonal antibodies against porphyromonas gingivalis in patients with periodontitis. Infect Immun. 1996;64:422–7.
63. Booth V, Lehner T. Characterization of the porphyromonas gingivalis antigen recognized by a monoclonal antibody which prevents colonization by the organism. J Periodontal Res. 1997;32:54–60.
64. Kelly CG, Booth V, Kendal H, Slaney JM, Curtis MA, Lehner T. The relationship between colonization and haemagglutination inhibiting and B cell epitopes of porphyromonas gingivalis. Clin Exp Immunol. 1997;110:285–91.
65. O'brien-Simpson NM, Holden JA, Lenzo JC, Tan Y, Brammar GC, Walsh KA, Singleton W, Orth RKH, Slakeski N, Cross KJ, Darby IB, Becher D, Rowe T, Morelli AB, Hammet A, Nash A, Brown A, Ma B, Vingadassalom D, Mccluskey J, Kleanthous H, Reynolds EC. A therapeutic porphyromonas gingivalis gingipain vaccine induces neutralising Igg1 antibodies that protect against experimental periodontitis. npj Vaccines. 2016;1:16022.
66. Hasan A, Sadoh D, Palmer R, Foo M, Marber M, Lehner T. The immune responses to human and microbial heat shock proteins in periodontal disease with and without coronary heart disease. Clin Exp Immunol. 2005;142:585–94.
67. Ebersole JL. Humoral immune responses in gingival crevice fluid: local and systemic implications. Periodontol 2000. 2003;31:135–66.
68. Kinane DF, Mooney J, Ebersole JL. Humoral immune response to actinobacillus actinomycetemcomitans and porphyromonas gingivalis in periodontal disease. Periodontol 2000. 1999;20:289–340.
69. Ohlrich EJ, Cullinan MP, Seymour GJ. The immunopathogenesis of periodontal disease. Aust Dent J. 2009;54(Suppl 1):S2–10.
70. Cheng W-C, Hughes FJ, Taams LS. The presence, function and regulation of Il-17 and Th17 cells in periodontitis. J Clin Periodontol. 2014;41:541–9.
71. Cheng WC, Van Asten SD, Burns LA, Evans HG, Walter GJ, Hashim A, Hughes FJ, Taams LS. Periodontitis-associated pathogens *P. gingivalis* and *A. actinomycetemcomitans* activate human Cd14(+) monocytes leading to enhanced Th17/Il-17 responses. Eur J Immunol. 2016;46:2211–21.
72. Abusleme L, Moutsopoulos NM. Il-17: overview and role in oral immunity and microbiome. Oral Dis. 2016;23:854–65.
73. Hajishengallis G, Moutsopoulos NM. Role of bacteria in leukocyte adhesion deficiency-associated periodontitis. Microb Pathog. 2016;94:21–6.
74. Kinane D, Bouchard P. Periodontal diseases and health: consensus report of the sixth european workshop on periodontology. J Clin Periodontol. 2008;35:333–7.
75. Persson GR, Persson RE. Cardiovascular disease and periodontitis: an update on the associations and risk. J Clin Periodontol. 2008;35:362–79.

76. Sahingur SE, Yeudall WA. Chemokine function in periodontal disease and oral cavity cancer. Front Immunol. 2015;6:214.
77. Ebersole JL, Kirakodu SS, Novak MJ, Orraca L, Martinez JG, Cunningham LL, Thomas MV, Stromberg A, Pandruvada SN, Gonzalez OA. Transcriptome analysis of B cell immune functions in periodontitis: mucosal tissue responses to the oral microbiome in aging. Front Immunol. 2016;7:272.
78. Ebersole JL, Dawson DR 3rd, Morford LA, Peyyala R, Miller CS, Gonzalez OA. Periodontal disease immunology: 'double indemnity' in protecting the host. Periodontol 2000. 2013;62:163–202.
79. Soames J, Southam J. Oral pathology. Oxford, UK: Oxford University Press; 2008.
80. Koelle DM, Corey L. Recent progress in herpes simplex virus immunobiology and vaccine research. Clin Microbiol Rev. 2003;16:96–113.
81. Kim DY, Cho S, Choi MJ, Sohn S, Lee ES, Bang D. Immunopathogenic role of herpes simplex virus in Behcet's disease. Genet Res Int. 2013;638273:24.
82. Hatemi G, Yazici H. Behçet's syndrome and micro-organisms. Best Pract Res Clin Rheumatol. 2011;25:389–406.
83. Young C, Lehner T, Barnes CG. Cd4 and Cd8 cell responses to herpes simplex virus in Behçet's disease. Clin Exp Immunol. 1988;73:6–10.
84. Khanna KM, Bonneau RH, Kinchington PR, Hendricks RL. Herpes simplex virus-specific memory Cd8+ T cells are selectively activated and retained in latently infected sensory ganglia. Immunity. 2003;18:593–603.
85. Pollara G, Katz DR, Chain BM. The host response to herpes simplex virus infection. Curr Opin Infect Dis. 2004;17:199–203.
86. Slots J, Saygun I, Sabeti M, Kubar A. Epstein-Barr virus in oral diseases. J Periodontal Res. 2006;41:235–44.
87. Seoudi N, Bergmeier LA, Hagi-Pavli E, Bibby D, Fortune F. The seroprevalence and salivary shedding of herpesviruses in Behçet's syndrome and recurrent aphthous stomatitis. J Oral Microbiol. 2015;7:27156.
88. Seoudi N, Bergmeier LA, Drobniewski F, Paster B, Fortune F. The oral mucosal and salivary microbial community of Behcet's syndrome and recurrent aphthous stomatitis. J Oral Microbiol. 2015;7:27150.
89. Seoudi N, Bergmeier LA, Hagi-Pavli E, Bibby D, Curtis MA, Fortune F. The role of Tlr2 and 4 in Behcet's disease pathogenesis. Innate Immun. 2014;20:412–22.
90. Rohner E, Wyss N, Heg Z, Faralli Z, Mbulaiteye SM, Novak U, Zwahlen M, Egger M, Bohlius J. HIV and human herpesvirus 8 co-infection across the globe: systematic review and meta-analysis. Int J Cancer. 2016;138:45–54.
91. Rohner E, Wyss N, Trelle S, Mbulaiteye SM, Egger M, Novak U, Zwahlen M, Bohlius J. Hhv-8 seroprevalence: a global view. Syst Rev. 2014;3:11.
92. Huhns M, Simm G, Erbersdobler A, Zimpfer A. HPV infection, but not EBV or HHV-8 infection, is associated with salivary gland tumours. Biomed Res Int. 2015;2015:829349.
93. Rautava J, Syrjanen S. Human papillomavirus infections in the oral mucosa. J Am Dent Assoc. 2011;142:905–14.
94. Syrjanen S, Lodi G, Von Bultzingslowen I, Aliko A, Arduino P, Campisi G, Challacombe S, Ficarra G, Flaitz C, Zhou HM, Maeda H, Miller C, Jontell M. Human papillomaviruses in oral carcinoma and oral potentially malignant disorders: a systematic review. Oral Dis. 2011;17(Suppl 1):58–72.
95. Cleveland JL, Junger ML, Saraiya M, Markowitz LE, Dunne EF, Epstein JB. The connection between human papillomavirus and oropharyngeal squamous cell carcinomas in the United States: implications for dentistry. J Am Dent Assoc. 2011;142:915–24.
96. Greenspan JS. Sentinels and signposts: the epidemiology and significance of the oral manifestations of HIV disease. Oral Dis. 1997;3(Suppl 1):S13–7.
97. Lifson AR, Hilton JF, Westenhouse JL, Canchola AJ, Samuel MC, Katz MH, Buchbinder SP, Hessol NA, Osmond DH, Shiboski S, Et A. Time from HIV seroconversion to oral candidiasis or hairy leukoplakia among homosexual and bisexual men enrolled in three prospective cohorts. AIDS. 1994;8:73–9.
98. Challacombe SJ, Naglik JR. The effects of HIV infection on oral mucosal immunity. Adv Dent Res. 2006;19:29–35.
99. Kohli A, Islam A, Moyes DL, Murciano C, Shen C, Challacombe SJ, Naglik JR. Oral and vaginal epithelial cell lines bind and transfer cell-free infectious HIV-1 to permissive cells but are not productively infected. PLoS One. 2014;9:E98077.
100. Moyes DL, Islam A, Kohli A, Naglik JR. Oral epithelial cells and their interactions with HIV-1. Oral Dis. 2016;22(Suppl 1):66–72.
101. Wood LF, Chahroudi A, Chen H-L, Jaspan HB, Sodora DL. The oral mucosa immune environment and oral transmission of HIV/SIV. Immunol Rev. 2013;254:34–53.
102. Moyes DL, Wilson D, Richardson JP, Mogavero S, Tang SX, Wernecke J, Hofs S, Gratacap RL, Robbins J, Runglall M, Murciano C, Blagojevic M, Thavaraj S, Forster TM, Hebecker B, Kasper L, Vizcay G, Iancu SI, Kichik N, Hader A, Kurzai O, Luo T, Kruger T, Kniemeyer O, Cota E, Bader O, Wheeler RT, Gutsmann T, Hube B, Naglik JR. Candidalysin is a fungal peptide toxin critical for mucosal infection. Nature. 2016;532:64–8.
103. Naglik JR, Richardson JP, Moyes DL. Candida albicans pathogenicity and epithelial immunity. PLoS Pathog. 2014;10:E1004257.
104. Moyes DL, Naglik JR. Mucosal immunity and candida albicans infection. Clin Dev Immunol. 2011;2011:346307.
105. Boyle JO, Gümüs ZH, Kacker A, Choksi VL, Bocker JM, Zhou XK, Yantiss RK, Hughes DB,

Du B, Judson BL, Subbaramaiah K, Dannenberg AJ. Effects of cigarette smoke on the human oral mucosal transcriptome. Cancer Prev Res (Phila). 2010;3:266–78.
106. Cruchley AT, Williams DM, Farthing PM, Speight PM, Lesch CA, Squier CA. Langerhans cell density in normal human oral mucosa and skin: relationship to age, smoking and alcohol consumption. J Oral Pathol Med. 1994;23:55–9.
107. Hovav AH. Dendritic cells of the oral mucosa. Mucosal Immunol. 2014;7:27–37.
108. Thomas MS, Parolia A, Kundabala M, Vikram M. Asthma and oral health: a review. Aust Dent J. 2010;55:128–33.
109. Lehner T, Challacombe SJ, Ivanyi L, Wilton JM. The relationship between serum and salivary antibodies and cell-mediated immunity in oral disease in man. Adv Exp Med Biol. 1974;45:485–95.
110. Lehner T, Wilton JM, Ivanyi L. Immunodeficiencies in chronic muco-cutaneous candidosis. Immunology. 1972;22:775–87.
111. Conti HR, Peterson AC, Brane L, Huppler AR, Hernández-Santos N, Whibley N, Garg AV, Simpson-Abelson MR, Gibson GA, Mamo AJ, Osborne LC, Bishu S, Ghilardi N, Siebenlist U, Watkins SC, Artis D, Mcgeachy MJ, Gaffen SL. Oral-resident natural Th17 cells and Γδ T cells control opportunistic candida albicans infections. J Exp Med. 2014;211:2075–84.
112. Conti HR, Shen F, Nayyar N, Stocum E, Sun JN, Lindemann MJ, Ho AW, Hai JH, Yu JJ, Jung JW, Filler SG, Masso-Welch P, Edgerton M, Gaffen SL. Th17 Cells and Il-17 receptor signaling are essential for mucosal host defense against oral candidiasis. J Exp Med. 2009;206:299–311.
113. Conti HR, Bruno VM, Childs EE, Daugherty S, Hunter JP, Mengesha BG, Saevig DL, Hendricks MR, Coleman BM, Brane L, Solis N, Cruz JA, Verma AH, Garg AV, Hise AG, Richardson JP, Naglik JR, Filler SG, Kolls JK, Sinha S, Gaffen SL. Il-17 receptor signaling in oral epithelial cells is critical for protection against oropharyngeal candidiasis. Cell Host Microbe. 2016;20:606–17.
114. Song X, He X, Li X, Qian Y. The roles and functional mechanisms of interleukin-17 family cytokines in mucosal immunity. Cell Mol Immunol. 2016;13:418–31.
115. Gaffen SL, Herzberg MC, Taubman MA, Van Dyke TE. Recent advances in host defense mechanisms/therapies against oral infectious diseases and consequences for systemic disease. Adv Dent Res. 2014;26:30–7.
116. Gaffen SL, Jain R, Garg AV, Cua DJ. The Il-23-Il-17 immune axis: from mechanisms to therapeutic testing. Nat Rev Immunol. 2014;14:585–600.
117. Cua DJ, Tato CM. Innate Il-17-producing cells: the sentinels of the immune system. Nat Rev Immunol. 2010;10:479–89.
118. Hernandez-Santos N, Gaffen SL. Th17 cells in immunity to candida albicans. Cell Host Microbe. 2012;11:425–35.
119. Hernandez-Santos N, Huppler AR, Peterson AC, Khader SA, Mckenna KC, Gaffen SL. Th17 cells confer long-term adaptive immunity to oral mucosal candida albicans infections. Mucosal Immunol. 2013;6:900–10.

11 Non-infectious Diseases of the Oral Mucosa: The Importance of Immune Functions

Lesley Ann Bergmeier and Farida Fortune

11.1 Introduction

The non-microbial diseases of the oral mucosa constitute a heterogenous group of disorders from rare life-threatening autoimmune disorders such as *Pemphigus vulgaris* to more common conditions like *recurrent aphthous stomatitis* (RAS). However, they have some features in common; a genetic susceptibility or association in some; and immune dysregulation or immune-driven pathology in many. In this chapter, the diseases are briefly described in terms of any genetic susceptibility/association, and the contribution of the immune response to pathology. Many of the diseases are managed by using *immunosuppressive* drugs which also include specific monoclonal antibodies, the so-called biologics. The mode of action of some of the drugs most commonly used is outlined in Table 11.1. An extensive description of diagnostic criteria and detailed clinical management is beyond the scope of this chapter but is referenced in a literature review of current clinical management of oral mucosal disease in Chap. 12.

Table 11.1 An outline of the mode of action of some immunosuppressant drugs commonly used in the management of oral mucosal disease. See Chapter 12 for a literature review of clinical management

Drug	Immune system target
Cyclosporin and tacrolimus	Inhibits IL-2: action and production Specific effect on T helper cells
Corticosteroids	Inhibits expression of cytokine genes
Azathioprine	Inhibits purine synthesis
Cyclophosphamide	Acts as an alkylating agent: Binds and cross-links DNA preventing replication and gene transcription
Methotrexate	Interferes with thymidine synthesis and therefore DNA synthesis through competitive inhibition of dihydrofolate reductase
Mycophenolate mofetil	Blocks the synthesis of guanine
Monoclonal antibodies	Antibodies with a single specificity. Many available specific for cytokines and cell surface molecules such as receptor molecules

See Chap. 12 for more detail of clinical management

11.2 Recurrent Aphthous Stomatitis

Recurrent aphthous stomatitis (RAS) or recurrent oral ulceration (ROU) is characterised by ulcers which occur singly or in crops lasting from 7 to

L.A. Bergmeier (✉) • F. Fortune
Centre for Oral Immunology and Regenerative Medicine, Institute of Dentistry, Barts and The London School of Medicine and Dentistry, London, UK
e-mail: l.a.bergmeier@qmul.ac.uk; f.fortune@qmul.ac.uk

L.A. Bergmeier (ed.), *Oral Mucosa in Health and Disease*,
https://doi.org/10.1007/978-3-319-56065-6_11

21 days before resolving and healing spontaneously (Table 11.2). They occur as either minor, major or herpetiform (Table 11.2) and have been carefully characterised to distinguish them from non-aphthous ulcers [1]. Clinical classification and severity-scoring systems have recently been developed [2]. About 10% of the population present with RAS although a wide range is noted in the literature. The age of onset is usually in the first decade of life, peaking in the second or third decade.

Genetics: There is somewhat conflicting evidence of genetic associations or indeed non-association in RAS which tend to depend on the ethnic origin of the study cohort [3]. A prevalence of HLA-A2 and B12 (B44) has been reported for some groups [4, 5]. HLA-B5 seems to be decreased in Sicilian patients but is similar to healthy controls in Turkish patients. Jurge et al. suggest that there is no consistent association between RAS and HLA haplotypes apart from HLA-B51. This is also an important association in Behçet's disease (see below). However, it is further suggested that a strong association was present in RAS for the alleles for IL-1 β-15 and IL-6-174.

There is some evidence of genetic susceptibility with a family history in about 40% of patients, a high concordance in twins.

The aetiopathogenesis of RAS is unclear with many factors implicated including hypersensitivity risk, endocrine factors, psychology, microbial agents, foods and socio-economic factors but the consensus is that an unidentified organism or environmental agent triggers an autoimmune-type reaction in susceptible individuals. Autoantibodies and T cells sensitised to the oral mucosa have been reported [6]. In most patients, cytotoxic antibodies specific for oral epithelial cells have also been demonstrated [7].

There is little evidence of a viral aetiology in RAS. However, there are increased plasma levels of IL-2 in active RAS and altered NK cell activity was also observed between periods of ulcer activity and remission [8]. Since NK cells have activities against varicella zoster virus (VSV) and CMV, a viral aetiology was explored but the evidence is somewhat confused as some studies failed to find any differences between NK subsets in RAS and healthy controls [3]. Elevated levels of antibodies to CMV have been seen in some RAS patients [9].

Cross-reactivity between oral bacterial antigens and oral tissues is an attractive hypothesis and there is very strong evidence that this occurs in RAS. Hasan et al. [10] have demonstrated a cross-reactivity between a peptide shared by the 65 kDa heat-shock protein (HSP) of *Mycobacterium sp.* and *Strep. sanguis*, its human mitochondrial 60 kDa equivalent. Both antibodies and autoreactive T cell clones have been demonstrated [11]. Importantly, this cross-reactivity has also been linked with immune responses to HSP60 in periodontal disease and cardiovascular disease [12], again linking the oral cavity with systemic disease.

More recently it has been noted that disturbances in the oral microbiota can be detected at both the genus and phylum levels in the ulcer sites in RAS compared with non-ulcerated sites [13].

Table 11.2 Characteristics of three types of oral ulcers

	Minor aphthous	Major aphthous	Herpetiform
Depth	• Shallow	• Deeper	• Shallow, pin-point
Size	• 2–7.5 mm	• >10 mm	• 0.3–0.5
Site	• Usually on non-keratinised section of oral mucosa including lateral sides and ventral surface of tongue, floor of the month, buccal sulcus but rarely on gingiva, palate and dorsum of the tongue	• Usually in any region of the month	• Usually in any site of the oral cavity
Numbers	• 1–10: number varies but can be as many as or more than 6 at times	• 1–5: few ulcers occur at a time	• 5–20: ulcers occurring in coalescing clusters
Healing	• 1–2 weeks: without scarring	• 2–8 weeks: with scarring	• 1–2 weeks: can heal with scarring

11.3 Behçet's Disease

Behçet's disease (BD) is a rare chronic immune-mediated inflammatory, multisystem disorder of unknown aetiology. More than 90% of patients present with oral ulceration and the presence of genital ulcers forms part of the clinical diagnosis, first established by the Behçet's International Study Group [14]. A subsequent validation study performed on multinational BD patients of 27 countries exhibited improved sensitivity over the ISG criteria and was proposed for adoption both as a guide for diagnosis and classification of BD [15].

The ulcers present as minor, major and herpetiform, as in RAS, but occur at much greater frequency and numbers and are slower to heal, sometimes with scarring not seen in RAS (Table 11.2).

A recent study has developed and validated a new genital ulcer severity-scoring system [16] which complements the scoring system for oral ulcers adapted from the RAS scoring system [2].

Other clinical manifestations include uveitis, and skin lesions, central nervous system and vascular complications which are severe and cause considerable morbidities and potential mortalities where the disease progresses to involve the large blood vessels, central nervous system or gastrointestinal tract (Fig. 11.1).

While it has a worldwide occurrence, it was traditionally known as the 'silk road' disease due to the prevalence along this ancient trading route [17].

The aetiopathogenesis of BD remains unknown, but both innate and adaptive immune mechanisms are involved. An initial reaction triggered by an infectious or environmental agent in a genetically susceptible individual is the most widely accepted hypothesis, supported by numerous investigations providing evidence of severe immune dysregulation [18–20].

Genetics: The aetiology of BD remains largely unknown, with a consensus that the pathogenesis is likely to be triggered by an environmental agent in a genetically susceptible host [21, 22]. HLA-B51, one of the splice variants of HLA-B5, has been found to be the most strongly associated genetic marker for BD to date from 1982 when it

International Study Group (ISG) criteria 1990 for Behçet's Disease

Recurrent oral ulceration	minor aphthous, major aphthous or Herpetiform ulceration with at least 3 times recurrence over a period of 12 months	Observed by patients or physicians
Plus-any 2 of the following		
Recurrent genital ulceration	aphthous Ulceration or scarring	Observed by patients or physicians
Eye Lesions	anterior uveitis, posterior uveitis or cells in vitreous on slit lamp examination or retinal vasculitis	Observed by ophthalmologist
Skin Lesions	erythema-nodosum, pseudo-folliculitis or papulo-pustular lesions or acne-form nodules observed in post- adolescent patients	Observed by physicians and/or patients
Positive Pathergy test	by oblique intra-cutaneous insertion of a 20 gauges or smaller needle under sterile conditions on forearm, read 24-48 hours later	Read by physicians

Clinical manifestation of Behçet's Disease

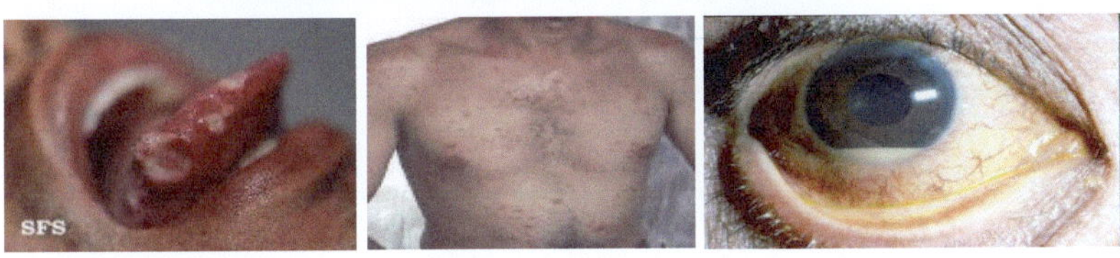

Oral Ulcers Skin lesions Hypopyon

Fig. 11.1 ISG diagnostic criteria for Behçet's disease with examples of some typical presentations

was first reported among the Japanese BD population [23, 24]. However, it accounts for less than 20% of the genetic risk and this indicates that other genetic factors might be involved. Several other studies showed a link between BD and MHC class I chain-related gene (MIC-A and MIC-B) and regarded the MIC-A allele (MICA*009) as a candidate for BD genetic susceptibility [25]. MIC antigens are expressed on the surface of various cells including fibroblasts, gastric epithelium and endothelial cells and these are also ligands for NKG2D, an activating natural killer receptor found on gamma delta (γδ) and CD8$^+$ αβT cells (Figs. 11.2 and 11.3).

The immunopathogenesis of BD has been subject to intense investigations over the last half century and new techniques such as genome-wide association studies (GWAS) have helped to elucidate many aspects of this complex disorder as outlined in Fig. 11.2 [26, 27].

The disease is increased approximately sixfold in patients with HLA-B51/B5 genetic polymorphisms [28] and is widely regarded as an auto-inflammatory condition although autoimmune responses to certain specific antigens have been described in the disease. These include retinal S antigen, heat-shock proteins and cytoskeletal proteins (Figs. 11.2 and 11.3).

Viral infection with Epstein–Barr virus (EBV) and herpes simplex virus (HSV) was also thought to be important in both initiating and triggering acute exacerbations of affected systems in BD [29, 30]. High levels of EBV shedding have been observed in both BD and RAS and a lower level of CMV IgG was observed in BD. The expression of unusual splice variants of TLR2 and TLR4 in BD suggested a defect in the crosstalk between innate and adaptive immune responses. A significant reduction in the response to cognate agonists of TLR1/2 heterodimer and TLR4 was also observed in BD. The TLR1/2 heterodimer is the initial receptor for sensing CMV [31]. Thus, viral associations are probably not causative but reflective of immune defects in BD.

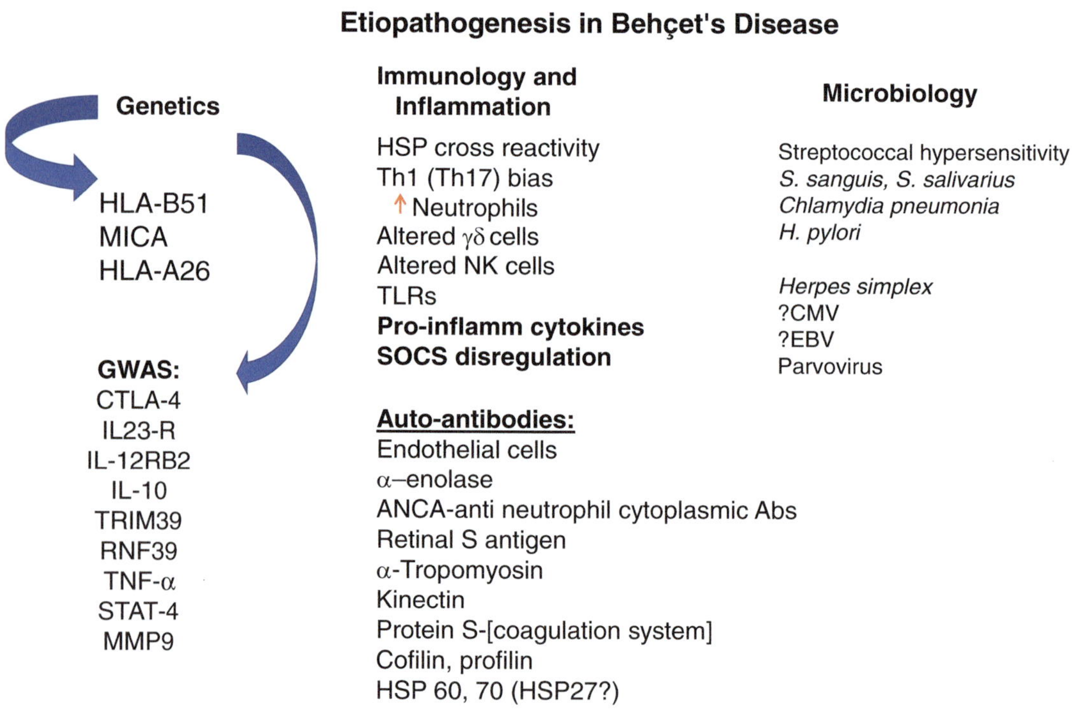

Fig. 11.2 Aetiopathogenesis of Behçet's disease showing genetic associations; infective and environmental agents and innate and adaptive immune drivers of pathogenesis

Like RAS, immune response HSPs have been demonstrated in BD and provided evidence of immunopathogenesis. The peptide involved in BD is different from that in RAS and an animal model for uveitis that mimics BD has been developed in the Lewis rat [32, 33]. Antibodies and T and B cell epitopes have been mapped within mycobacterial HSP65 [34–36]. Most importantly, the uveitis model could be ameliorated using *oral tolerisation*, by linking the uveitogenic peptide to the mucosal adjuvant *CHOLERA TOXIN B subunit* [32]. This protocol was successfully used in a phase I/II clinical trial which reduced uveitis relapse in BD patients [37].

There is a large body of evidence suggesting that loss of normal immune regulation plays a key role in BD pathology [25]. Neutrophil activation and recruitment to the site of inflammatory lesions [38, 39], Th1/Th17 cytokine polarisation of $CD4^+$ T cells and increased IFN-γ, TNF-α, IL-8 and IL-17 levels have been correlated with BD activity [40]. Conversely, a reduction in T_{regs} and the suppressive cytokine, IL-10, has also been described in the disease [41, 42]. While BD lesions are dominated by neutrophils and $CD4^+$ T cells, innate lymphoid cells including γδ T cells and conventional NK cells are also found in BD lesions and may play a significant role in driving the $CD4^+$ Th1 response characteristic of BD lesions [43–46]. Recent studies have shown a marked decrease in circulating NK cells in BD patients [47].

Proinflammatory cytokines are a key feature of the disease, which is usually described as having a Th1 profile [48, 49]. However, the triggers for cytokine induction are not well understood and/or controversial. Suppressor of cytokine signalling (SOCS) proteins that negatively regulate the JAK–STAT signalling pathway of cytokine induction have been shown to be differentially expressed in BD [50, 51]. The expression of SOCS1 and 3 mRNA and protein was studied in peripheral blood mononuclear cells (PBMCs) and neutrophils of patients with BD and compared with healthy controls (HCs) and patients with recurrent aphthous stomatitis (RAS) using RT-PCR, Western blot and immunohistochemistry. SOCS1 and 3 mRNA was also measured in buccal mucosal cells (BMC) of patients with BD and HCs. SOCS1 and 3 mRNA was significantly upregulated in PBMCs of patients with BD compared with HCs. In addition, there were subtle differences between expression in active and symptom-free BD (quiescent BD). SOCS1

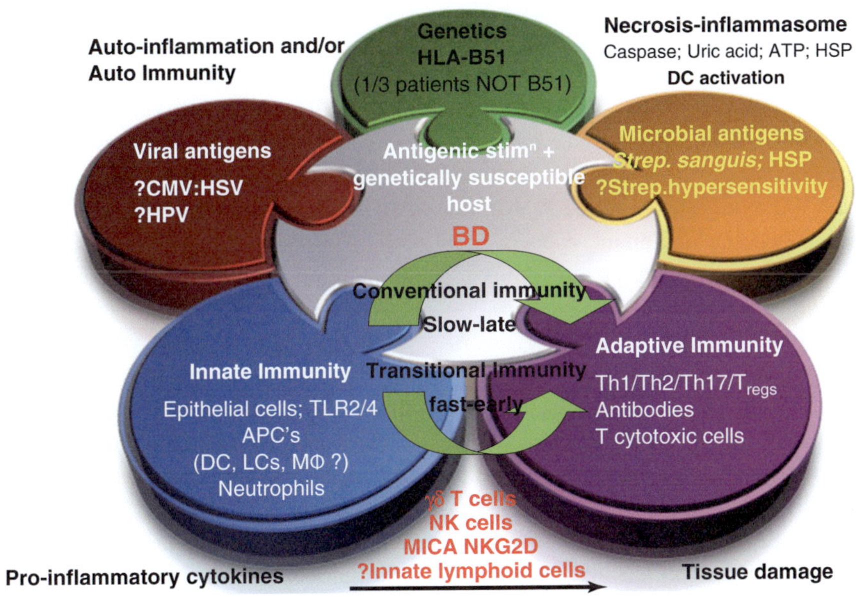

Fig. 11.3 Complexity of immune involvement in Behçet's disease

and SOCS3 were also significantly upregulated in buccal mucosal cells from oral ulcers of BD compared with HCs. These observations suggest a differential expression of these important regulators, not only between patients with BD and healthy controls, but also between mucosal and systemic tissues.

BD demonstrates many immune-disordered features but of interest is the differential expression of SOCS in the oral tissues compared with the circulating cells. Given that more than 90% of patients present with oral ulceration and the clinical observations that the control of oral symptoms frequently results in better control of systemic disease, it would seem appropriate that much greater attention should be paid to restoration of immune homeostasis in the oral cavity (Fig. 11.3).

11.4 Lichen Planus (LP)

Oral lichen planus (OLP) is a mucocutaneous disease that usually presents in the mouth with a bilateral symmetry of white patches or striae on the mucosa. In this form, it is quite common (1–2% of the adult population), but can also present as a bullous, ulcerative or erosive condition [52]. In about 10% of patients a skin manifestation occurs which presents in the flexor surfaces of the arms. Additionally, 25% of female patients present with genital LP. This can either be restricted to the vulva, or in a small subgroup may present additionally in the vagina and the gingiva, the VVG syndrome [53]. OLP also has the potential to become malignant, especially if other risk factors are involved such as smoking [54]. In a recent meta-analysis of about 20,000 patient data a small subset of about 1.1% of patients were at risk of developing oral squamous cell carcinoma (OSCC). A higher incidence of OLP was noted in patients with additional risk factors that include smoking, alcohol consumption and HCV infection, although the authors felt that these associations required further investigation [55–57]. HPV infection has also been cited as precipitating factors along with stress, anxiety and autoimmune thyroid disease [58].

OLP is regarded as a T cell-mediated autoimmune disease where $CD8^+$ cytotoxic T cell triggers apoptosis of the oral epithelial cells [59, 60].

An intense lymphocytic infiltrate is characteristic of OLP with degeneration of the basal layer. Keratinocytes are the target probably due to the expression of foreign or altered-self proteins which are recognised as PAMPS or DAMPS by the innate immune system [61]. There is a suggestion that TLR-mediated signalling might be affected in OLP, especially TLR4, which recognises HSPs [62].

$CD4^+$ T helper cells and $CD8^+$ cytotoxic cells have been shown to drive pathology, and while the antigen is unknown the oligoclonal usage of a restricted set of T cell receptors suggests that an altered-self molecule or superantigens readily activate the T cells [63–65]. Langerhans cells in the mucosa of OLP are more activated than in normal mucosa and the trafficking between the draining lymph nodes and the oral mucosa is increased [66].

Trafficking of cells between the circulation, draining lymph nodes and mucosa is strongly influenced by cytokines, chemokines and adhesion molecules and it has been shown that TNF-α and IFN-γ are important in recruiting cells into the mucosa. TNF-α and IFN-γ induce the expression of the mucosal addressins, E-selectin and MAdCAM-1 on endothelial cells. Selective recruitment of cells expressing either CLA (skin homing marker) or $\alpha_e\beta_7$ (mucosal homing marker) results in infiltration of both $CD4^+$ and $CD8^+$ cells which are then exposed to the antigens on the damaged keratinocytes. This results in recirculation through the draining lymph nodes and further activation and recruitment of cells [67–69]. Cytotoxic T cells are then able to kill antigen-expressing basal cells.

Th17 cells have also been implicated in the pathology of OLP [64, 70] while the loss of T_{regs} and the downregulation of TLR signalling pathways have also been noted [62, 71, 72]. The role of novel Th subsets such as Th22, Th9 and Tfh (follicular helper cells) has yet to be fully elucidated [64].

Oral lichenoid lesions (OLL) are a hypersensitivity reaction that can arise in response to dental restoration materials such as amalgam. Lichenoid reactions also occur in response to NSAIDS and some angiotensin-converting enzyme inhibitors and are a major complication of graft-versus-host disease (GVHD) especially if the graft is allogeneic hematopoietic stem cells or bone marrow [58, 73]. In acute, but more especially chronic, GVHD, patients present with gingivitis, mucositis erythema and pain. Donor T cells are thought to react to major tissue antigens and activate type 1 interferons. It has been suggested that destruction of the thymus allows loss of central tolerance. However, both B cells and T cells are implicated in the general loss of tolerance with cytokines and antigen-presenting cells (DCs) also playing a role [73].

In summary, while the full picture of the aetiopathogenesis has yet to be fully elucidated for OLP, once again immune dysregulation plays a major role in this disease.

11.5 Dermatoses

Oral manifestations of dermatoses include those seen in lupus erythematosus (both discoid DLE and systemic SLE), pemphigus/pemphigoid and erythema multiforme. For all these conditions, there is strong evidence of a genetic involvement. In DLE/SLE lesions similar to lichen planus (LP) occur and there is an association with HLA-DR2 and DR3 as well as HLADRB1*0301 in DLE. Trucci et al. [72] have suggested a dysregulation in DLE similar to that of LP in terms of DC, TLR and type I interferon activity (Table 11.3).

In *erythema multiforme* (EM), while there is an association with HLA-B*1502 an aetiological agent has not been found although hypersensitivity to drugs and some viruses (Herpes sp.) and other infectious agents (*Mycoplasma, Histoplasma, Trichomonas*) have been postulated as causative agents [74]. Currently the most robust association is between HSV infection and EM—this has been termed HAEM (herpes-associated EM). HSV-DNA has been isolated from lesions in 36–80% of patients, although infectious HSV was absent. Drugs which have been shown to precipitate the disease include sulphonamides, NSAIDS, penicillin and anticonvulsants.

Table 11.3 Immune-mediated disorders frequently present with oral manifestations and represent considerable morbidities and mortalities

Prevalence of oral mucosal involvement in immune-mediated disorders		
Disease	No. of cases	References
Lichen planus (65%)	82	Carvalho et al. (2011)*
Pemphigus vulgaris (26.8%)		
Pemphigoid (7.3%)		
Lichen planus (70.2%)	309	Jaafari-Ashkavandi et al. (2011)*
Pemphigus vulgaris (24.9%)		
Pemphigoid (3.3%)		
Erythema multiforme (1.3%)		
Lupus erythematosus (0.33%)		
Lichen planus (51%)	88	Goncalves et al. (2010)*
Lupus erythematosus (20%)		
Erythema multiforme (20%)		
Pemphigus vulgaris (9%)		
Lichen planus (76.56%)	64	Arisawa et al. (2008)*
Pemphigoid (9.37%)		
Erythema multiforme (7.82%)		
Pemphigus vulgaris (6.25%)		
Lichen planus (70.5%)	187	Leo et al. (2008)*
Pemphigoid (14%)		
Pemphigus vulgaris (13%)		
Linear IgA disease (1.6%)		

From Mustafa et al. [58] with permission of the publishers: *Investigations of Immune mediated disorders reviewed in Mustafa et al. [58]

There are several different forms of EM dependent on the extent of mucosal involvement. EM minor affects usually only one mucosa and sometimes presents with symmetrical skin lesions on the extremities. EM major is a more serious condition involving more than one mucosal membrane and variable skin involvement. Stevens-Johnson syndrome (SJS), while presenting with similar skin involvement to EM, is more extensive and more serious with a mortality rate of 5–15%. Both EM and SJS can be associated with systemic symptoms and toxic epidermal necrolysis was thought to be a form of EM but is now considered to be a different disease.

Genetics: Recurrent EM has been associated with HLA-B15, -B35, -A33, -DR53 and HLADQB1*0301. Patients with extensive mucosal involvement may have a rare allele, HLA-DQB1*0402.

EM induces painful erythematous/ulcerative lesions of the mucous membranes and the presence of apoptotic keratinocytes, possibly caused by infiltrating cytotoxic lymphocytes ($CD8^+$ T cells and macrophages). It has been suggested that immune complex disease might be responsible for the histological changes that are observed: lymphohistiocytic infiltration of the lamina propria and infiltration of eosinophils in the degenerating oral epithelium. The recognition of HSV-DNA-expressing keratinocytes by specific $CD4^+$Th1 cells induces the production of IFNγ which in turn upregulates pro-inflammatory cytokines and chemokines. These cells then recruit autoreactive T cells to the oral mucosa where the cells are damaged by cytotoxic T cells, NK cells and chemokines. There appears to be a difference in the pathogenesis when EM is induced by drug sensitivity. Here, there is no IFNγ but TNFα is induced and the tissue damage is because of apoptotic cell death. In SJS the cell death has been suggested to be induced by Fas-FasL interaction which is mediated through caspases [74].

A recent review of drug-induced exfoliative dermatitis (including EM) indicated a strong association with altered cell death mechanisms [75].

Pemphigus is a group of potentially life-threatening autoimmune diseases characterised by cutaneous and/or mucosal blistering. They are classical antibody-driven autoimmune disorders where the target antigens are intracellular substances in the suprabasilar epithelium. Serum IgG and IgM antibodies can be demonstrated and salivary IgA is also present in patients with oral manifestations. The titre of antibody in serum is strongly correlated with the severity of disease and the antibody titre decreases as lesions heal [76, 77].

Genetics: There is a robust genetic association with PV and certain ethnic groups such as Ashkenazi Jews and those of Mediterranean and South Asian origin. The major association is with MHC class II alleles such as HLA-DR4 (DQB1*0503), DRw14. In Japanese patients, the association lies in HLA-B15. These alleles are critical for the recognition of Dsg3 by T lymphocytes which drive the class switching to pathogenic IgG subclasses. There is direct evidence that IgG antibodies to Dsg3 are critical for pathogenesis. IgG1 antibodies are seen in remission but IgG4 predominates in active PV. IgG4 is a very interesting molecule as it can interchange its Fab arms so there is the potential for the antibody to be *monovalent* as seen in myasthenia gravis [78]—or indeed bi-specific. This has been exploited in the design of cancer therapies where a bi-specific antibody can target two antigens [79].

The molecular mechanism of cell-cell interaction and the effect of disruption of the desmosomes in these conditions are discussed in more detail in Chap. 3. Different profiles of antibody specificity have been described, with IgG anti-Dsg3 associated with pemphigus vulgaris (PV) [80], whereas IgG anti-Dsg1 is associated with pemphigus foliaceus—this form of the disease does not usually affect the oral mucosa [81]. However, disease severity can be correlated with the balance between both of these antibodies, present in about 50% of PV patients. Anti-Dsg3 predominates in oral manifestations. Where IgA antibodies occur in the saliva, it is thought that this is a serum exudate.

A third variant of the disease, paraneoplastic pemphigus, also gives rise to mucosal presentations and is caused by both antibody- and cell-mediated autoimmune responses. Patients can also develop autoreactive IgG antibodies to other cytoplasmic proteins including epiplakin, plectin and desmoplakins [77] (Table 11.4).

Mucous membrane pemphigoid (MMP) is a collection of disparate autoimmune conditions,

Table 11.4 Major autoantigens in disorders affecting the oral mucosa

Disease	Autoantigen
Pemphigus diseases	
Pemphigus vulgaris	Desmoglein 3, Desmoglein 1
Paraneoplastic pemphigus	Desmoglein 3, Desmoglein 1, Desmoplakin, Periplakin, Envoplakin
	Plectin, Desmocollins 1–3. BP230
	Alpha-2-macroglobuline-like-1
Pemphigus vegetans	Desmoglein 3, Desmoglein 1
Pemphigus foliaceus	Desmoglein 1
Pemphigoid diseases	
Mucous membrane pemphigoid	Collagen XVII/BP180, BP230, Laminin 332, α6β4 integrin
Linear IgA disease	LAD-1 (120 kDa), LABD97(97 kDa), 285 kDa, 180 kDa
Epidermolysis bullosa acquisita	Collagen VII
Bullous pemphigoid	Collagen XVII/BP180, BP230
Dermatitis herpetiformis	Tissue/epidermal transglutaminase
Chronic ulcerative stomatitis	deltaNp63alpha
Lichen planus	Not known
Erythema multiforme	Not known, Desmoplakin 1 and II (?)
Systemic lupus erythematosus	Nuclear antigens

From Mustafa et al. [58] with permission of the publishers

now referred to as immune-mediated subepithelial blistering diseases (IMSEBD). The group includes bullous pemphigoid (BP) and pemphigoid (herpes) gestationis which generally present with skin but not oral lesions. Cicatricial pemphigoid involves the oral mucous membranes and the eyes [82]. Two types occur in the oral mucosa: those that are ulcerative and involve the non-keratinised (and occasionally) and the keratinised mucosa and that which induces a desquamative gingivitis. They are characterised by deposition of antibodies (IgG) and complement (C3) in the basement membrane, which results in a bullous-type lesion and separation of the epithelium [83–85]. The autoantibodies have been shown to initiate a signal transduction cascade that increases the secretion of IL-6 and IL-8. This results in the recruitment of leucocytes (neutrophils) which on release of cytolytic enzymes effect the detachment of the basal cells from the basement membrane zone (BMZ). Other cytokines and chemokines are also thought to be involved including RANTES, TNF-α and IFN-γ.

SIgA antibodies specific for BP antigens of the hemidesmosomes, at the gap junctions in the epithelia, are characteristic of the MMP.

Genetics: An association with HLA DQB1*0301 has been noted, especially in patients with higher clinical scores or those with ocular disease [86]. Other studies suggest an involvement of HLA DQ7. There is a predominantly female presentation with an age of onset between 51 and 62 years.

Circulating IgG and/or IgA antibodies are induced to a variety of basement membrane zone antigens because of loss of tolerance to these self-proteins in BP but these are rarely detectable in MMP if skin substrates are used (Table 11.4). However, using mucosal substrates such as monkey oesophagus, antibodies are detected in variable degrees. In CP, circulating antibodies specific for extracellular and intracellular domains of BP180 are detectable. The specificity of autoantibodies has been used to distinguish CP and BP. The evidence is growing that the different presentation of pemphigoid variants reflects not just the specificity of the autoantibodies but also the mechanisms of pathogenesis (reviewed in [82]). When both classes of antibody (IgG and IgA) are present in the circulation the disease tends to be more severe and persistent than when IgG serum antibodies are present alone.

A variety of autoantigens including the bullous pemphigoid antigen 1 (BPAg1) (a 230 kDa protein, BP230), the bullous pemphigoid antigen 2 (BPAg2) (a 180 kDa protein, BP180), integrin subunits α6/β4, laminin-332 (also known as epiligrin and laminin-5), laminin-6 and collagen type I have been identified. BPAg1 is an intracellular protein, whereas BPAg2 and α6/β4 integrins are transmembrane proteins. The most frequently targeted autoantigen in MMP is BPAg2. Laminin-5 is thought to be the major ligand between the transmembrane proteins and the anchoring filaments. Anchoring fibrils, composed of type VII collagen, are located deeper in the *lamina densa*. These autoantigens are not exclusive to MMP and anti-BPAg1 and anti-BPAg2 have been demonstrated in BP, although anti-BPAg2 is more common, and autoantibodies to type VII collagens are also found in *epidermolysis bullosa acquisita* [87].

Linear IgA disease is a variant of pemphigoid that is distinguishable by the deposition of IgA at the epidermal basement membrane and oral presentation occurs in about 80% of cases with multiple painful ulcers occurring following the rupture of blisters. Desquamative gingivitis is frequently observed.

11.6 Sjögren's Syndrome (SS)

Primary *Sjögren's syndrome (SS)* is a chronic autoimmune disorder characterised by lymphocytic infiltration of the exocrine glands, especially the salivary glands and lacrimal glands resulting in mucosal dryness. A broad clinical presentation of exocrine involvement may also extend to the vagina but can also include systemic disorders (musculoskeletal, pulmonary, gastric, renal and nervous systems) and lymphoproliferative conditions (lymphoma).

Viral infections have been associated with SS, including EBV, HHV-6 and HTLV-1, and the salivary glands can act as sites of latency for these viruses [88]. In a Japanese cohort, high titres of HTLV-1 were described along with salivary IgA specific for HTLV-1 antigens [89] and cross-reactivity between the La/SSB protein and viral domains of EBV, HHV-6 and HIV-1 has been suggested [90]. Activation of TLR3 has been observed and there is evidence of dysregulation in elements of both the innate and adaptive immune responses [91, 92] (Fig. 11.4).

Pro-inflammatory cytokines, IL-1, TNF and IL6, are upregulated in salivary gland tissues in SS and IL-33 (an IL-1 family member) acts synergistically with IL-23 and IL-12 to upregulate the production of IFNγ by NK and NKT cells [93]. Recent investigations have shown that IL-17 is dysregulated along with the SOCS and JAK-STAT signalling pathways [94]. In the B cell compartment, there is hyperactivity and the formation of so-called ectopic germinal centres, which are now recognised as fundamental to the diagnosis of SS [95, 96]. The activation of type I IFN and the TLR signalling pathways are reminiscent of the observations made in lichen planus and lupus [72] and suggest that these pathways are central to oral mucosal homeostasis and that the genetic background, antigenic and/or environmental exposure that disrupt these pathways do so in an individualist manner to produce disparate presentations of common dysregulations (Fig. 11.4).

SS patients are at increased risk of developing lymphoma especially those associated with mucosal associated lymphoid tissues (MALT). The two main features of SS, namely, the chronic exposure to autoantigens and immune activation, are key in the pathogenesis of lymphoma. The ectopic germinal centres described in SS are defined as B cell aggregates surrounded by T cells and are indicative of inflammation in secondary lymphoid organs such as the lymph nodes draining the salivary glands and are dependent on IL-21 and IL-22 for their formation.

Th17 cells are a major source of these cytokines [97–101].

Genetics: Genetic association is not as clearly defined compared with other autoimmune diseases such as SLE or RA, but is expected to be highly complex. There is known familial association and over a 600-fold higher risk of developing SS in an unaffected twin if a twin sibling has the disease [102, 103]. GWAS

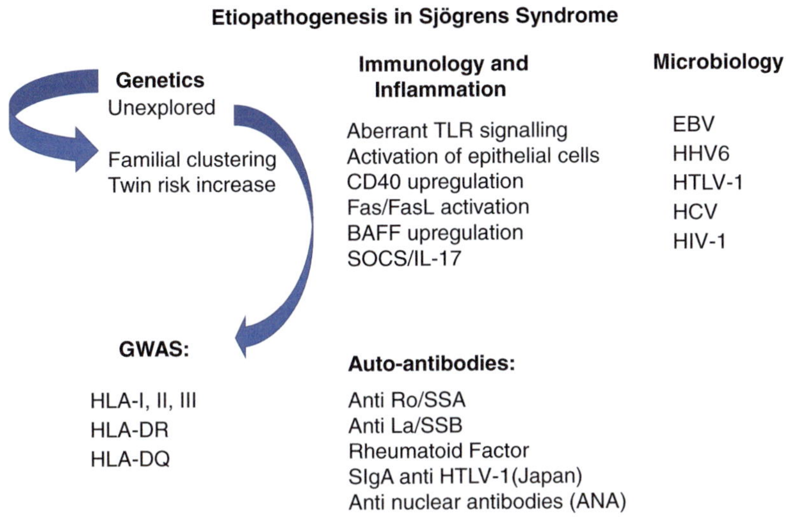

Fig. 11.4 Genetic associations, infection and dysregulation of both innate and adaptive immune responses can be demonstrated in SS

studies have been carried out and several strong associations have been noted in HLA class I, II and III as well as HLA-DR and DQ loci [104], indicating that both the innate and adaptive immune responses contribute to disease pathogenesis. Other associations include genes which control expression of type I and II IFN signalling (*IRF5, IL12A* and *STAT4*), NF-κB signalling (*CXCR5*) and genes involved in activation and differentiation pathways in B cells (*BLK*) [105].

11.7 Tumours

In 2013 oral cancer resulted in 135,000 deaths in the USA, an increase from 84,000 deaths in 1990 [106]. The 5-year survival rates in the USA are 63%. Oncogenes are activated because of DNA mutations. Risk factors that predispose a person to oral cancer have been identified in epidemiological studies. Smoking, alcohol and chewing betel, paan and areca are known to be strong risk factors for developing oral cancer. In India where such practices are common, oral cancer represents up to 40% of all cancers, compared to just 4% in the UK.

Around 75% of oral cancers are linked to these modifiable behaviours. Other factors include poor oral hygiene, irritation caused by ill-fitting dentures and other rough surfaces on the teeth, poor nutrition and some chronic infections caused by fungi, bacteria or viruses [107]. If oral cancer is diagnosed in its earliest stages, treatment is generally very effective.

Oral cancer often presents as a non-healing ulcer (shows no sign of healing after 2 weeks). In the USA, oral cancer accounts for about 8% of all malignant growths. Men are affected twice as often as women, particularly men older than 40.

Head and neck squamous cell carcinoma (HNSCC) is the most common in the oral cavity. It is an aggressive epithelial malignancy but has an 80% 5-year survival rate when treated in the early stages. This drops to 19% for late-stage disease. The rate of second primary tumours in the oral cavity has been reported as 3–7% which is higher than any other malignancy. This observation has led to the concept of 'field characterisation' where it is postulated that multiple primary tumours develop independently in the aero-digestive tract because of continuous exposure to carcinogens. It is now known that at least 50% of oropharyngeal cancers, especially those involving the tonsils and the base of the tongue and oropharynx, are infected with oncogenic variants of HPV [108, 109]. It has however been noted that patients with HPV-positive HNSCC do better than HPV-negative HNSCC. The success of the HPV16 vaccine for cervical carcinoma gives hope that other vaccine strategies might be used against at least some of these malignancies.

The molecular mechanisms that predominate in oral carcinomas and the effect on homeostasis are explored in more detail by Professor Parkinson in Chap. 8.

11.8 Oral Manifestations of Systemic Disease

Many systemic diseases present with oral manifestations which significantly impact on the quality of life either due to loss or compromised function in the oral cavity or due to the pain that frequently accompanies such alterations to normal homeostasis (Table 11.2). There is a very extensive literature [110–112] suggesting that the mouth is an excellent 'mirror' or 'window' for studying the oral presentations of systemic disease and in some cases may occur before systemic disease is obvious. This is certainly the case for some of the oral manifestations of HIV, at the transition from the asymptomatic phase to full-blown AIDS (Table 11.5). Tables 11.3 and 11.6 give an indication of the many diseases that have oral manifestations which are beyond the scope of this chapter.

Pemphigus/pemphigoid, Sjögren's syndrome and erythema multiforme have been described above and the final section of this chapter deals with the gastrointestinal disease which has oral manifestations.

11.9 Crohn's Disease and Orofacial Granulomatosis

Crohn's disease (CD) in the oral cavity (OCD) is also described as orofacial granulomatosis (OFG) in the literature and presents as a non-caseating granuloma with a lymphocytic infiltration and swelling at various sites in the oral cavity. OFG can occur without gastrointestinal involvement and in some cases is associated with established Crohn's and can also occur as a manifestation of sarcoidosis. Oral presentation is usually that of thickened rubbery lips and cheeks and swollen gingival mucosa. Ulcers and tags in the buccal mucosa are frequent [113, 114]. Serum and salivary IgA and IgA2 responses to *Streptococcus cerevisiae* have been demonstrated in OFG and CD suggesting a method by which the two

Table 11.5 Clinical presentations of oral manifestations of some systemic diseases

Systemic diseases	Oral manifestations
Scarlet fever	Fiery red tongue, prominent papillae (raspberry tongue), white-coated tongue with projected papillae (strawberry tongue)
Measles	Spotty enanthema in the oral cavity, often precedes skin rash, ulcerated buccal mucosa Koplik spots
Infectious mononucleosis (EBV)	Acute pharyngitis and tonsillitis occasionally with grey-white exudative membrane, enlarged lymph nodes, palatal petechia
Diphtheria	Characterised by dirty-white, fibrinosuppurative, tough, inflammatory membranes over the tonsils and retropharynx
HIV	Opportunistic oral infections, Candida, herpes viruses, Kaposi sarcoma, hairy leukoplakia
Lichen planus	Reticulate, lace-like white keratotic lesions; bullous and ulcerated (rare); seen in ~50% or patients with cutaneous LP; usually with other systemic manifestations
Pemphigus	Vesicles and bullae-prone to rupture. Produce hyperaemic erosion covered with exudate
Bullous pemphigoid	Oral lesions similar to pemphigus: histologically distinct
Erythema multeforme (Stevens–Johnson syndrome)	Maculopapular, vesiculobullous eruptions. Sometimes following infection, ingestion of drugs, cancer development, collagen vascular disease

diseases might be distinguished [115]. CD has been associated with several infectious agents in the past, including *Mycobacterium paratuberculosis* and Saccharomyces [116]. More recently the effect of signalling through the CD40-CD40L pathway using HSP70 peptides that inhibit the production of TNFα has demonstrated a potential route of amelioration of inflammation on CD and possibly OFG [117], again demonstrating the power of immune intervention in these inflammatory diseases. More recently a review of the histology of biopsies and intestinal pattern of disease has suggested that OCD and OFG are two distinct diseases [118].

There is also increasing evidence that the interactions of the host with the microbiome are critical in the maintenance or loss of homeostasis, suggesting that a normal commensal flora might become dysbiotic and give rise to a selection of pathobionts in CD, similar to the dysbiotic changes that occur in periodontitis [119, 120].

There is some overlap between the presentation and disease associations in Crohn's disease and Behçet's disease, especially in terms of gastrointestinal symptoms, and it is important that good differential diagnosis of these two conditions is established [121] (Fig. 11.5).

Celiac disease tends to present in the oral cavity as oral ulceration. However, oral ulceration does not predispose to the development of celiac disease. Celiac disease is a classical hypersensitivity reaction to wheat gliadin which results in a flattening of the villi of the small intestine and a reduced surface area for nutrient absorption. The loss of absorptive surface area in the gut of celiac patients leads to low levels of iron and folate which are easily remedied and this results in resolution of the oral ulcers. More recently the crucial role of B cells in celiac disease has been reviewed [122] and HSP autoantigens have now been suggested to be involved in this disease [123].

Ulcerative colitis also presents with oral ulceration, but in addition to aphthous ulcers patients may also present with more severe types of ulcers including pyostomatitis necrotica, pyostomatitis vegetans or haemorrhagic ulcers. These ulcers do not appear in the oral cavity in the absence of bowel symptoms, in contrast to Crohn's disease

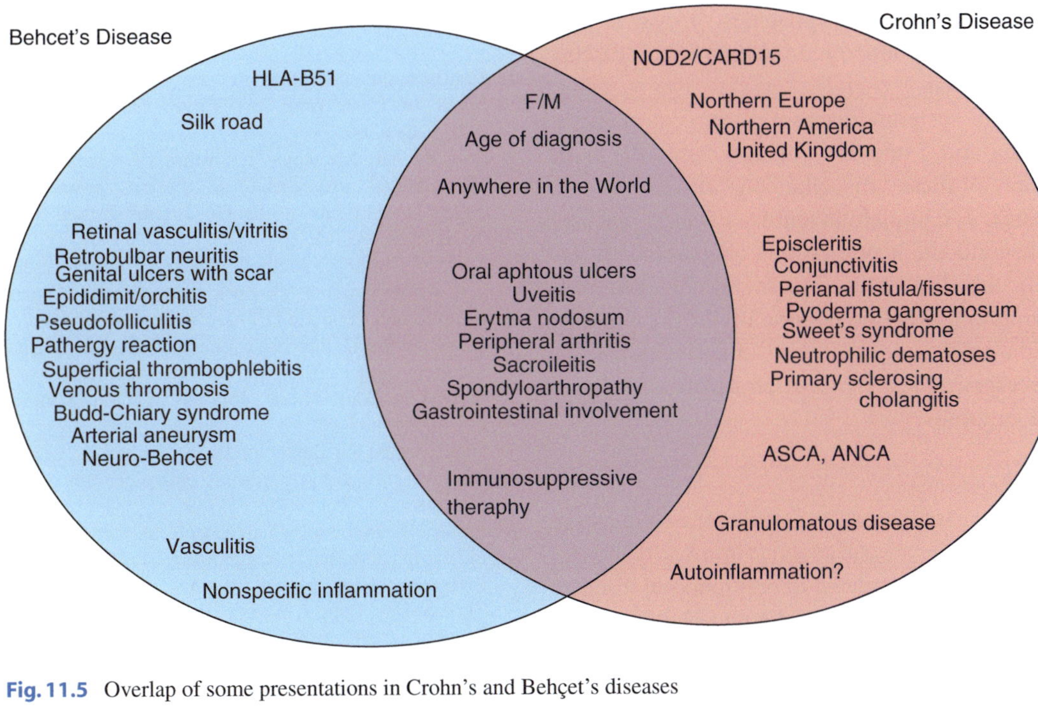

Fig. 11.5 Overlap of some presentations in Crohn's and Behçet's diseases

Table 11.6 Many systemic diseases present with oral manifestations

Miscellaneous disease with oral manifestations	
• **Haematological**	• **Graft vs. host disease**
• Anaemia	
• Leukaemia	
• Multiple myeloma	
• **Rheumatological**	• **Autoinflammatory syndromes**
• Scleroderma	• Periodic fever
• Lupus	• Familial Mediterranean fever
• Rheumatoid arthritis	• Hyperimmunoglobulineamia D
• **Oncological**	• Mevalonate aciduria
• Metastatic disease	• TNF receptor-associated periodic fever
• Histiocytosis	• Pyogenic sterile arthritis
• Mucositis (as a result of cancer therapy)	
• **Endocrine disorders**	• **Adverse pregnancy outcomes**
• Diabetes (Periodontitis)	
• Hypo and hyperthyroidism	• **Coronary heart disease**
• Hyperadrenocortisism	• Infective endocarditis
• **Renal disease**	• **Aspiration pneumonia**
• Uremic stomatitis	

Adapted from: Scully et al. [112], Long et al. [111]

where oral presentation can occur without gut symptoms. Interestingly, cell necrosis frequently releases HSPs and if these are taken up by resident DCs or Langerhans cells in the oral mucosa it might precipitate autoimmune responses to these chaperone molecules. However, the involvement of mucosal immunity in this disease is not proven.

Many systemic diseases show oral manifestations and Table 11.6 demonstrates the wide variety of these 'miscellaneous' *inside-out* conditions. A complete inventory of these diseases is beyond the scope of this chapter but hopefully this list reflects an increased awareness within the scientific community of the capacity of the oral cavity to reveal underlying disease processes and to signpost protocols and potential therapies.

Conclusions

This chapter seeks to describe the elements of both the innate and adaptive immune responses that are dysregulated in oral mucosal diseases and to try to point out the underlying mechanisms of disease presentation. This is by no means an exhaustive examination of all aspects of oral mucosal disease but seeks to emphasise the role that the immune system has in both resolution of disease and disease progression.

References

1. Challacombe SJ, Shirlaw PJ. Oral ulceration: when to treat, refer or ignore. Dental Update. 1991;18(9):368–73.
2. Tappuni AR, Kovacevic T, Shirlaw PJ, Challacombe SJ. Clinical assessment of disease severity in recurrent aphthous stomatitis. J Oral Pathol Med. 2013;42(8):635–41.
3. Jurge S, Kuffer R, Scully C, Porter SR. Mucosal disease series. Number VI. Recurrent aphthous stomatitis. Oral Dis. 2006;12(1):1–21.
4. Challacombe SJ, Barkhan P, Lehner T. Haematological features and differentiation of recurrent oral ulceration. The British Journal of Oral Surgery. 1977a;15(1):37–48.
5. Challacombe SJ, Batchelor JR, Kennedy LA, Lehner T. HLA antigens in recurrent oral ulceration. Arch Dermatol. 1977b;113(12):1717–9.
6. Lehner T. Immunological aspects of recurrent oral ulceration and Behcet's syndrome. Journal of Oral Pathology. 1978;7(6):424–30.
7. Thomas DW, Bagg J, Walker DM. Characterisation of the effector cells responsible for the in vitro cytotoxicity of blood leucocytes from aphthous ulcer patients for oral epithelial cells. Gut. 1990;31(3):294–9.

8. Sun A, Chu CT, Wu YC, Yuan JH. Mechanisms of depressed natural killer cell activity in recurrent aphthous ulcers. Clin Immunol Immunopathol. 1991;60(1):83–92.
9. Pedersen A, Hornsleth A. Recurrent aphthous ulceration: a possible clinical manifestation of reactivation of varicella zoster or cytomegalovirus infection. J Oral Pathol Med. 1993;22(2):64–8.
10. Hasan A, Childerstone A, Pervin K, Shinnick T, Mizushima Y, Van der Zee R, Vaughan R, Lehner T. Recognition of a unique peptide epitope of the mycobacterial and human heat shock protein 65-60 antigen by T cells of patients with recurrent oral ulcers. Clin Exp Immunol. 1995;99(3):392–7.
11. Hasan A, Shinnick T, Mizushima Y, van der Zee R, Lehner T. Defining a T-cell epitope within HSP 65 in recurrent aphthous stomatitis. Clin Exp Immunol. 2002;128(2):318–25.
12. Hasan A, Sadoh D, Palmer R, Foo M, Marber M, Lehner T. The immune responses to human and microbial heat shock proteins in periodontal disease with and without coronary heart disease. Clin Exp Immunol. 2005;142(3):585–94.
13. Hijazi K, Lowe T, Meharg C, Berry SH, Foley J, Hold GL. Mucosal microbiome in patients with recurrent aphthous stomatitis. J Dent Res. 2015;94(3 Suppl):87s–94s.
14. International Study Group for Behcet's Disease. Criteria for diagnosis of Behcet's disease. Lancet. 1990;**335**(8697):1078–80.
15. Davatchi F, Assaad-Khalil S, Calamia KT, Crook JE, Sadeghi-Abdollahi B, Schirmer M, Tzellos T, Zouboulis CC, Akhlagi M, Al-Dalaan A, Alekberova ZS, Ali AA, Altenburg A, Arromdee E, Baltaci M, Bastos M, Benamour S, Ghorbel I, Boyvat A, Carvalho L, Chen W, Ben-Chetrit E, Chams-Davatchi C, Correia J, Crespo J, Dias C, Dong Y, Paixao-Duarte F, Elmuntaser K, Elonakov AV, Grana Gil J, Haghdoost AA, Hayani RM, Houman H, Isayeva AR, Jamshidi AR, Kaklamanis P, Kumar A, Kyrgidis A, Madanat W, Nadji A, Namba K, Ohno S, Olivieri I, Vaz Patto J, Pipitone N, de Queiroz MV, Ramos F, Resende C, Rosa CM, Salvarani C, Serra MJ, Shahram F, Shams H, Sharquie KE, Sliti-Khanfir M, Tribolet de Abreu T, Vasconcelos C, Vedes J, Wechsler B, Cheng YK, Zhang Z, Ziaei N. The International criteria for Behcet's disease (ICBD): a collaborative study of 27 countries on the sensitivity and specificity of the new criteria. J Eur Acad Dermatol Venereol. 2014;28(3):338–47.
16. Senusi A, Seoudi N, Bergmeier LA, Fortune F. Genital ulcer severity score and genital health quality of life in Behçet's disease. Orphanet Journal of Rare Diseases. 2015;10:117.
17. Verity DH, Marr JE, Ohno S, Wallace GR, Stanford MR. Behcet's disease, the Silk Road and HLA-B51: historical and geographical perspectives. Tissue Antigens. 1999;54(3):213–20.
18. Lehner T. Immunopathogenesis of Behcet's disease. Ann Med Interne (Paris). 1999;150(6):483–7.
19. Lehner T, Stanford MR, Phipps PA, Sun JB, Xiao BG, Holmgren J, Shinnick T, Hasan A, Mizushima Y. Immunopathogenesis and prevention of uveitis with the Behcet's disease-specific peptide linked to cholera toxin B. Adv Exp Med Biol. 2003;528:173–80.
20. Takeuchi M, Kastner DL, Remmers EF. The immunogenetics of Behçet's disease: a comprehensive review. J Autoimmun. 2015;64:137–48.
21. Deuter CM, Kotter I, Wallace GR, Murray PI, Stubiger N, Zierhut M. Behcet's disease: ocular effects and treatment. Prog Retin Eye Res. 2008;27(1):111–36.
22. Marshall SE. Behcet's disease. Best Pract Res Clin Rheumatol. 2004;18(3):291–311.
23. Jenkins MK, Chen CA, Jung G, Mueller DL, Schwartz RH. Inhibition of antigen-specific proliferation of type 1 murine T cell clones after stimulation with immobilized anti-CD3 monoclonal antibody. J Immunol. 1990;144(1):16–22.
24. Kaya TI. Genetics of Behcet's disease. Pathol Res Int. 2012;2012:912589.
25. Pineton de Chambrun M, Wechsler B, Geri G, Cacoub P, Saadoun D. New insights into the pathogenesis of Behcet's disease. Autoimmun Rev. 2012a;11(10):687–98.
26. Kirino Y, Bertsias G, Ishigatsubo Y, Mizuki N, Tugal-Tutkun I, Seyahi E, Ozyazgan Y, Sacli FS, Erer B, Inoko H, Emrence Z, Cakar A, Abaci N, Ustek D, Satorius C, Ueda A, Takeno M, Kim Y, Wood GM, Ombrello MJ, Meguro A, Gül A, Remmers EF, Kastner DL. Genome-wide association analysis identifies new susceptibility loci for Behçet's disease and epistasis between HLA-B*51 and ERAP1. Nat Genet. 2013;45(2):202–7.
27. Remmers EF, Cosan F, Kirino Y, Ombrello MJ, Abaci N, Satorius C, Le JM, Yang B, Korman BD, Cakiris A, Aglar O, Emrence Z, Azakli H, Ustek D, Tugal-Tutkun I, Akman-Demir G, Chen W, Amos CI, Dizon MB, Kose AA, Azizlerli G, Erer B, Brand OJ, Kaklamani VG, Kaklamanis P, Ben-Chetrit E, Stanford M, Fortune F, Ghabra M, Ollier WE, Cho YH, Bang D, O'Shea J, Wallace GR, Gadina M, Kastner DL, Gül A. Genome-wide association study identifies variants in the MHC class I, IL10, and IL23R-IL12RB2 regions associated with Behçet's disease. Nat Genet. 2010;42(8):698–702.
28. Elliott GC, Zeidler MP. MsSOCS expression indicates a potential role for JAK/STAT signalling in the early stages of Manduca sexta spermatogenesis. Insect Mol Biol. 2008;17(5):475–83.
29. Kageyama R, Yoshiura S, Masamizu Y, Niwa Y. Ultradian oscillators in somite segmentation and other biological events. Cold Spring Harb Symp Quant Biol. 2007;72:451–7.
30. Liu ZJ, Liu XL, Zhao J, Shi YJ, Yan LN, Chen XF, Li XH, You HB, Xu FL, Gong JP. The effects of SOCS-1 on liver endotoxin tolerance development induced by a low dose of lipopolysaccharide are related to dampen NF-kappaB-mediated pathway. Dig Liver Dis. 2008;40(7):568–77.

31. Seoudi N, Bergmeier LA, Hagi-Pavli E, Bibby D, Curtis MA, Fortune F. The role of TLR2 and 4 in Behcet's disease pathogenesis. Innate Immunity. 2014;20(4):412–22.
32. Phipps PA, Stanford MR, Sun JB, Xiao BG, Holmgren J, Shinnick T, Hasan A, Mizushima Y, Lehner T. Prevention of mucosally induced uveitis with a HSP60-derived peptide linked to cholera toxin B subunit. Eur J Immunol. 2003;33(1):224–32.
33. Stanford MR, Kasp E, Whiston R, Hasan A, Todryk S, Shinnick T, Mizushima Y, Dumonde DC, van der Zee R, Lehner T. Heat shock protein peptides reactive in patients with Behcet's disease are uveitogenic in Lewis rats. Clin Exp Immunol. 1994;97(2):226–31.
34. Direskeneli H, Hasan A, Shinnick T, Mizushima R, van der Zee R, Fortune F, Stanford MR, Lehner T. Recognition of B-cell epitopes of the 65 kDa HSP in Behcet's disease. Scand J Immunol. 1996;43(4):464–71.
35. Pervin K, Childerstone A, Shinnick T, Mizushima Y, van der Zee R, Hasan A, Vaughan R, Lehner T. T cell epitope expression of mycobacterial and homologous human 65-kilodalton heat shock protein peptides in short term cell lines from patients with Behçet's disease. J Immunol. 1993;151(4):2273–82.
36. Uchio E, Stanford M, Hasan A, Satoh S, Ohno S, Shinnick T, van der Zee R, Mizushima Y, Lehner T. HSP-derived peptides inducing uveitis and IgG and IgA antibodies. Exp Eye Res. 1998;67(6):719–27.
37. Stanford M, Whittall T, Bergmeier LA, Lindblad M, Lundin S, Shinnick T, Mizushima Y, Holmgren J, Lehner T. Oral tolerization with peptide 336-351 linked to cholera toxin B subunit in preventing relapses of uveitis in Behcet's disease. Clin Exp Immunol. 2004;137(1):201–8.
38. Masuhiro Y, Kayama K, Fukushima A, Baba K, Soutsu M, Kamiya Y, Gotoh M, Yamaguchi N, Hanazawa S. SOCS-3 inhibits E2F/DP-1 transcriptional activity and cell cycle progression via interaction with DP-1. J Biol Chem. 2008;283(46):31575–83.
39. Ogawa Y, Duru EA, Ameredes BT. Role of IL-10 in the resolution of airway inflammation. Curr Mol Med. 2008;8(5):437–45.
40. Beverley PC, Daser A, Michie CA, Wallace DL. Functional subsets of T cells defined by isoforms of CD45. Biochem Soc Trans. 1992;20(1):184–7.
41. Fei X, Zhang Y, Gu X, Qiu R, Mao Y, Ji C. Crystallization and preliminary X-ray analysis of the splice variant of human ankyrin repeat and suppressor of cytokine signaling box protein 9 (hASB9-2). Protein Pept Lett. 2008;15(6):647–9.
42. Lv L, Zhang J, Huang X, Zhao Y, Zhou Z, Zhang H. Lentivirus-mediated RNA interference targeting STAT4 inhibits the proliferation of vascular smooth muscle cells. Arch Med Res. 2008; 39(6):582–9.
43. Hasan A, Fortune F, Wilson A, Warr K, Shinnick T, Mizushima Y, van der Zee R, Stanford MR, Sanderson J, Lehner T. Role of gamma delta T cells in pathogenesis and diagnosis of Behcet's disease. Lancet. 1996;347(9004):789–94.
44. Hasan MS, Bergmeier LA, Petrushkin H, Fortune F. Gamma delta (gammadelta) T cells and their involvement in Behcet's disease. J Immunol Res. 2015;2015:705831.
45. Kaneko F, Takahashi Y, Muramatsu R, Adachi K, Miura Y, Nakane A, Minagawa T. Natural killer cell numbers and function in peripheral lymphoid cells in Behcet's disease. Br J Dermatol. 1985;113(3):313–8.
46. Yamaguchi Y, Takahashi H, Satoh T, Okazaki Y, Mizuki N, Takahashi K, Ikezawa Z, Kuwana M. Natural killer cells control a T-helper 1 response in patients with Behcet's disease. Arthritis Research & Therapy. 2010;12(3):R80.
47. Hasan MS, Ryan PL, Bergmeier LA, Fortune F. Circulating NK cells and their subsets in Behcet's disease. Clin Exp Immunol. 2017;188:311–22.
48. Ben Ahmed M, Houman H, Miled M, Dellagi K, Louzir H. Involvement of chemokines and Th1 cytokines in the pathogenesis of mucocutaneous lesions of Behçet's disease. Arthritis Rheum. 2004;50(7):2291–5.
49. Curnow SJ, Pryce K, Modi N, Knight B, Graham EM, Stewart JE, Fortune F, Stanford MR, Murray PI, Wallace GR. Serum cytokine profiles in Behçet's disease: is there a role for IL-15 in pathogenesis? Immunol Lett. 2008;121(1):7–12.
50. Hamedi M, Bergmeier LA, Hagi-Pavli E, Vartoukian SR, Fortune F. Differential expression of suppressor of cytokine signalling proteins in Behçet's disease. Scand J Immunol. 2014;80(5):369–76.
51. Tulunay A, Dozmorov MG, Ture-Ozdemir F, Yilmaz V, Eksioglu-Demiralp E, Alibaz-Oner F, Ozen G, Wren JD, Saruhan-Direskeneli G, Sawalha AH, Direskeneli H. Activation of the JAK/STAT pathway in Behcet's disease. Genes Immun. 2015;16:176.
52. Eisen D, Carrozzo M, Bagan Sebastian JV, Thongprasom K. Number V oral lichen planus: clinical features and management. Oral Dis. 2005;11(6):338–49.
53. Setterfield JF, Neill S, Shirlaw PJ, Theron J, Vaughan R, Escudier M, Challacombe SJ, Black MM. The vulvovaginal gingival syndrome: a severe subgroup of lichen planus with characteristic clinical features and a novel association with the class II HLA DQB1*0201 allele. J Am Acad Dermatol. 2006;55(1):98–113.
54. Fitzpatrick SG, Hirsch SA, Gordon SC. The malignant transformation of oral lichen planus and oral lichenoid lesions: a systematic review. Journal of the American Dental Association (1939). 2014;145(1):45–56.
55. Aghbari SMH, Abushouk AI, Attia A, Elmaraezy A, Menshawy A, Ahmed MS, Elsaadany BA, Ahmed EM. Malignant transformation of oral lichen planus

and oral lichenoid lesions: a meta-analysis of 20095 patient data. Oral Oncol. 2017;68:92–102.
56. Lodi G, Scully C, Carrozzo M, Griffiths M, Sugerman PB, Thongprasom K. Current controversies in oral lichen planus: report of an international consensus meeting. Part 1. Viral infections and etiopathogenesis. Oral Surg Oral Med Oral Pathol Oral Radiol Endod. 2005;100(1):40–51.
57. Porter SR, Lodi G, Chandler K, Kumar N. Development of squamous cell carcinoma in hepatitis C virus-associated lichen planus. Oral Oncol. 1997b;33(1):58–9.
58. Mustafa MB, Porter SR, Smoller BR, Sitaru C. Oral mucosal manifestations of autoimmune skin diseases. Autoimmun Rev. 2015;14(10):930–51.
59. Eversole LR, Dam J, Ficarra G, Hwang CY. Leukocyte adhesion molecules in oral lichen planus: a T cell-mediated immunopathologic process. Oral Microbiol Immunol. 1994;9(6):376–83.
60. Porter SR, Kirby A, Olsen I, Barrett W. Immunologic aspects of dermal and oral lichen planus: a review. Oral Surg Oral Med Oral Pathol Oral Radiol Endod. 1997a;83(3):358–66.
61. Walsh LJ, Savage NW, Ishii T, Seymour GJ. Immunopathogenesis of oral lichen planus. J Oral Pathol Med. 1990;19(9):389–96.
62. Sinon SH, Rich AM, Parachuru VP, Firth FA, Milne T, Seymour GJ. Downregulation of toll-like receptor-mediated signalling pathways in oral lichen planus. J Oral Pathol Med. 2016;45(1):28–34.
63. Thornhill MH. Immune mechanisms in oral lichen planus. Acta Odontol Scand. 2001;59(3):174–7.
64. Wang H, Zhang D, Han Q, Zhao X, Zeng X, Xu Y, Sun Z, Chen Q. Role of distinct CD4(+) T helper subset in pathogenesis of oral lichen planus. J Oral Pathol Med. 2016;45(6):385–93.
65. Zhou XJ, Savage NW, Sugerman PB, Walsh LJ, Aldred MJ, Seymour GJ. TCR V beta gene expression in lesional T lymphocyte cell lines in oral lichen planus. Oral Dis. 1996;2(4):295–8.
66. Walton LJ, Macey MG, Thornhill MH, Farthing PM. Intra-epithelial subpopulations of T lymphocytes and Langerhans cells in oral lichen planus. J Oral Pathol Med. 1998;27(3):116–23.
67. Brown DW, Furness J, Speight PM, Thomas GJ, Li J, Thornhill MH, Farthing PM. Mechanisms of binding of cutaneous lymphocyte-associated antigen-positive and alphaebeta7-positive lymphocytes to oral and skin keratinocytes. Immunology. 1999;98(1):9–15.
68. Li J, Farthing PM, Thornhill MH. Cytokine regulation of major histocompatibility complex antigen expression by human oral and skin keratinocytes. Arch Oral Biol. 1996a;41(6):533–8.
69. Li J, Ireland GW, Farthing PM, Thornhill MH. Epidermal and oral keratinocytes are induced to produce RANTES and IL-8 by cytokine stimulation. The Journal of Investigative Dermatology. 1996b;106(4):661–6.
70. Monteiro BV, Pereira Jdos S, Nonaka CF, Godoy GP, da Silveira EJ, Miguel MC. Immunoexpression of Th17-related cytokines in oral lichen planus. Appl Immunohistochem Mol Morphol. 2015;23(6):409–15.
71. Firth FA, Friedlander LT, Parachuru VPB, Kardos TB, Seymour GJ, Rich AM. Regulation of immune cells in oral lichen planus. Arch Dermatol Res. 2015;307(4):333–9.
72. Trucci VM, Salum FG, Figueiredo MA, Cherubini K. Interrelationship of dendritic cells, type 1 interferon system, regulatory T cells and toll-like receptors and their role in lichen planus and lupus erythematosus: a literature review. Arch Oral Biol. 2013;58(10):1532–40.
73. Margaix-Muñoz M, Bagán JV, Jiménez Y, Sarrión MG, Poveda-Roda R. Graft-versus-host disease affecting oral cavity. A review. J Clin Exp Dent. 2015;7(1):e138–45.
74. Farthing P, Bagan JV, Scully C. Mucosal disease series. Number IV. Erythema multiforme. Oral Dis. 2005;11(5):261–7.
75. Yacoub M-R, Berti A, Campochiaro C, Tombetti E, Ramirez GA, Nico A, Di Leo E, Fantini P, Sabbadini MG, Nettis E, Colombo G. Drug induced exfoliative dermatitis: state of the art. Clin Mol Allergy. 2016;14(1):9.
76. Black M, Mignogna MD, Scully C. Number II. Pemphigus vulgaris. Oral Dis. 2005;11(3): 119–30.
77. Kasperkiewicz M, Ellebrecht CT, Takahashi H, Yamagami J, Zillikens D, Payne AS, Amagai M. Pemphigus. Nat Rev Dis Primers. 2017;3(17026):26.
78. Koneczny I, Stevens JA, De Rosa A, Huda S, Huijbers MG, Saxena A, Maestri M, Lazaridis K, Zisimopoulou P, Tzartos S, Verschuuren J, van der Maarel SM, van Damme P, De Baets MH, Molenaar PC, Vincent A, Ricciardi R, Martinez-Martinez P, Losen M. IgG4 autoantibodies against muscle-specific kinase undergo Fab-arm exchange in myasthenia gravis patients. J Autoimmun. 2017;77:104–15.
79. Bernhard H, Karbach J, Strittmatter W, Meyer zum Buschenfelde KH, Knuth A. Induction of tumor-cell lysis by bi-specific antibody recognizing ganglioside GD2 and T-cell antigen CD3. Int J Cancer. 1993;55(3):465–70.
80. Harman KE, Seed PT, Gratian MJ, Bhogal BS, Challacombe SJ, Black MM. The severity of cutaneous and oral pemphigus is related to desmoglein 1 and 3 antibody levels. Br J Dermatol. 2001;144(4):775–80.
81. Amagai M, Matsuyoshi N, Wang ZH, Andl C, Stanley JR. Toxin in bullous impetigo and staphylococcal scalded-skin syndrome targets desmoglein 1. Nat Med. 2000;6(11):1275–7.
82. Bagan J, Lo Muzio L, Scully C. Mucosal disease series. Number III. Mucous membrane pemphigoid. Oral Dis. 2005;11(4):197–218.

83. Dabelsteen E. Molecular biological aspects of acquired bullous diseases. Crit Rev Oral Biol Med. 1998;9(2):162–78.
84. Setterfield J, Challacombe SJ, Black MM. Bullous pemphigoid localized to the mouth. Br J Dermatol. 1997;137(5):825.
85. Setterfield J, Shirlaw PJ, Kerr-Muir M, Neill S, Bhogal BS, Morgan P, Tilling K, Challacombe SJ, Black MM. Mucous membrane pemphigoid: a dual circulating antibody response with IgG and IgA signifies a more severe and persistent disease. Br J Dermatol. 1998;138(4):602–10.
86. Setterfield J, Theron J, Vaughan RW, Welsh KI, Mallon E, Wojnarowska F, Challacombe SJ, Black MM. Mucous membrane pemphigoid: HLA-DQB1*0301 is associated with all clinical sites of involvement and may be linked to antibasement membrane IgG production. Br J Dermatol. 2001;145(3):406–14.
87. Xu HH, Werth VP, Parisi E, Sollecito TP. Mucous membrane pemphigoid. Dent Clin N Am. 2013;57(4):611–30.
88. Lucchesi D, Pitzalis C, Bombardieri M. EBV and other viruses as triggers of tertiary lymphoid structures in primary Sjogren's syndrome. Expert Rev Clin Immunol. 2014;10(4):445–55.
89. Terada K, Katamine S, Eguchi K, Moriuchi R, Kita M, Shimada H, Yamashita I, Iwata K, Tsuji Y, Nagataki S, et al. Prevalence of serum and salivary antibodies to HTLV-1 in Sjogren's syndrome. Lancet. 1994;344(8930):1116–9.
90. Haaheim LR, Halse AK, Kvakestad R, Stern B, Normann O, Jonsson R. Serum antibodies from patients with primary Sjogren's syndrome and systemic lupus erythematosus recognize multiple epitopes on the La(SS-B) autoantigen resembling viral protein sequences. Scand J Immunol. 1996;43(1):115–21.
91. Kyriakidis NC, Kapsogeorgou EK, Gourzi VC, Konsta OD, Baltatzis GE, Tzioufas AG. Toll-like receptor 3 stimulation promotes Ro52/TRIM21 synthesis and nuclear redistribution in salivary gland epithelial cells, partially via type I interferon pathway. Clin Exp Immunol. 2014a;178(3):548–60.
92. Kyriakidis NC, Kapsogeorgou EK, Tzioufas AG. A comprehensive review of autoantibodies in primary Sjogren's syndrome: clinical phenotypes and regulatory mechanisms. J Autoimmun. 2014b;51:67–74.
93. Awada A, Nicaise C, Ena S, Schandene L, Rasschaert J, Popescu I, Gangji V, Soyfoo MS. Potential involvement of the IL-33-ST2 axis in the pathogenesis of primary Sjogren's syndrome. Ann Rheum Dis. 2014;73(6):1259–63.
94. Vartoukian SR, Tilakaratne WM, Seoudi N, Bombardieri M, Bergmeier L, Tappuni AR, Fortune F. Dysregulation of the suppressor of cytokine signalling 3-signal transducer and activator of transcription-3 pathway in the aetiopathogenesis of Sjögren's syndrome. Clin Exp Immunol. 2014;177(3):618–29.
95. Ferro F, Marcucci E, Orlandi M, Baldini C, Bartoloni-Bocci E. One year in review 2017: primary Sjogren's syndrome. Clin Exp Rheumatol. 2017;35(2):179–91.
96. Mavragani CP. Mechanisms and new strategies for primary Sjogren's syndrome. Annu Rev Med. 2017;68:331–43.
97. Alunno A, Carubbi F, Bistoni O, Caterbi S, Bartoloni E, Mirabelli G, Cannarile F, Cipriani P, Giacomelli R, Gerli R. T regulatory and T helper 17 cells in primary Sjogren's syndrome: facts and perspectives. Mediat Inflamm. 2015;243723(10):28.
98. Barone F, Bombardieri M, Manzo A, Blades MC, Morgan PR, Challacombe SJ, Valesini G, Pitzalis C. Association of CXCL13 and CCL21 expression with the progressive organization of lymphoid-like structures in Sjögren's syndrome. Arthritis Rheum. 2005;52(6):1773–84.
99. Barone F, Bombardieri M, Rosado MM, Morgan PR, Challacombe SJ, De Vita S, Carsetti R, Spencer J, Valesini G, Pitzalis C. CXCL13, CCL21, and CXCL12 expression in salivary glands of patients with Sjogren's syndrome and MALT lymphoma: association with reactive and malignant areas of lymphoid organization. J Immunol. 2008;180(7):5130–40.
100. Barone F, Nayar S, Campos J, Cloake T, Withers DR, Toellner KM, Zhang Y, Fouser L, Fisher B, Bowman S, Rangel-Moreno J, Garcia-Hernandez Mde L, Randall TD, Lucchesi D, Bombardieri M, Pitzalis C, Luther SA, Buckley CD. IL-22 regulates lymphoid chemokine production and assembly of tertiary lymphoid organs. Proc Natl Acad Sci U S A. 2015;112(35):11024–9.
101. Risselada AP, Looije MF, Kruize AA, Bijlsma JW, van Roon JA. The role of ectopic germinal centers in the immunopathology of primary Sjogren's syndrome: a systematic review. Semin Arthritis Rheum. 2013;42(4):368–76.
102. Kuo CF, Grainge MJ, Valdes AM, See LC, Luo SF, Yu KH, Zhang W, Doherty M. Familial aggregation of systemic lupus erythematosus and coaggregation of autoimmune diseases in affected families. JAMA Intern Med. 2015;175(9):1518–26.
103. Nezos A, Mavragani CP. Contribution of genetic factors to Sjogren's syndrome and Sjogren's syndrome related lymphomagenesis. J Immunol Res. 2015;754825(10):15.
104. Lessard CJ, Li H, Adrianto I, Ice JA, Rasmussen A, Grundahl KM, Kelly JA, Dozmorov MG, Miceli-Richard C, Bowman S, Lester S, Eriksson P, Eloranta ML, Brun JG, Goransson LG, Harboe E, Guthridge JM, Kaufman KM, Kvarnstrom M, Jazebi H, Cunninghame Graham DS, Grandits ME, Nazmul-Hossain AN, Patel K, Adler AJ, Maier-Moore JS, Farris AD, Brennan MT, Lessard JA, Chodosh J, Gopalakrishnan R, Hefner KS, Houston GD, Huang AJ, Hughes PJ, Lewis DM, Radfar L, Rohrer MD, Stone DU, Wren JD, Vyse TJ, Gaffney PM, James JA, Omdal R,

Wahren-Herlenius M, Illei GG, Witte T, Jonsson R, Rischmueller M, Ronnblom L, Nordmark G, Ng WF, Mariette X, Anaya JM, Rhodus NL, Segal BM, Scofield RH, Montgomery CG, Harley JB, Sivils KL. Variants at multiple loci implicated in both innate and adaptive immune responses are associated with Sjogren's syndrome. Nat Genet. 2013;45(11):1284–92.

105. Brito-Zeron P, Baldini C, Bootsma H, Bowman SJ, Jonsson R, Mariette X, Sivils K, Theander E, Tzioufas A, Ramos-Casals M. Sjogren syndrome. Nat Rev Dis Primers. 2016;2:16047. https://doi.org/10.1038/nrdp.2016.47.
106. GBD 2013 Mortality and Causes of Death Collaborators. Global, regional, and national age-sex specific all-cause and cause-specific mortality for 240 causes of death, 1990–2013: a systematic analysis for the Global Burden of Disease Study 2013. Lancet. 2015;385(9963):117–71.
107. Srinivasprasad V, Dineshshankar J, Sathiyajeeva J, Karthikeyan M, Sunitha J, Ragunathan R. Liaison between micro-organisms and oral cancer. Journal of Pharmacy & Bioallied Sciences. 2015;7(Suppl 2):S354–60.
108. Chaturvedi AK, D'Souza G, Gillison ML, Katki HA. Burden of HPV-positive oropharynx cancers among ever and never smokers in the U.S. population. Oral Oncol. 2016;60:61–7.
109. Chaturvedi AK, Engels EA, Anderson WF, Gillison ML. Incidence trends for human papillomavirus-related and -unrelated oral squamous cell carcinomas in the United States. J Clin Oncol. 2008;26(4):612–9.
110. Epstein JB. The mouth: a window on systemic disease. Can Fam Physician. 1980;26:953–7.
111. Long RG, Hlousek L, Doyle JL. Oral manifestations of systemic diseases. Mt Sinai J Med. 1998;65(5–6):309–15.
112. Scully C, Hodgson T, Lachmann H. Auto-inflammatory syndromes and oral health. Oral Dis. 2008;14(8):690–9.
113. Campbell H, Escudier M, Patel P, Nunes C, Elliott TR, Barnard K, Shirlaw P, Poate T, Cook R, Milligan P, Brostoff J, Mentzer A, Lomer MC, Challacombe SJ, Sanderson JD. Distinguishing orofacial granulomatosis from crohn's disease: two separate disease entities? Inflamm Bowel Dis. 2011a;17(10):2109–15.
114. Campbell HE, Escudier MP, Patel P, Challacombe SJ, Sanderson JD, Lomer MC. Review article: cinnamon- and benzoate-free diet as a primary treatment for orofacial granulomatosis. Aliment Pharmacol Ther. 2011b;34(7):687–701.
115. Savage NW, Barnard K, Shirlaw PJ, Rahman D, Mistry M, Escudier MP, Sanderson JD, Challacombe SJ. Serum and salivary IgA antibody responses to Saccharomyces cerevisiae, Candida albicans and Streptococcus mutans in orofacial granulomatosis and Crohn's disease. Clin Exp Immunol. 2004;135(3):483–9.
116. Barnes RM, Allan S, Taylor-Robinson CH, Finn R, Johnson PM. Serum antibodies reactive with Saccharomyces cerevisiae in inflammatory bowel disease: is IgA antibody a marker for Crohn's disease? International Archives of Allergy and Applied Immunology. 1990;92(1):9–15.
117. Whittall T, Wang Y, Kelly CG, Thompson R, Sanderson J, Lomer M, Soon SY, Bergmeier LA, Singh M, Lehner T. Tumour necrosis factor-alpha production stimulated by heat shock protein 70 and its inhibition in circulating dendritic cells and cells eluted from mucosal tissues in Crohn's disease. Clin Exp Immunol. 2006;143(3):550–9.
118. Zbar AP, Ben-Horin S, Beer-Gabel M, Eliakim R. Oral Crohn's disease: is it a separable disease from orofacial granulomatosis? A review. Journal of Crohn's & Colitis. 2012;6(2):135–42.
119. Butto LF, Haller D. Dysbiosis in Crohn's disease: joint action of stochastic injuries and focal inflammation in the gut. Gut Microbes. 2017;8(1):53–8.
120. Butto LF, Schaubeck M, Haller D. Mechanisms of microbe-host interaction in Crohn's disease: dysbiosis vs. pathobiont selection. Front Immunol. 2015;6:555.
121. Yazisiz V. Similarities and differences between Behcet's disease and Crohn's disease. World J Gastrointest Pathophysiol. 2014;5(3):228–38.
122. Stamnaes J, Sollid LM. Celiac disease: autoimmunity in response to food antigen. Semin Immunol. 2015;27(5):343–52.
123. Tukaj S, Gorog A, Kleszczynski K, Zillikens D, Karpati S, Kasperkiewicz M. Autoimmunity to heat shock proteins and vitamin D status in patients with celiac disease without associated dermatitis herpetiformis. J Steroid Biochem Mol Biol. 2017; 173:23–7.

12 Clinical Management of Oral Mucosal Disease: A Literature Review

Lesley Ann Bergmeier and Farida Fortune

An extensive description of diagnostic criteria and detailed clinical management is beyond the scope of this chapter and it should be borne in mind that these are not clinical recommendations but represent the consensus from the literature of successful management strategies that not only control oral manifestations but also contribute to the control of systemic manifestations of the diseases. New understanding of the pathologies of diseases, especially in terms of signalling pathways, has revealed potentially new approaches to clinical management with new classes of drugs.

Many of the diseases are managed by using *immunosuppressive* drugs which also include specific monoclonal antibodies, the so-called biologics. The mode of action of some of the drugs most commonly used is outlined in Table 12.1.

12.1 Recurrent Aphthous Stomatitis (RAS)

Diagnosis of RAS is based on history and clinical observations and it is important to distinguish systemic causes and iron and mineral deficiencies. There are few specific agents for RAS but Table 12.2 shows a list of therapies based on data from Jurge et al. [10]. As with many of the diseases described in this chapter management is based around corticosteroid therapies and immune-modulating drugs.

A systematic review of clinical trials undertaken as part of a Cochrane review could find no evidence of a single systemic treatment for RAS [11] and topical corticosteroids remain the main treatment for RAS [12] as indicated in Table 12.2. These topical treatments vary in their efficacy and can be administered as mouth rinses, ointments and creams or in adhesive vehicles.

The present consensus for treatment remains that of reducing physical trauma in the oral mucosa and controlling any inflammatory responses that drive ulceration.

12.2 Behçet's Disease (BD)

In the last decade, multidisciplinary clinics have been firmly established in the UK that are now funded by the NHS as Behçet's Centres of Excellence (BCE). These centres have provided a step change in the treatment of this multisystem disease with improved time to diagnosis and improved clinical management [13]. The pathway used at the London BCE is outlined in Fig. 12.1.

L.A. Bergmeier (✉) • F. Fortune
Centre for Oral Immunology and Regenerative Medicine, Institute of Dentistry, Queen Mary School of Medicine and Dentistry, London, UK
e-mail: l.a.bergemeier@qmul.ac.uk; f.fortune@qmul.ac.uk

L.A. Bergmeier (ed.), *Oral Mucosa in Health and Disease*,
https://doi.org/10.1007/978-3-319-56065-6_12

Table 12.1 Modes of action of some immunosuppressive drugs in common use for oral mucosal diseases

Medication		Immune system targets	Mode of action
Corticosteroids	For example: Prednisolone	General effects: production *or* inhibition of the transcription of anti-inflammatory proteins. Stimulates lipocortin (anti-inflammatory protein) production which also inhibits the production of various pro-inflammatory cytokines including IFN-γ and TNF-α Reduces T cell numbers and suppress their functions; *reviewed by Bijlsma and Jacobs* [1]	Anti-inflammatory
Anti-inflammatory agents	Colchicine	Inhibition of polymerization of microtubules results in decreased cytokine secretion; inhibits phagocytosis and mitosis. Inhibitory effect on neutrophils; inhibits chemotaxis and downregulates numerous functions; *reviewed by Slobodnick et al.* [2]	
Immunosuppressants	Azathioprine	Thought to act by increasing apoptosis of *activated* T cells; *reviewed by Sahasranaman et al.* [3] Controls T cell apoptosis by modulating a costimulatory signal (CD28) into an apoptotic signal [4]	Reduces immune activity
	MMF	Inhibits the proliferation of T and B lymphocytes. Inhibits infiltration of lymphocytes and monocytes into the site of inflammation Decreases the production of pro-inflammatory cytokines including TNF-α and IL-lβ; *reviewed by Allison* [5]	
"Biologics"		*Infliximab*—binds to TNF-α and TNF-β and lyses TNF-producing cells which neutralises their activity *Etanercept*—a recombinant dimer of human TNF receptor proteins fused and bound to human IgG1, preventing the binding of TNF to its cell surface receptor *Adalimumab*—a monoclonal fully human anti-TNF-α antibody which binds to TNF-α with high affinity; *reviewed by Nash and Florin, Silva et al.* [6, 7] *Rituximab*: anti-CD20. *Anakinra:* anti-IL-1 *Daclizumab*: anti IL-2. *Tocilizumab*: anti-IL-6. *Secukinumab*: anti-IL-17 *Anti-IFN-α* therapy which acts by targeting various immune cells that produce type I interferons; *reviewed by Nava et al. and Saleh and Arayssi* [8, 9]	Reduces inflammation by specifically inhibiting different molecules

The treatment for BD patients consists of several immunomodulatory agents including corticosteroids, immunosuppressants and biological therapies but they are not a cure [9, 14, 15]. The treatment is often symptomatic and the main goal is to induce and maintain disease remission. In BD, high-quality therapeutic trials have been lacking and the choice of therapy is largely based on case reports and several randomised control trials (RCT). However, the results have been inconsistent and the efficacy of the medications used was found to be organ specific. For example, while immunosuppressive therapies prevented relapse of ocular lesions, they were often

Table 12.2 RAS therapies

Local physical treatment	Surgical removal
	Debridement
	Laser ablation
	Low dense ultrasound
	Chemical cautery (e.g. silver nitrate sticks)
	Physical barriers (e.g. cyanoacrylate adhesives)
Antimicrobials	Chlorhexidine gluconate (mouthrinse)
	Triclosan (mouthrinse)
	Topical tetracyclines (e.g. aureomycin, chlortetracycline, tetracycline)
Topical corticosteroids	Hydrocortisone hemisuccinate (pellets)
	Triamcinolone acetonide (in adhesive paste)
	Flucinonide (cream)
	Betamethasone valerate (mouthrinse)
	Betamethasone-17-benzoate (mouthrinse)
	Betamethasone-17-valerate (mouthrinse)
	Flumethasone pivolate (spray)
	Beclomethasone diproprionate (spray)
	Clobetasol proprionate (cream)
	Mometasone furoate (cream)
Topical analgesics	Benzydamine hydrochloride (spray or mouthrinse)
	Topical anaesthetics (gel)
Other topical anti-inflammatory agents	Amlexanox
	Sodium cromoglycate (lozenges)
	Carbenoxolone sodium mouthrinse
	Azalestine
	Human alpha-2-interferon (cream)
	Ciclosporin (mouthrinse)
	Deglycirrhizinated liquorice
	Topical 5-aminosalicylic acid
	Prostaglandin E2 (gel)
	Topical granulocyte-macrophage colony-stimulating factor
	Aspirin mouthrinse
	Diclofenac in hyaluronase
	Sucralfate
Systemic immunosuppression	Prednisolone; Azathioprine; Levamisole
	Colchicine; Thalidomide; Pentoxifylline
	Dapsone
	Cimetidine

These range from topical preparations to systemic anti-inflammatory drugs and their use is dependent on individual severity scores and potential systemic disease which might result in RAS development (from Jurge et al. [10] with permission of the publishers)
NOTE: many of the drug therapies used in RAS are also used in other diseases described below because of their immune modulating mechanisms

ineffective in repeated mucocutaneous lesions [16]. This difference in the efficacy of therapies in different organs is thought to be due to different antigenic stimuli for different manifestations in BD. The complexity of the clinical management decisions can be appreciated when examining the drug pathways illustrated in Fig. 12.1.

The severity of oral ulceration in BD varies from mild infrequent symptoms requiring no treatment to frequent moderate and severe painful symptoms that interfere with quality of life (QoL). The treatment is therefore personalised according to disease severity in individual patients. The drug pathway guidelines recommend that mild symptoms of oral ulceration are treated with topical colchicine and topical steroid and non-steroid (anti-inflammatory) mouthwash and/or triple-therapy mouthwash (1 tablet betamethasone + 1 tablet doxycycline + 1 mL nystatin dissolved in 10 mL of water). Clinicians' observations suggest that control of the oral symptoms in BD frequently leads to better control of the systemic symptoms. However, this is a complex disease and although more than 90% of patients first present with oral ulceration clinical management is a challenge as a result of other systemic manifestations that arise as the disease progresses.

In 2008 the European League Against Rheumatology (EULAR) published a recommendation guideline for the management of BD [17].

A recent review of new therapeutic agents for BD noted that IL-1 inhibitors currently represent the most studied agents among the latest treatment

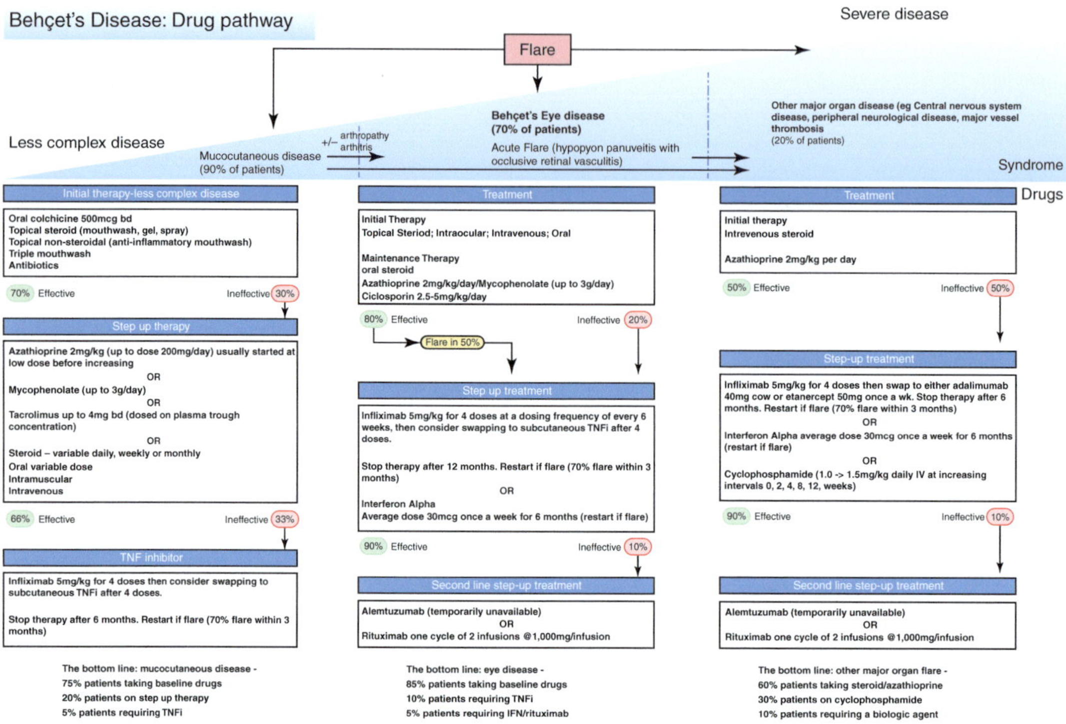

Fig. 12.1 The London Behçet's Centre of Excellence (BCE) BD management protocol

options for BD [18]. These are proving to be effective, safe and with an acceptable retention on treatment. However, BD is a difficult disorder to manage where some symptoms respond to certain treatments that in turn can make other manifestations worse. Identifying new treatment options for patients unresponsive to the current drug regimens relies on the type of drug pathway outlined in Fig. 12.1.

12.3 Autoimmune Skin Diseases with Oral Manifestations

The most frequent site of mucosal manifestations of autoimmune skin disorders is the oral cavity and in the next section the more frequently encountered of these diseases is briefly described in terms of clinical management. In a recent review, the therapeutic options were outlined and the "*usual suspects*" of corticosteroids, immune-modulating drugs and biologics were listed along with adverse reactions [19] (Table 12.3).

12.4 Lichen Planus (LP)

The treatment aim is to eliminate ulcerative and atrophic lesions and as far as possible reduce the risk of malignant transformation. Mechanical trauma or irritants such as ill-fitting dentures should be managed appropriately and where lichenoid reactions occur hypersensitivity to amalgam restorations might be implicated, and these should be replaced with appropriate materials.

There are no medications that have been specifically developed for oral lichen planus (OLP) and the drugs most frequently used are those with immunosuppressive activities. Topical corticosteroids are a first-line medication for OLP and are used widely to suppress T cell responses. Some patients may require systemic medications and these have included azathioprine, cyclosporine and mycophenolate mofetil [20, 21]. In OLP resulting from graft-versus-host disease (GVHD), where autoantibodies are demonstrated, anti-CD20 (rituximab) has been

Table 12.3 Autoimmune skin diseases have oral manifestations and are treated with immune-modulating drugs as well as corticosteroids

Therapeutic options for oral lesions associated with autoimmune skin diseases	
Therapy (indications and properties)	Drugs
Topical corticosteroids	
First line therapy for localised mild or chronic disease	For example, mometasome furoate, triamcinolone acetonide, clobetasol proprionate
Maintenance therapy after short course of systemic corticosteroids	
Calcineurin inhibitors	
Second-line therapy, for patients who fail to respond to corticosteroids	Tacrolimus, Pimercrolimus
Systemic corticosteroids	
Cornerstone therapy with rapid onset and high effectiveness	Prednisolone
Immunosuppressant drugs	
Slower in onset than corticosteroids	Azathioprine
Used in conjunction with corticosteroids for their steroid-sparing actions	Mycophenolate mofetil
Can be used alone to maintain remission after corticosteroids withdrawal	Cyclophosphamide Cyclosporine

Adapted from Mustafa et al. [19]

suggested as a therapy. More recently the development of JAK/STAT pathway inhibitors has been investigated in skin diseases including LP [22] while the effects of dexamethasone and cyclosporine A in OLP have been shown to act through modulation of the TLR4/NF-κB pathway [23]. Total glucosides of paeony have recently been investigated as suppressive agents of the NF-κB pathway for OLP [24].

Novel therapies such as low-level laser therapy have been used as alternatives to corticosteroids and were shown to be effective in the management of symptomatic OLP [25]. However, this systematic review noted that due to a variety of methods and substantial variations in laser parameters among these studies, more randomised clinical trials with large sample sizes are warranted. This was also the conclusion from another systematic review which investigated new therapies such as biologics and nutraceuticals [26]. In other words, until more RTCs with larger sample sizes and longer treatment periods can be carried out and evaluated the management of OLP remains steroids and the immune-modulating drugs reviewed by Eisen et al. in 2005 [20].

12.5 Erythema Multiforme (EM)

There is no specific treatment of EM but supportive care is of great importance, with intravenous hydration and liquid diets often necessary. Acyclovir is a successful treatment in many patients even when a clear viral association (HSV) is not established, while the use of antimicrobials, such as tetracycline, is a successful strategy when EM is associated with *Mycoplasma pneumoniae.* Corticosteroids can be used in EM as they are effective in reducing the amount of keratinocyte cell death and reduction in caspase activity. Tapering regimes of prednisolone have proved useful as well as azathioprine. Other immunomodulating drugs and biologics such as the anti-TNF drugs have also been shown to be efficacious [27]. Thalidomide has controlled previously resistant disease, in one report, and one patient with long-standing recalcitrant CP responded rapidly and lastingly to therapy with the TNF-a antagonist, *etanercept* (reviewed in Farthing et al. [28]). In a recent paper the guidelines for treatment of Steven-Johnson syndrome and toxic epidermal necrolysis were outlined [29] along with differential diagnosis of this disease.

12.6 Pemphigus

Current treatment is based on systemic immunosuppression using corticosteroids and additionally azathioprine or cyclophosphamide has been used (Fig. 12.2). Cyclosporine has also been effective for some patients. However, adverse effects to

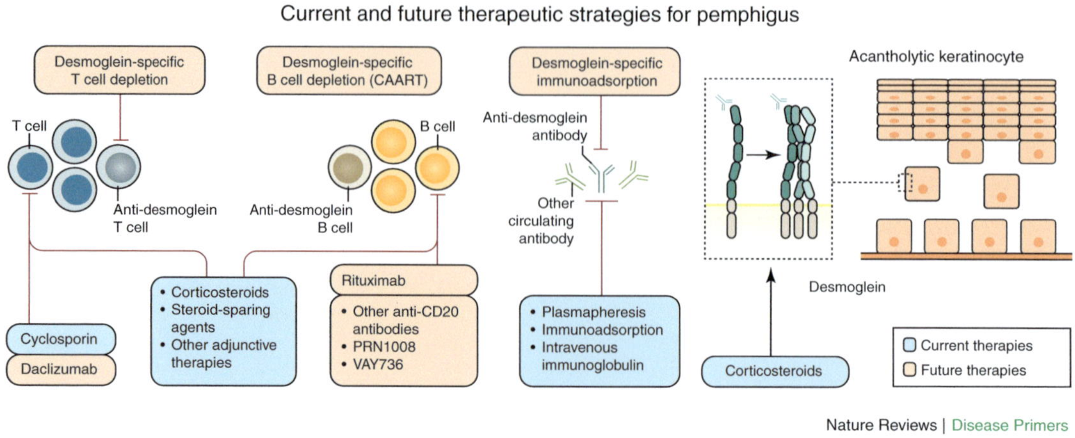

Fig. 12.2 Treatment pathways for pemphigus. Current treatments are in blue; future treatments are in orange. From [32]. Reproduced with publishers' permission

these drugs are recorded (reviewed in [30]). The use of biologics, namely rituximab—an anti-CD20 reagent which targets B cell—has also been effective [31, 32]. In a recent review [33] the potential for therapies that absorb autoantibodies was discussed and this can be regarded as an update of the plasmapheresis that has been used in the past [30]. Relapse after *rituximab* therapy of pemphigus vulgaris may be due to incomplete B cell depletion and recurrence of the same anti-DSG3 B cells observed during initial disease, a model best supported by the current data. Alternatively, disease relapse may be triggered by the appearance of a new anti-DSG3 B cell repertoire after complete B cell depletion. Production of anti-DSG3 antibodies by long-lived plasma cells, which are CD20—and hence not targeted by rituximab, appears not to play a significant role for many patients, given the serologic remissions of disease observed after rituximab therapy.

One of the most exciting potential therapies is the use of T cells that have been engineered to express a chimeric immunoreceptor consisting of the DSG3 extracellular domain fused to the T cell receptor cytoplasmic signalling and costimulatory domains [34]. These experiments have been carried out in mice but have enormous potential as the DSG3 chimeric autoantibody receptor T cells (CAARTs) specifically kill anti-DSG3-specific B cells. There is also potential for the development of long-term memory CAARTs, which could lead to long-term remission but with no global immunosuppressive effects.

12.7 Mucous Membrane Pemphigoid (MMP)

Patients presenting with oral lesions alone are best treated with topical anti-inflammatory reagents such as corticosteroids. Topical tacrolimus has been used to treat CP. Patients with recalcitrant pemphigus or involvement of the skin or large oral lesion requires systemic therapy which depends on the extent and severity of disease. The drugs in use include prednisolone, azathioprine, methotrexate and etanercept [35]. Calcineurin antagonists, such as topical cyclosporine and tacrolimus, have been useful. However, the FDA (USA) has discouraged the use of tacrolimus as it is a potential carcinogen [35].

An outline of treatments for oral lesions is presented in Table 12.4 based on data from Mustafa et al. [19, 36] (Fig. 12.3).

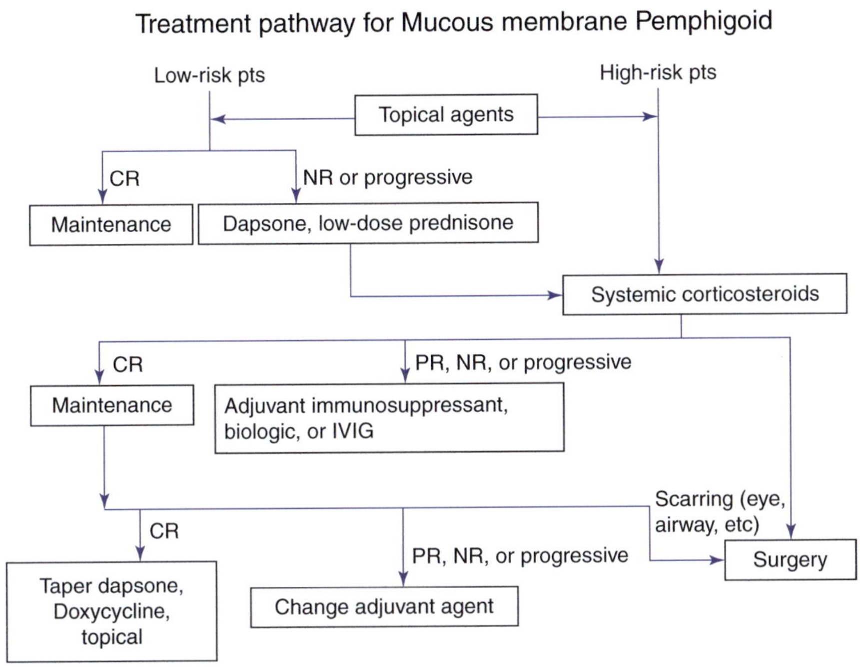

Fig. 12.3 Treatment pathway for MMP: *CR* complete response, *NR* no response, *PR* partial response. Reproduced from Xu et al. 2013 with permission of the publishers

12.8 Sjögren's Syndrome (SS)

Sicca symptoms are managed with eye drops and artificial saliva along with mucolytic drugs. Good oral hygiene along with maintaining moistening of the epithelium and fluorides to prevent or control caries is important. Systemic secretagogues, such as pilocarpine and cevimeline, can be used but have significant adverse effects. Treatment of the non-glandular systemic presentations depends on the system involvement. Corticosteroids and hydroxychloroquine are frequently used for systemic disease; for example low-dose prednisolone is useful for skin and arthritic symptoms. Hydroxychloroquine inhibits TLR signalling and therefore downregulates innate immune responses, especially to viral infections. This drug works well in patients with concomitant RA [37]. However, a recent randomised clinical trial suggested that this drug was no better than placebo in patients with SS [38].

Biological therapies targeting the pathogenic B cells have shown very promising results. Rituximab (anti-CD20) reduced some symptoms such as fatigue and increased salivary flow in two small clinical trials [39]. However, there are increased risks of patients developing a serum sickness type of response and developing antibodies to rituximab [40].

Primary SS patients, along with two other common autoimmune diseases (rheumatoid arthritis and SLE), are at high risk of developing lymphoma. In fact, pSS have the highest risk factors of all three diseases. In a retrospective study carried out in 2011 Pollard et al. [41] found that an initial high SS disease activity was likely to result in an adverse prognosis for the progression of lymphoma and/or SS. These patients required treatment for both MALT lymphoma and SS. But in patients with only localised asymptomatic MALT lymphoma and low SS activity a so-called watchful waiting strategy seems to give good outcomes.

There remains a lack of targeted therapy against the glandular and extra-glandular manifestations [42], but therapies under investigation include inhibitors of cathepsin S, B7-related molecules and CD40, abatacept, BAFF, CD20 and CD22 blocking agents, PI3Kδ and lymphotoxin-β receptor repressors [43].

12.9 Oral Crohn's Disease

Oral symptoms in Crohn's disease (OCD) are often asymptomatic and may spontaneously resolve over time. These granulomatous lesions need to be differentiated from those of oral facial granulomatous disease (OFG) and a recent review clearly suggests that these two presentations are indeed distinct diseases with different treatment pathways as indicated in Fig. 12.4 [44]. It is essential that there is expert evaluation by multidisciplinary teams that consist of oral medicine, gastroenterologists, dermatologists and coloproctologists in order that accurate diagnosis is affected.

There are also similarities between the aphthous ulcers of Behçet's disease and the ulcers seen in OCD and the similarities and differences have recently been reviewed [45] and the different treatment modalities are outlined in Table 12.4.

Once again there are very similar treatment options open to clinicians for these diseases, all of which are based around reducing inflamma-

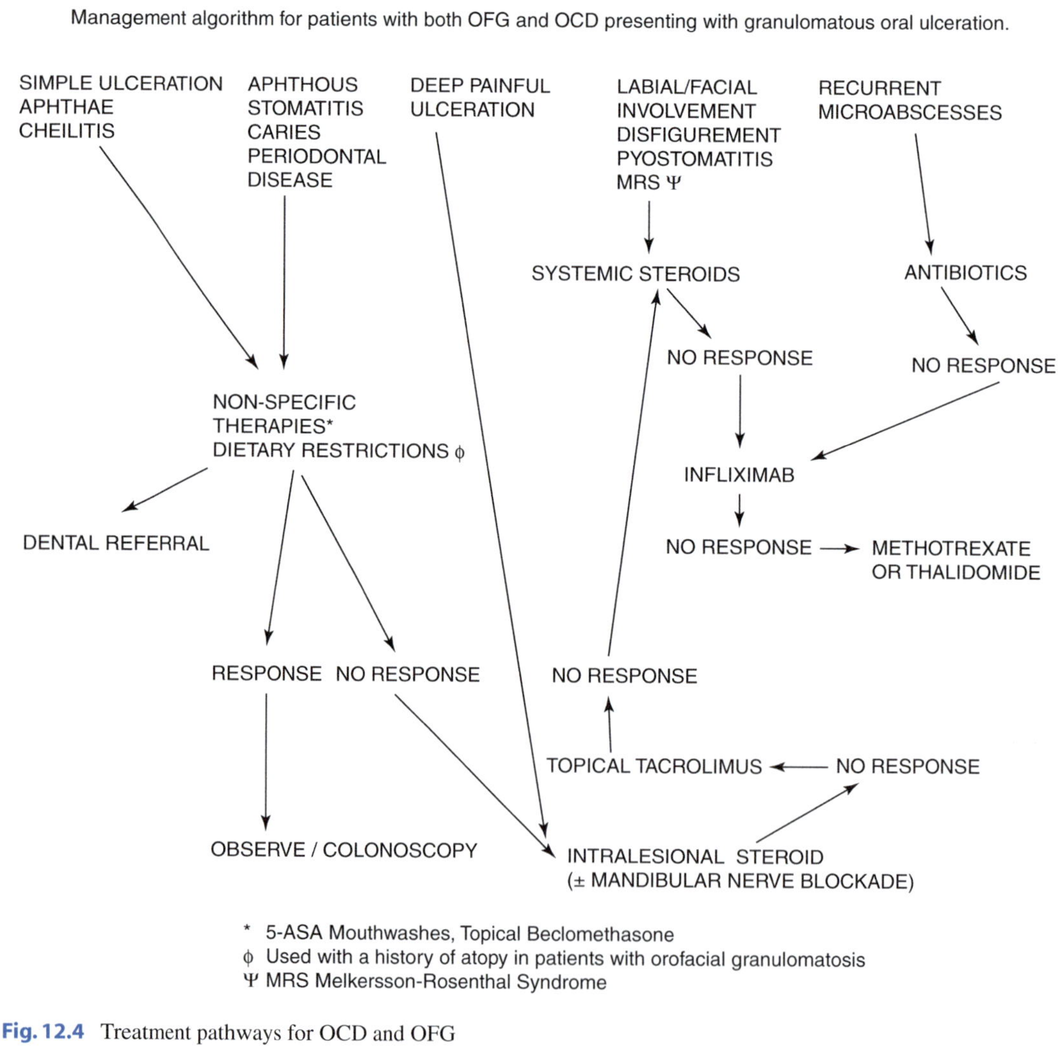

Fig. 12.4 Treatment pathways for OCD and OFG

Table 12.4 Comparison of the treatments used in Crohn's and Behçet's diseases

	Behçet's disease		Crohn's disease	
	Non-GIT BD	GIT BD	Non-GIT CD	GIT CD
Colchicine	S, **M**, A	–	–	–
Corticosteroids	All manifestations	+	All manifestations	+
Azathioprine	S, **M**, O, V, N	+	S	+
6-Mercaptopurine	–	??	–	+
Cyclosporine A	O	–	–	–
Interferon-alpha	O,N	–	–	–
Mycophenolate mofetil	O	–	–	–
Cyclophosphamide	O, V, N	–	–	–
Methotrexate	A, N	+	A, S	+
Sulphasalazine	A		A	
Mesalazine	–	+	–	+
Anti-TNF agents	A, O, N	+	A, S, O	+

A Arthritis, *S* Skin, *M* mucosal (GIT), *O* ocular, *V* vascular, *N* neurological, (+) effective, (−) not effective, *BD* Behçet's disease. *CD* Crohn's disease

tion and modulating the immune responses, either responsible for the pathology or unable to resolve following inflammation-driven damage.

12.10 Graft-Versus-Host Disease

The great success of modern medical advances has regrettably produced some iatrogenic oral mucosal diseases, especially in the field of transplantation. This is especially true for haematopoietic stem cell transplantation. Graft-versus-host disease (GVDH) is a leading cause of late mortality. Presentation in the mouth is common with gingivitis, mucositis and erythema common to both acute and chronic GVHD. In chronic GVHD (CGVHD) further development of lichen planus like lesions, hyperkeratotic plaques and a restriction of the mouth opening is frequent [46].

Once again, the options for treatment largely depend on corticosteroids and immunomodulating drugs (Table 12.5).

12.11 Tumours

The clinical management of tumours still resides with the surgical team along with radiotherapy and chemotherapy.

While more recent vaccination studies suggest that this avenue of endeavour might be promising for oral cancers and some head and neck cancers the detailed investigation is beyond the scope of this chapter but is eluded to in Chap. 8.

Conclusion

One of the most important elements of treatment of oral mucosal diseases is the accurate diagnosis and monitoring of severity in these diseases. While many of the treatment options are still based around corticosteroids and immunosuppressive drugs, one of the step changes in treatment has been the establishment of the multidisciplinary clinics for patient management of these multisystem complex conditions.

The understanding of thc contribution of the oral microbiota to health and/or dysbiosis is at the frontier of new developments and the search for early biomarkers will make considerable impacts on the way oral healthcare is regulated in the future.

However, exploiting the *immunology toolbox* with the development and use of monoclonal antibody therapy and the potential for designing other biologics may change the landscape of clinical management in the years to come.

Table 12.5 Topical management of oral mucosal cGVHD

Therapeutical options			Instructions for use
Corticosteroids	Solution	Dexamethasone 0.1 mg/mL (5 mL)	Keep solution in mouth for 4–6 min without swallowing
		Budesonide 0.3–0.6 mg/mL (10 mL)	Wait 10–15 min before eating/drinking
		Prednisolone 3 mg/mL (5 mL)	Repeat up to 4–6 times per day
		Triamcinolone 1% (5 mL)	
	Gel, cream, and ointment	Fluocinonide 0.05%	Apply it directly over the lesions 2–4 times per day
		Clobetasol 0.05%	
		Triamcinolone 0.1–0.5%	
Calcineurin inhibitors	Solution	Tacrolimus 0.1 mg/mL (5 mL)	Keep solution in mouth for 4–6 min without swallowing.
		Cyclosporine	Repeat up to 4–6 times per day
	Ointment	Tacrolimus 0.1%	Apply it directly over the lesions 2–4 times per day
Oral phototherapy	Methoxypsoralen 3 mg/kg + UVA light 0.5 J/cm^2		3–4 times per week
Antimetabolite and immunosupressive agents	Azathioprine (solution and gel) 5 mg/cm^3		
	Thalidomide (solution and ointment)		

References

1. Bijlsma JW, Jacobs JW. Glucocorticoids in the treatment of rheumatoid arthritis: still used after 65 years. Ann N Y Acad Sci. 2014;18:27–31.
2. Slobodnick A, Shah B, Pillinger MH, Krasnokutsky S. Colchicine: old and new. Am J Med. 2015;128(5):461–70.
3. Sahasranaman S, Howard D, Roy S. Clinical pharmacology and pharmacogenetics of thiopurines. Eur J Clin Pharmacol. 2008;64(8):753–67.
4. Tiede I, Fritz G, Strand S, Poppe D, Dvorsky R, Strand D, Lehr HA, Wirtz S, Becker C, Atreya R, Mudter J, Hildner K, Bartsch B, Holtmann M, Blumberg R, Walczak H, Iven H, Galle PR, Ahmadian MR, Neurath MF. CD28-dependent Rac1 activation is the molecular target of azathioprine in primary human CD4+ T lymphocytes. J Clin Invest. 2003;111(8):1133–45.
5. Allison AC. Mechanisms of action of mycophenolate mofetil. Lupus. 2005;14(1):s2–8.
6. Nash PT, Florin TH. Tumour necrosis factor inhibitors. Med J Aust. 2005;183(4):205–8.
7. Silva LC, Ortigosa LC, Benard G. Anti-TNF-alpha agents in the treatment of immune-mediated inflammatory diseases: mechanisms of action and pitfalls. Immunotherapy. 2010;2(6):817–33.
8. Nava F, Ghilotti F, Maggi L, Hatemi G, Del Bianco A, Merlo C, Filippini G, Tramacere I. Biologics, colchicine, corticosteroids, immunosuppressants and interferon-alpha for Neuro-Behcet's Syndrome. Cochrane Database Syst Rev. 2014;12(10):18.
9. Saleh Z, Arayssi T. Update on the therapy of Behcet disease. Ther Adv Chronic Dis. 2014;5(3):112–34.
10. Jurge S, Kuffer R, Scully C, Porter SR. Mucosal disease series. Number VI. Recurrent aphthous stomatitis. Oral Dis. 2006;12(1):1–21.
11. Spivakovsky S. Inconclusive evidence on systemic treatments for recurrent aphthous stomatitis. Evid Based Dent. 2012;13(4):6400901.
12. Porter S, Scully C. Aphthous ulcers (recurrent). Clin Evid. 2005;13(94):1687–94.
13. Moots RJ, Fortune F, Situnayake D. The Behcet's centres of excellence: a new paradigm for care delivery in complex diseases. Rheumatology (Oxford). 2017. https://doi.org/10.1093/rheumatology/kex037.
14. Ambrose NL, Haskard DO. Differential diagnosis and management of Behcet syndrome. Nat Rev Rheumatol. 2013;9(2):79–89.
15. Mendes D, Correia M, Barbedo M, Vaio T, Mota M, Goncalves O, Valente J. Behcet's disease—a contemporary review. J Autoimmun. 2009;32(3–4):178–88.
16. Direskeneli H. Behcet's disease: infectious aetiology, new autoantigens, and HLA-B51. Ann Rheum Dis. 2001;60(11):996–1002.
17. Hatemi G, Silman A, Bang D, Bodaghi B, Chamberlain AM, Gul A, Houman MH, Kotter I, Olivieri I, Salvarani C, Sfikakis PP, Siva A, Stanford MR, Stubiger N, Yurdakul S, Yazici H. EULAR recommendations for the management of Behcet disease. Ann Rheum Dis. 2008;67(12):1656–62.
18. Vitale A, Rigante D, Lopalco G, Emmi G, Bianco MT, Galeazzi M, Iannone F, Cantarini L. New therapeutic solutions for Behcet's syndrome. Expert Opin Investig Drugs. 2016;25(7):827–40.
19. Mustafa MB, Porter SR, Smoller BR, Sitaru C. Oral mucosal manifestations of autoimmune skin diseases. Autoimmun Rev. 2015;14(10):930–51.

20. Eisen D, Carrozzo M, Bagan Sebastian JV, Thongprasom K. Number V Oral lichen planus: clinical features and management. Oral Dis. 2005; 11(6):338–49.
21. Wang H, Zhang D, Han Q, Zhao X, Zeng X, Xu Y, Sun Z, Chen Q. Role of distinct CD4(+) T helper subset in pathogenesis of oral lichen planus. J Oral Pathol Med. 2016;45:385–93.
22. Alves de Medeiros AK, Speeckaert R, Desmet E, Van Gele M, De Schepper S, Lambert J. JAK3 as an emerging target for topical treatment of inflammatory skin diseases. PLoS One. 2016;11(10):e0164080.
23. Ge Y, Xu Y, Sun W, Man Z, Zhu L, Xia X, Zhao L, Zhao Y, Wang X. The molecular mechanisms of the effect of Dexamethasone and Cyclosporin A on TLR4/NF-kappaB signaling pathway activation in oral lichen planus. Gene. 2012;508(2):157–64.
24. Wang Y, Zhang H, Du G, Cao T, Luo Q, Chen J, Chen F, Tang G. Total glucosides of paeony (TGP) inhibits the production of inflammatory cytokines in oral lichen planus by suppressing the NF-kappaB signaling pathway. Int Immunopharmacol. 2016;36:67–72.
25. Al-Maweri SA, Kalakonda B, Al-Soneidar WA, Al-Shamiri HM, Alakhali MS, Alaizari N. Efficacy of low-level laser therapy in management of symptomatic oral lichen planus: a systematic review. Lasers Med Sci. 2017;24(10):017–2233.
26. Gupta S, Ghosh S. Interventions for the management of oral lichen planus: a review of the conventional and novel therapies. Oral Dis. 2017;5(10):12634.
27. Yacoub M-R, Berti A, Campochiaro C, Tombetti E, Ramirez GA, Nico A, Di Leo E, Fantini P, Sabbadini MG, Nettis E, Colombo G. Drug induced exfoliative dermatitis: state of the art. Clinical and Molecular Allergy: CMA. 2016;14(1):9.
28. Farthing P, Bagan JV, Scully C. Mucosal disease series. Number IV. Erythema multiforme. Oral Dis. 2005;11(5):261–7.
29. Creamer D, Walsh SA, Dziewulski P, Exton LS, Lee HY, Dart JK, Setterfield J, Bunker CB, Ardern-Jones MR, Watson KM, Wong GA, Philippidou M, Vercueil A, Martin RV, Williams G, Shah M, Brown D, Williams P, Mohd Mustapa MF, Smith CH. U.K. guidelines for the management of Stevens-Johnson syndrome/toxic epidermal necrolysis in adults 2016. Br J Dermatol. 2016;174(6):1194–227.
30. Black M, Mignogna MD, Scully C. Number II. Pemphigus vulgaris. Oral Dis. 2005;11(3):119–30.
31. Eming R, Nagel A, Wolff-Franke S, Podstawa E, Debus D, Hertl M. Rituximab exerts a dual effect in pemphigus vulgaris. J Invest Dermatol. 2008;128(12):2850–8.
32. Kasperkiewicz M, Ellebrecht CT, Takahashi H, Yamagami J, Zillikens D, Payne AS, Amagai M. Pemphigus. Nat Rev Dis Primers. 2017;3(17026):26.
33. Ellebrecht CT, Payne AS. Setting the target for pemphigus vulgaris therapy. JCI Insight. 2017;2(5):92021.
34. Ellebrecht CT, Bhoj VG, Nace A, Choi EJ, Mao X, Cho MJ, Di Zenzo G, Lanzavecchia A, Seykora JT, Cotsarelis G, Milone MC, Payne AS. Reengineering chimeric antigen receptor T cells for targeted therapy of autoimmune disease. Science. 2016;353(6295):179–84.
35. Scully C, Lo Muzio L. Oral mucosal diseases: mucous membrane pemphigoid. Br J Oral Maxillofac Surg. 2008;46(5):358–66.
36. Schifter M, Yeoh SC, Coleman H, Georgiou A. Oral mucosal diseases: the inflammatory dermatoses. Aust Dent J. 2010;55(Suppl 1):23–38.
37. Kruize AA, Hene RJ, Kallenberg CG, van Bijsterveld OP, van der Heide A, Kater L, Bijlsma JW. Hydroxychloroquine treatment for primary Sjogren's syndrome: a two year double blind crossover trial. Ann Rheum Dis. 1993;52(5):360–4.
38. Gottenberg JE, Ravaud P, Puechal X, Le Guern V, Sibilia J, Goeb V, Larroche C, Dubost JJ, Rist S, Saraux A, Devauchelle-Pensec V, Morel J, Hayem G, Hatron P, Perdriger A, Sene D, Zarnitsky C, Batouche D, Furlan V, Benessiano J, Perrodeau E, Seror R, Mariette X. Effects of hydroxychloroquine on symptomatic improvement in primary Sjogren syndrome: the JOQUER randomized clinical trial. JAMA. 2014;312(3):249–58.
39. Dass S, Bowman SJ, Vital EM, Ikeda K, Pease CT, Hamburger J, Richards A, Rauz S, Emery P. Reduction of fatigue in Sjogren syndrome with rituximab: results of a randomised, double-blind, placebo-controlled pilot study. Ann Rheum Dis. 2008;67(11):1541–4.
40. Meijer JM, Meiners PM, Vissink A, Spijkervet FK, Abdulahad W, Kamminga N, Brouwer E, Kallenberg CG, Bootsma H. Effectiveness of rituximab treatment in primary Sjogren's syndrome: a randomized, double-blind, placebo-controlled trial. Arthritis Rheum. 2010;62(4):960–8.
41. Pollard RP, Pijpe J, Bootsma H, Spijkervet FK, Kluin PM, Roodenburg JL, Kallenberg CG, Vissink A, van Imhoff GW. Treatment of mucosa-associated lymphoid tissue lymphoma in Sjogren's syndrome: a retrospective clinical study. J Rheumatol. 2011;38(10):2198–208.
42. Ferro F, Marcucci E, Orlandi M, Baldini C, Bartoloni-Bocci E. One year in review 2017: primary Sjogren's syndrome. Clin Exp Rheumatol. 2017;35(2):179–91.
43. Mavragani CP. Mechanisms and New Strategies for Primary Sjogren's Syndrome. Annu Rev Med. 2017;68:331–43.
44. Zbar AP, Ben-Horin S, Beer-Gabel M, Eliakim R. Oral Crohn's disease: is it a separable disease from orofacial granulomatosis? A review. J Crohns Colitis. 2012;6(2):135–42.
45. Yazisiz V. Similarities and differences between Behcet's disease and Crohn's disease. World J Gastrointest Pathophysiol. 2014;5(3):228–38.
46. Margaix-Muñoz M, Bagán JV, Jiménez Y, Sarrión MG, Poveda-Roda R. Graft-versus-host disease affecting oral cavity. A review. J Clin Exp Dent. 2015;7(1):e138–45.

Future Directions of Research in the Oral Mucosa

13

Lesley Ann Bergmeier and E. Kenneth Parkinson

From the initial observations of dental plaque, by Anton van Leeuwenhoek in the 1680s (Fig. 13.1), to the first draft of the human genome [1, 2], investigation of the oral cavity has been central to many discoveries in human biology. It is now common for individuals interested in their ancestry to have their genomic DNA analysed from buccal mucosal swabs or saliva. Many of the key observations that underpin the understanding of the mucosal immune system were made by groups working in dental institutes investigating oral diseases such as dental caries [3].

13.1 Introduction

The knowledge gained from the new technologies that have unravelled the genetic background of humans in health and disease has generated new methodologies that enable in-depth analysis of genetic predispositions to disease and the effects of gain or loss of function of genes in the context of disease mechanisms. Genetic analysis by genome-wide association studies (GWAS) has enabled the dissection of genetic anomalies in a variety of oral mucosal diseases, for example the large GWAS studies carried out on Behçet's disease patients that revealed polymorphisms in key genes of the innate immune system as well as confirming the association of HLA genotypes with the disease [4, 5]. GWAS studies of the periodontal pathogens have also been carried out [6–8]. However, the limitations of these studies lie in their potential to either overestimate associations or indeed mask associations (Fig.13.2).

However, a list of the genes is not enough. Post-translational modification has profound effects on function and signalling pathways and how the proteins within and between the cells of the oral mucosa communicate and interact. Interactions of these proteins, and with those of the oral microbiome, in health and disease have generated *Big Data* and have taken oral mucosal research from the reductionist approach that gave rise to the "omics" into the realms of systems biology (Fig. 13.3).

In a recent review of a symposium held at the 94th General Session of the IADR 2016 ("How the OMICS are contributing to the understanding of caries") the consensus was that these powerful technologies are expected to reveal novel caries biomarkers and next-generation diagnostics and therapies [9]. This approach will also inform other diseases of the oral mucosa.

In the UK, the 1000 Genome Project reported in 2015 and mapped the structural variation of 2504 human genomes [10]. Investigations of populations with high degrees of consanguinity have enabled the investigation of "human knockouts",

L.A. Bergmeier (✉) • E.K. Parkinson
Centre for Oral Immunology and Regenerative Medicine, Institute of Dentistry, Barts and The London School of Medicine and Dentistry, London, UK
e-mail: l.a.bergmeier@qmul.ac.uk

L.A. Bergmeier (ed.), *Oral Mucosa in Health and Disease*,
https://doi.org/10.1007/978-3-319-56065-6_13

Fig. 13.1 The observations on dental plaque by van Leeuwenhoek

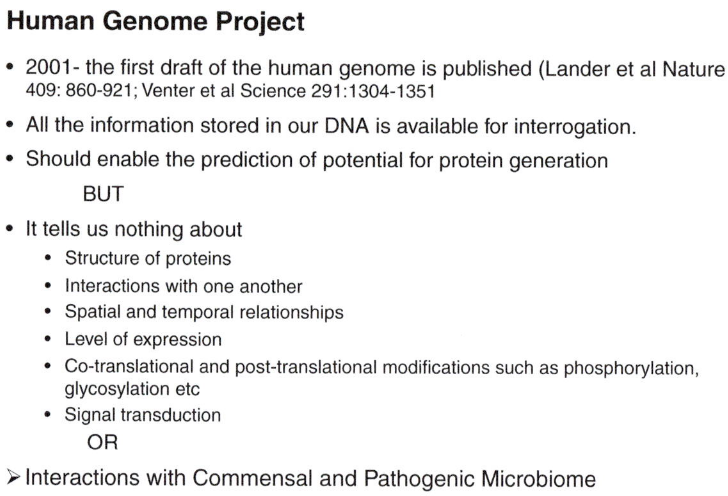

Fig. 13.2 Human genome project and beyond

where loss-of-function mutations can disrupt both copies of a given gene. Phenotypic analysis of such "human knockouts" can provide insight into gene function. Consanguineous unions are more likely to result in offspring carrying homozygous loss-of-function mutations [11].

13.2 Big Data

Human Microbiome Project (HMP) was established in 2008, with the mission of generating resources that would enable the comprehensive characterisation of the human microbiome and analysis of its role in human health and disease, http://hmpdacc.org/. There are now large comprehensive databases that can be mined for information, including the human *oral microbiome* (HTTP://www.homd.org). The interactions of proteins within and between different cell types of the human body (protein–protein interactions—PPI) are being mapped and yield enormous amounts of information that has given rise to the concept of the *interactome* (http://interactome.baderlab.org/). The first computational/predictive model of the *human-microbial oral interactome* was published in 2014 [12] and has the potential to reveal drug targets and

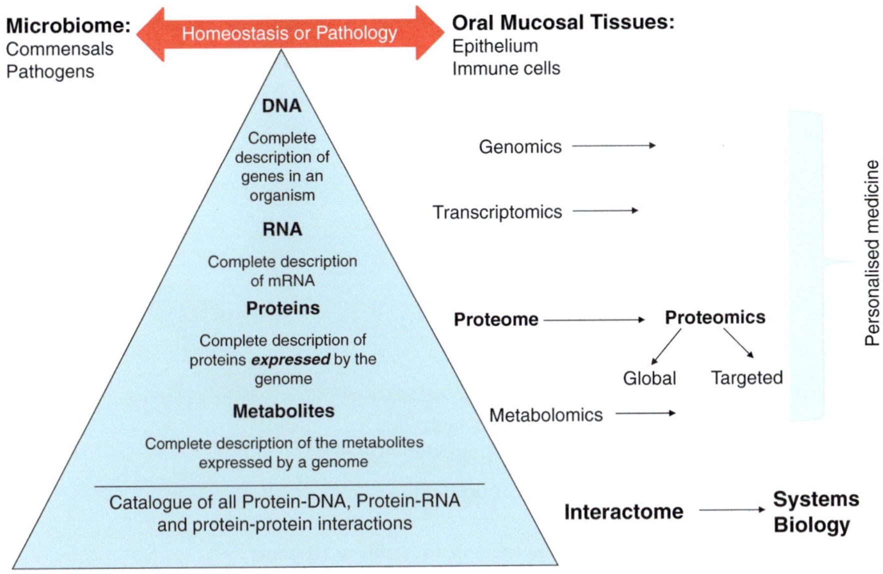

Fig. 13.3 From "omics" to systems biology

interactions that are key to the maintenance of homeostasis within the complex ecology of the oral cavity. This work is based on the development of the OralCard [13]. The tools that were developed became a key resource for understanding the molecular foundations of the biology and disease mechanisms within the oral cavity in the context of analysis of the oral proteome associated with type 2 diabetes and other conditions (http://bioinformatics.ua.pt/oralcard).

In a recent paper an orthogonal analysis was carried out in 77 individuals vaccinated against herpes zoster, a common oral pathogen. The authors constructed a "multiscale, multifactorial response network (MMRN)", using datasets from peripheral blood transcriptomics, flow cytometry of blood cells, plasma cytokine analysis and metabolomics to identify molecular networks induced by vaccination [14]. The authors could show striking relationships between metabolomic and transcriptomic signatures and were further able to distinguish significant differences in the response of young compared with older individuals. This type of integrated approach to dissecting and understanding biological networks signifies a potential paradigm shift in the way investigations are carried out in understanding health and disease.

13.3 Oral Microbiome: Commensals, Pathogens and Homeostasis

The microbiome evolves throughout life and is continuously influenced by the environment. At birth the colonisation by bacterial species is influenced by the route of delivery with the microbiome of vaginally delivered infants represented by the maternal vaginal and gut microbiome, while caesarean-delivered newborns exhibit a maternal skin-derived microbiome [15–17]. The oral cavity is host to more than 700 species of commensal bacteria, of which about 60% are cultivatable (HTTP://www.homd.org), but despite this high level of colonisation and exposure to allergens in foods there is relatively little acute inflammation or allergic reactions in the normal oral mucosa.

Periodontitis is the result of a chronic inflammation associated with a dramatic change from a symbiotic community of mostly facultative organisms to a dysbiotic community consisting of anaerobic organisms that have evolved to thrive in an inflammatory environment including the acquisition of numerous virulence factors [18].

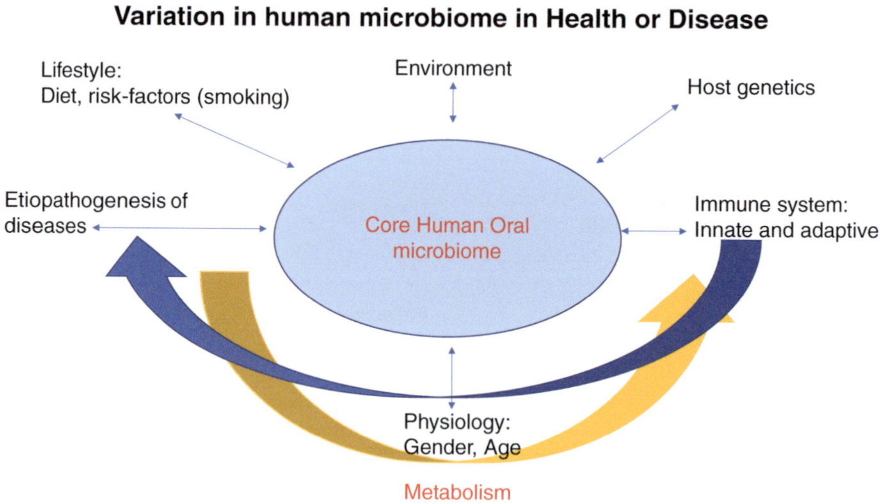

Fig. 13.4 Potential interactions of the human microbiome in health and disease

Recent research has built on the knowledge that certain "keystone" organisms have a disproportionate effect on their environment relative to their abundance [19]. *P. gingivalis* is a prime example of an organism that co-operates with others to remodel its microenvironment into a dysbiotic and disease-provoking microbiota. These new communities can subvert the normally tolerogenic immune system to maintain an inflammatory environment that promotes their survival in the face of a robust immune response [19–21]. The interaction of the host responses was discussed in Chap. 10.

In a prescient review of oral microbiology "Past, Present and Future", He and colleagues [22] drew attention to the "metagenomics" analysis which would be required for understanding communities of bacteria [23]. The study of biofilms has enhanced the understanding of the complexities of the oral microbiome and its interaction with host tissues. The use of multispecies oral biofilms, and investigating their ability to generate chemical messengers, such as cytokines and chemokines from immortalised epithelial cell lines, can elicit information about the way in which these microbial communities contribute to inflammatory processes in the oral mucosa [24–26].

Similarly, the construction of in vitro 3D models of the oral mucosa is essential in exploring the potential for oral cancer studies as well as drug delivery, infection models [27] and biocompatibility studies [28]. These integrated cellular structures offer modelled ecosystems of cells, biomaterials and microorganisms and the ability to manipulate the proteomic and metabolomic environment to understand disease mechanisms (Fig.13.4).

13.4 Immunology in the Oral Cavity: New Kids on the Block

In the general literature, there is still a tendency to assume that the rules and observations that govern and inform the biology of the gastrointestinal tract also apply to the oral mucosa. The oral mucosa resembles the skin in some respects but shares greater similarities with the oesophagus, cervix and vagina (Chap. 2). However, in the last decade the unique features of the oral mucosa have become better understood and amenable to investigation.

New techniques such as sublingual immunotherapy for type 1 hypersensitivity offer the potential for desensitisation from life-threatening allergies and a recent paper has begun to elaborate the mechanisms involved, where particular phenotypes of dendritic cells in the sublingual mucosa migrate to the submandibular lymph nodes where they induce T regulatory cells that modulate responses [29].

Multi-coloured flow cytometry (up to 15 colours) has enabled the dissection of many of the cells of the immune response, both new cell types and subpopulations with diverse functions reflected by unique phenotypes. *Immunophenotyping* has been explored in numerous papers with several recent investigations showing discrete phenotypic subpopulations with different functions in NK cells [30], γδT cells [31] and mucosal dendritic cells [32, 33]. Immunophenotyping has also been proposed as a method for monitoring autoimmune diseases [34]. Villani et al. [35] have used single-cell RNA-seq to reveal new types of human blood dendritic cells, monocytes and progenitor cells. These techniques have the potential to build an immune cell atlas that would map responses in a similar manner to the MMRN methods used by the Pulendran group [14].

The relatively new discovery of *innate lymphoid cells* has expanded the understanding of how the immune system is regulated and how tissue homeostasis is maintained in the face of inflammatory responses [36]. These cells populate barrier tissues and are therefore very important in homeostasis and defence from pathogens [37]. To date these cells have been identified in gingival tissue [38], show functional changes during SIV infection [39] and are important in responses against fungal infections [40]. There is however scope for investigation of these cells in the buccal mucosa especially since it has been reported that these cells are regulated by the microbiome [41]. The immune system has evolved and responds within the context of different microbiota depending on the tissue. Eberl [42] has suggested that the divisions between symbionts and pathogens are too strict and that an equilibrium exists between the host and microbiome that creates a "superorganism". This is echoed in two recent reviews on the subversion of the immune response by the microbiota, generation of dysbiosis and chronic inflammation [43, 44].

The recognition that microorganisms can drive the activation of appropriate immune responses and are important in maintaining homeostasis [45–47] has also revealed that the types of responses induced by commensals are very different from those of pathogenic bacteria [48].

The enormous potential of the immune response for dealing with pathogens, and the host response contribution to disease states (e.g. periodontitis, allergy and autoimmunity), emphasises the importance of homeostasis (see Chap. 5). In a recent review, an equilibrium model has been posed for immunity [49]. It was suggested that four types of response coexist in balance, where a type 1 response deals with intracellular bacteria, viruses and tumours; type 2 responds to helminth infections; type 3 deals with extracellular organisms and type 4 excludes microorganisms. The oral immune responses fit within several arms of this model and understanding the oral equilibrium would be a very important area of research.

13.5 The Ageing Oral Mucosa

Ageing is a growing issue in most countries where a high proportion of populations live to relatively old age and as fewer individuals are edentulous there is an opportunity to investigate the ageing process in a system that lends itself to easy and non-traumatic investigation through the salivary proteome and metabolome. However, dissecting the contributions of the oral bacteria and the different cell types in the oral cavity will require other lines of investigation to test hypotheses and translate these findings clinically.

Chronic inflammation increases with age and is a predisposing factor for many cancers and other diseases. Recent evidence indicates that senescent cells are responsible for many diseases associated with chronological ageing [50, 51] and have been demonstrated to contribute to the side effects of cancer therapy [52]. Senescent cells are also responsible for secreting a variety of cytokines known as the inflammasome that in healthy individuals targets the senescent cells for clearance by the innate and adaptive immune system [53]. Furthermore, chronic inflammation can induce senescence in mouse models [54] and senescent cells are associated with chronic inflammatory conditions of the oral cavity such as oral submucous fibrosis [55]. As humans age the efficiency of their immune system declines, a process known as immunosenescence, and this may, or may not, be responsible for the accumulation of senescent cells with age in many human diseases. Several strategies are now being explored to selectively deplete tissues of senescent cells including boosting the host immune system and development of selective drugs known collectively as senolytics [56, 57]. Senolytics work very well in mice [57, 58] but are known to have unacceptable side effects in humans [56] and so whether the mouse models will be translatable into the clinic remains to be seen.

In addition, the molecular changes that are responsible for these changes identified by the many omic studies described above are not fully understood. Divergence of cytokine profiles, mucosal stability and microbial sensing via TLRs, NLRs and RIGIs are part of the innate immune system and there is increasing evidence that perturbations of these functions with age have profound consequences. Ontological transcriptomic analysis of the oral mucosa in non-human primates suggests that changes in the expression of sets of genes associated with the inflammasome (NLRs) with age might impair the recognition of microbes at the mucosal surfaces and establish inflammation [59]. Inflammatory disorders such as cardiovascular disease, chronic diabetic wounds and periodontal disease result from the amplification of inflammation provided by the influx of neutrophils and phagocytic cells—often as a direct result of initial microbial interactions with pattern recognition molecules [60–62] and to what extent this is linked to the accumulation of senescent cells and/or chronic inflammation remains to be investigated. The effect of post-translational modification on the protein expression of these molecules, the existence of splice variants [63] and indeed a greater understanding of the "spliceosome" [64, 65] will be of great value in the interpretation of the patterns of response and feeds into the ever-evolving field of systems biology.

13.6 Biomarker Discovery

The NIH has defined biomarkers as "quantifiable biological parameters that are measurable and evaluated as an indicator of normal biological, pathogenic, or pharmacological responses to a therapeutic intervention" [66]. Indeed, the recent disclosure that oral cancer-driver mutations and human papilloma virus DNA can be detected in as little as 5 mL saliva from established cancers [67] suggests that modern sequencing techniques may offer some means of detecting oral cancer early which previously has been a serious barrier to successful and cost-effective treatment. However, where such definitive markers such as a disease mutation are not available other approaches need to be developed.

The salivary proteomes and metabolomes are an ideal media for the investigation of disease markers as collecting saliva is a non-invasive technique that can be used in many of the *multiplex* platforms. However, dissecting the roles of the microbiome, inflammation and cellular senescence is likely to be very difficult in diseases such as oral cancer and so far has produced no consistent biomarker candidates.

No longer are scientists obliged to investigate one protein at a time, but the advent of microarrays for both RNA and proteins allows for the analysis of multiple analytes. These methodologies have enormous potential from the analysis of inflammatory markers in periodontal disease [68, 69] to the diagnosis of oral cancer [70] and many other oral mucosal diseases as discussed in Chap. 10. Furthermore, the ability to analyse sets of markers contextualises these substances in a way that single sample analysis cannot.

Data mining of established databases has shown that about 30% of the proteins found in saliva are also present in serum and this overlap could be exploited for biomarker discovery for disease. Various groups have drawn connections between the salivary proteome and diseases as diverse as cystic fibrosis, diabetes, periodontal disease, caries and AIDS [71].

Proteomics also has a role to play in understanding the network of immune cells. In a recent *Resource*, Rieckmann et al. [72] investigated more than 10,000 proteins from the total and secreted proteomes of 28 cell populations comparing activated and steady-state conditions. The results provide a framework for investigation of protein interactions but also a window into altered communication and signalling pathways associated with disease pathology (http://www.immprot.org/).

The investigation of biomarkers in the proteome has also been enriched by the knowledge of the microbiome and this has led to the realisation that the *metabolome* also has a major influence on the function of both the cells of the oral mucosa and the microbiome. The salivary proteome will, by definition, contain "metabolites" produced by the mucosal tissues *and* the commensal oral flora. Understanding the way metabolites influence the microorganisms of the oral biofilm but also their effects on the mucosal epithelium and the immune cells that contribute to protection may reveal early markers of disease or disease progression.

In the crosstalk between microbiota and immune system, metabolites have an important role [73]. It is well recognised that the microorganisms in the oral cavity use metabolic pathways that differ from other microorganisms such as gut *E. coli* [74–76].

Nutrients such as vitamin D have been shown to have an influence on the aetiopathogenesis of oral disease [77] including the ability to prevent periodontitis in deficiency diseases such as X-linked hypophosphatemia [78].

The tumour environment is also under investigation for the competition that occurs for nutrients between tumour cells and the infiltrating immune cells [79, 80]. Again, here there is a link with ageing as senescent cells are found to be associated with cancer-prone conditions such as oral submucous fibrosis [55] and the oral cancer environment [81]. Furthermore, oral cancer cells have been show to induce senescence in normal cells [81]. The metabolic pathways that tumour cells manipulate, such as those of lipid metabolism, glucose and amino acid metabolism, have significant impact on immune cells. Increased glycolysis by tumour cells and neighbouring senescent cells in the cancer environment [81, 82] can deprive infiltrating T lymphocytes of glucose which decreases their ability to signal, produce IFNγ and carry out cytotoxic functions. The presence of short-chain fatty acids and vitamins helps maintain barrier functions (in the gut) by supporting the development and survival of T regulatory cells and innate lymphoid cells. These mechanisms have recently been extensively reviewed [79].

Changes in metabolism in immune cells are also thought to contribute to the immune dysfunction of autoimmune diseases. Metabolites such as reactive oxygen species and changes to cholesterol glucose and amino acid catabolism have been shown to be similar in multiple sclerosis, autoimmune arthritis and other diseases with autoimmune components such as type 1 diabetes [83].

One of the key questions that might be asked is this: "What does normal look like"? In the HIV field the observation of correlates of protection has established some parameters for the development of vaccine candidates [84].

Perhaps it is possible to take a leaf out of the HIV field and ask what the correlates of "normality" are in the context of oral homeostasis and resolution of disease.

The databases generated by the "omics" platforms may indeed lead to such a profile but it is likely to be at a highly individual level that might open new avenues of research for drug discovery. At the very least it might lead to recognising markers that indicate if a patient can respond to a drug, thus saving the trial-and-error approach of medication.

It would be easy to be overwhelmed by the sheer amount of data generated but a "step back and look" approach [79] might bring targets into sharper focus.

This book has attempted to revisit the oral mucosa in the context of health and disease and while it does not claim to be an exhaustive study of the topic there is an interconnectedness of many distinct disciplines that meet in the oral cavity at the interface of the hard tissues and this unique mucosal tissue.

The new knowledge that is accumulating in the context of the microbiome and the nature of the cells that protect the oral mucosa opens many new avenues of research and this chapter has attempted to address some topics which could not be covered in the material presented in the main chapters and point to future directions of research. New technologies and methodologies have enabled a paradigm shift in the way oral mucosal biology research is carried out.

The map of the landscape of the oral mucosa, the "flora and fauna" that inhabit this ecosystem and the underlying structures that support the function of this gateway tissue are far more complex than van Leuwenhoek could have imagined when he first set eyes on the "little animalcules" in plaque.

References

1. Lander ES, Linton LM, Birren B, Nusbaum C, Zody MC, Baldwin J, Devon K, Dewar K, Doyle M, FitzHugh W, Funke R, Gage D, Harris K, Heaford A, Howland J, Kann L, Lehoczky J, LeVine R, McEwan P, McKernan K, Meldrim J, Mesirov JP, Miranda C, Morris W, Naylor J, Raymond C, Rosetti M, Santos R, Sheridan A, Sougnez C, Stange-Thomann Y, Stojanovic N, Subramanian A, Wyman D, Rogers J, Sulston J, Ainscough R, Beck S, Bentley D, Burton J, Clee C, Carter N, Coulson A, Deadman R, Deloukas P, Dunham A, Dunham I, Durbin R, French L, Grafham D, Gregory S, Hubbard T, Humphray S, Hunt A, Jones M, Lloyd C, McMurray A, Matthews L, Mercer S, Milne S, Mullikin JC, Mungall A, Plumb R, Ross M, Shownkeen R, Sims S, Waterston RH, Wilson RK, Hillier LW, McPherson JD, Marra MA, Mardis ER, Fulton LA, Chinwalla AT, Pepin KH, Gish WR, Chissoe SL, Wendl MC, Delehaunty KD, Miner TL, Delehaunty A, Kramer JB, Cook LL, Fulton RS, Johnson DL, Minx PJ, Clifton SW, Hawkins T, Branscomb E, Predki P, Richardson P, Wenning S, Slezak T, Doggett N, Cheng JF, Olsen A, Lucas S, Elkin C, Uberbacher E, Frazier M, Gibbs RA, Muzny DM, Scherer SE, Bouck JB, Sodergren EJ, Worley KC, Rives CM, Gorrell JH, Metzker ML, Naylor SL, Kucherlapati RS, Nelson DL, Weinstock GM, Sakaki Y, Fujiyama A, Hattori M, Yada T, Toyoda A, Itoh T, Kawagoe C, Watanabe H, Totoki Y, Taylor T, Weissenbach J, Heilig R, Saurin W, Artiguenave F, Brottier P, Bruls T, Pelletier E, Robert C, Wincker P, Smith DR, Doucette-Stamm L, Rubenfield M, Weinstock K, Lee HM, Dubois J, Rosenthal A, Platzer M, Nyakatura G, Taudien S, Rump A, Yang H, Yu J, Wang J, Huang G, Gu J, Hood L, Rowen L, Madan A, Qin S, Davis RW, Federspiel NA, Abola AP, Proctor MJ, Myers RM, Schmutz J, Dickson M, Grimwood J, Cox DR, Olson MV, Kaul R, Shimizu N, Kawasaki K, Minoshima S, Evans GA, Athanasiou M, Schultz R, Roe BA, Chen F, Pan H, Ramser J, Lehrach H, Reinhardt R, McCombie WR, de la Bastide M, Dedhia N, Blöcker H, Hornischer K, Nordsiek G, Agarwala R, Aravind L, Bailey JA, Bateman A, Batzoglou S, Birney E, Bork P, Brown DG, Burge CB, Cerutti L, Chen HC, Church D, Clamp M, Copley RR, Doerks T, Eddy SR, Eichler EE, Furey TS, Galagan J, Gilbert JG, Harmon C, Hayashizaki Y, Haussler D, Hermjakob H, Hokamp K, Jang W, Johnson LS, Jones TA, Kasif S, Kaspryzk A, Kennedy S, Kent WJ, Kitts P, Koonin EV, Korf I, Kulp D, Lancet D, Lowe TM, McLysaght A, Mikkelsen T, Moran JV, Mulder N, Pollara VJ, Ponting CP, Schuler G, Schultz J, Slater G, Smit AF, Stupka E, Szustakowki J, Thierry-Mieg D, Thierry-Mieg J, Wagner L, Wallis J, Wheeler R, Williams A, Wolf YI, Wolfe KH, Yang SP, Yeh RF, Collins F, Guyer MS, Peterson J, Felsenfeld A, Wetterstrand KA, Patrinos A, Morgan MJ, de Jong P, Catanese JJ, Osoegawa K, Shizuya H, Choi S, Chen YJ, Consortium IHGS. Initial sequencing and analysis of the human genome. Nature. 2001;409(6822):860–921.
2. Venter JC, Adams MD, Myers EW, Li PW, Mural RJ, Sutton GG, Smith HO, Yandell M, Evans CA, Holt RA, Gocayne JD, Amanatides P, Ballew RM, Huson DH, Wortman JR, Zhang Q, Kodira CD, Zheng XH, Chen L, Skupski M, Subramanian G, Thomas PD, Zhang J, Gabor Miklos GL, Nelson C, Broder S, Clark AG, Nadeau J, McKusick VA, Zinder N, Levine AJ, Roberts RJ, Simon M, Slayman C, Hunkapiller M, Bolanos R, Delcher A, Dew I, Fasulo D, Flanigan M, Florea L, Halpern A, Hannenhalli S, Kravitz S, Levy S, Mobarry C, Reinert K, Remington K, Abu-Threideh J, Beasley E, Biddick K, Bonazzi V, Brandon R, Cargill M, Chandramouliswaran I, Charlab R, Chaturvedi K, Deng Z, Di Francesco V, Dunn P, Eilbeck K, Evangelista C, Gabrielian AE, Gan W, Ge W, Gong F, Gu Z, Guan P, Heiman TJ, Higgins ME, Ji RR, Ke Z, Ketchum KA, Lai Z, Lei Y, Li Z, Li J, Liang Y, Lin X, Lu F, Merkulov GV, Milshina N, Moore HM, Naik AK, Narayan VA, Neelam B, Nusskern D, Rusch DB, Salzberg S, Shao W, Shue B, Sun J, Wang Z, Wang A, Wang X, Wang J, Wei M, Wides R, Xiao C, Yan C, Yao A, Ye J, Zhan M, Zhang W, Zhang H, Zhao Q, Zheng L, Zhong F, Zhong W, Zhu S, Zhao S, Gilbert D, Baumhueter S, Spier G, Carter C, Cravchik A, Woodage T, Ali F, An H, Awe A, Baldwin D, Baden H, Barnstead M, Barrow I, Beeson K, Busam D, Carver A, Center A, Cheng ML, Curry L, Danaher S, Davenport

L, Desilets R, Dietz S, Dodson K, Doup L, Ferriera S, Garg N, Gluecksmann A, Hart B, Haynes J, Haynes C, Heiner C, Hladun S, Hostin D, Houck J, Howland T, Ibegwam C, Johnson J, Kalush F, Kline L, Koduru S, Love A, Mann F, May D, McCawley S, McIntosh T, McMullen I, Moy M, Moy L, Murphy B, Nelson K, Pfannkoch C, Pratts E, Puri V, Qureshi H, Reardon M, Rodriguez R, Rogers YH, Romblad D, Ruhfel B, Scott R, Sitter C, Smallwood M, Stewart E, Strong R, Suh E, Thomas R, Tint NN, Tse S, Vech C, Wang G, Wetter J, Williams S, Williams M, Windsor S, Winn-Deen E, Wolfe K, Zaveri J, Zaveri K, Abril JF, Guigó R, Campbell MJ, Sjolander KV, Karlak B, Kejariwal A, Mi H, Lazareva B, Hatton T, Narechania A, Diemer K, Muruganujan A, Guo N, Sato S, Bafna V, Istrail S, Lippert R, Schwartz R, Walenz B, Yooseph S, Allen D, Basu A, Baxendale J, Blick L, Caminha M, Carnes-Stine J, Caulk P, Chiang YH, Coyne M, Dahlke C, Mays A, Dombroski M, Donnelly M, Ely D, Esparham S, Fosler C, Gire H, Glanowski S, Glasser K, Glodek A, Gorokhov M, Graham K, Gropman B, Harris M, Heil J, Henderson S, Hoover J, Jennings D, Jordan C, Jordan J, Kasha J, Kagan L, Kraft C, Levitsky A, Lewis M, Liu X, Lopez J, Ma D, Majoros W, McDaniel J, Murphy S, Newman M, Nguyen T, Nguyen N, Nodell M, Pan S, Peck J, Peterson M, Rowe W, Sanders R, Scott J, Simpson M, Smith T, Sprague A, Stockwell T, Turner R, Venter E, Wang M, Wen M, Wu D, Wu M, Xia A, Zandieh A, Zhu X. The sequence of the human genome. Science. 2001;291(5507):1304–51.

3. Kiyono H, Azegami T. The mucosal immune system: from dentistry to vaccine development. Proc Jpn Acad Ser B Phys Biol Sci. 2015;91(8):423–39.
4. Kirino Y, Bertsias G, Ishigatsubo Y, Mizuki N, Tugal-Tutkun I, Seyahi E, Ozyazgan Y, Sacli FS, Erer B, Inoko H, Emrence Z, Cakar A, Abaci N, Ustek D, Satorius C, Ueda A, Takeno M, Kim Y, Wood GM, Ombrello MJ, Meguro A, Gül A, Remmers EF, Kastner DL. Genome-wide association analysis identifies new susceptibility loci for Behçet's disease and epistasis between HLA-B*51 and ERAP1. Nat Genet. 2013;45(2):202–7.
5. Remmers EF, Cosan F, Kirino Y, Ombrello MJ, Abaci N, Satorius C, Le JM, Yang B, Korman BD, Cakiris A, Aglar O, Emrence Z, Azakli H, Ustek D, Tugal-Tutkun I, Akman-Demir G, Chen W, Amos CI, Dizon MB, Kose AA, Azizlerli G, Erer B, Brand OJ, Kaklamani VG, Kaklamanis P, Ben-Chetrit E, Stanford M, Fortune F, Ghabra M, Ollier WE, Cho YH, Bang D, O'Shea J, Wallace GR, Gadina M, Kastner DL, Gül A. Genome-wide association study identifies variants in the MHC class I, IL10, and IL23R-IL12RB2 regions associated with Behçet's disease. Nat Genet. 2010;42(8):698–702.
6. Divaris K, Monda KL, North KE, Olshan AF, Lange EM, Moss K, Barros SP, Beck JD, Offenbacher S. Genome-wide association study of periodontal pathogen colonization. J Dent Res. 2012;**91**(7 Suppl):21s–8s.
7. Divaris K, Monda KL, North KE, Olshan AF, Reynolds LM, Hsueh WC, Lange EM, Moss K, Barros SP, Weyant RJ, Liu Y, Newman AB, Beck JD, Offenbacher S. Exploring the genetic basis of chronic periodontitis: a genome-wide association study. Hum Mol Genet. 2013;22(11):2312–24.
8. Rhodin K, Divaris K, North KE, Barros SP, Moss K, Beck JD, Offenbacher S. Chronic periodontitis genome-wide association studies: gene-centric and gene set enrichment analyses. J Dent Res. 2014;93(9):882–90.
9. Nascimento MM, Zaura E, Mira A, Takahashi N, Ten Cate JM. Second era of OMICS in caries research: moving past the phase of disillusionment. J Dent Res. 2017;96:733–40.
10. Sudmant PH, Rausch T, Gardner EJ, Handsaker RE, Abyzov A, Huddleston J, Zhang Y, Ye K, Jun G, Hsi-Yang Fritz M, Konkel MK, Malhotra A, Stutz AM, Shi X, Paolo Casale F, Chen J, Hormozdiari F, Dayama G, Chen K, Malig M, Chaisson MJP, Walter K, Meiers S, Kashin S, Garrison E, Auton A, Lam HYK, Jasmine Mu X, Alkan C, Antaki D, Bae T, Cerveira E, Chines P, Chong Z, Clarke L, Dal E, Ding L, Emery S, Fan X, Gujral M, Kahveci F, Kidd JM, Kong Y, Lameijer E-W, McCarthy S, Flicek P, Gibbs RA, Marth G, Mason CE, Menelaou A, Muzny DM, Nelson BJ, Noor A, Parrish NF, Pendleton M, Quitadamo A, Raeder B, Schadt EE, Romanovitch M, Schlattl A, Sebra R, Shabalin AA, Untergasser A, Walker JA, Wang M, Yu F, Zhang C, Zhang J, Zheng-Bradley X, Zhou W, Zichner T, Sebat J, Batzer MA, McCarroll SA, The Genomes Project Consortium, Mills RE, Gerstein MB, Bashir A, Stegle O, Devine SE, Lee C, Eichler EE, Korbel JO. An integrated map of structural variation in 2,504 human genomes. Nature. 2015;526(7571):75–81.
11. Saleheen D, Natarajan P, Armean IM, Zhao W, Rasheed A, Khetarpal SA, Won HH, Karczewski KJ, O'Donnell-Luria AH, Samocha KE, Weisburd B, Gupta N, Zaidi M, Samuel M, Imran A, Abbas S, Majeed F, Ishaq M, Akhtar S, Trindade K, Mucksavage M, Qamar N, Zaman KS, Yaqoob Z, Saghir T, Rizvi SNH, Memon A, Hayyat Mallick N, Rasheed SZ, Memon FU, Mahmood K, Ahmed N, Do R, Krauss RM, MacArthur DG, Gabriel S, Lander ES, Daly MJ, Frossard P, Danesh J, Rader DJ, Kathiresan S. Human knockouts and phenotypic analysis in a cohort with a high rate of consanguinity. Nature. 2017;544(7649):235–9.
12. Coelho ED, Arrais JP, Matos S, Pereira C, Rosa N, Correia MJ, Barros M, Oliveira JL. Computational prediction of the human-microbial oral interactome. BMC Syst Biol. 2014;8(1):24.
13. Arrais JP, Rosa N, Melo J, Coelho ED, Amaral D, Correia MJ, Barros M, Oliveira JL. OralCard: a bioinformatic tool for the study of oral proteome. Arch Oral Biol. 2013;58(7):762–72.
14. Li S, Sullivan NL, Rouphael N, Yu T, Banton S, Maddur MS, McCausland M, Chiu C, Canniff J,

Dubey S, Liu K, Tran V, Hagan T, Duraisingham S, Wieland A, Mehta AK, Whitaker JA, Subramaniam S, Jones DP, Sette A, Vora K, Weinberg A, Mulligan MJ, Nakaya HI, Levin M, Ahmed R, Pulendran B. Metabolic phenotypes of response to vaccination in humans. Cell. 2017;169:862–77.

15. Dominguez-Bello MG, Costello EK, Contreras M, Magris M, Hidalgo G, Fierer N, Knight R. Delivery mode shapes the acquisition and structure of the initial microbiota across multiple body habitats in newborns. Proc Natl Acad Sci U S A. 2010;107(26):11971–5.
16. Johnson CL, Versalovic J. The human microbiome and its potential importance to pediatrics. Pediatrics. 2012;129(5):950–60.
17. Salvucci E. Microbiome, holobiont and the net of life. Crit Rev Microbiol. 2016;42(3):485–94.
18. Hajishengallis G, Liang S, Payne MA, Hashim A, Jotwani R, Eskan MA, McIntosh ML, Alsam A, Kirkwood KL, Lambris JD, Darveau RP, Curtis MA. Low-abundance biofilm species orchestrates inflammatory periodontal disease through the commensal microbiota and complement. Cell Host Microbe. 2011;10(5):497–506.
19. Hajishengallis G, Darveau RP, Curtis MA. The keystone-pathogen hypothesis. Nat Rev Microbiol. 2012;10(10):717–25.
20. Curtis MA, Zenobia C, Darveau RP. The relationship of the oral microbiotia to periodontal health and disease. Cell Host Microbe. 2011;10(4):302–6.
21. Hajishengallis G. Periodontitis: from microbial immune subversion to systemic inflammation. Nat Rev Immunol. 2015;15(1):30–44.
22. He XS, Shi WY. Oral microbiology: past, present and future. Int J Oral Sci. 2009;1(2):47–58.
23. Handelsman J, Rondon MR, Brady SF, Clardy J, Goodman RM. Molecular biological access to the chemistry of unknown soil microbes: a new frontier for natural products. Chem Biol. 1998;5(10):R245–9.
24. Ebersole JL, Dawson DR, Morford LA, Peyyala R, Miller CS, Gonzaléz OA. Periodontal disease immunology: 'double indemnity' in protecting the host. Periodontol. 2013;62(1):163–202.
25. Peyyala R, Ebersole JL. Multispecies biofilms and host responses: "discriminating the trees from the forest". Cytokine. 2013;61(1):15–25.
26. Peyyala R, Kirakodu SS, Novak KF, Ebersole JL. Oral epithelial cell responses to multispecies microbial biofilms. J Dent Res. 2013;92(3):235–40.
27. Moyes DL, Runglall M, Murciano C, Shen C, Nayar D, Thavaraj S, Kohli A, Islam A, Mora-Montes H, Challacombe SJ, Naglik JR. A biphasic innate immune MAPK response discriminates between the yeast and hyphal forms of Candida albicans in epithelial cells. Cell Host Microbe. 2010;8(3):225–35.
28. Moharamzadeh K, Colley H, Murdoch C, Hearnden V, Chai WL, Brook IM, Thornhill MH, Macneil S. Tissue-engineered oral mucosa. J Dent Res. 2012;91(7):642–50.
29. Tanaka Y, Nagashima H, Bando K, Lu L, Ozaki A, Morita Y, Fukumoto S, Ishii N, Sugawara S. Oral CD103-CD11b+ classical dendritic cells present sublingual antigen and induce Foxp3+ regulatory T cells in draining lymph nodes. Mucosal Immunol. 2017;10(1):79–90.
30. Hasan MS, Ryan PL, Bergmeier LA, Fortune F. Circulating NK cells and their subsets in Behcet's disease. Clin Exp Immunol. 2017;188(2):311–22.
31. Ryan PL, Sumaria N, Holland CJ, Bradford CM, Izotova N, Grandjean CL, Jawad AS, Bergmeier LA, Pennington DJ. Heterogeneous yet stable Vdelta2(+) T-cell profiles define distinct cytotoxic effector potentials in healthy human individuals. Proc Natl Acad Sci U S A. 2016;113(50):14378–83.
32. Hovav AH. Dendritic cells of the oral mucosa. Mucosal Immunol. 2014;7(1):27–37.
33. Nudel I, Elnekave M, Furmanov K, Arizon M, Clausen BE, Wilensky A, Hovav A-H. Dendritic cells in distinct oral mucosal tissues engage different mechanisms to prime CD8+ T cells. J Immunol. 2011;186(2):891–900.
34. Renaudineau Y. Immunophenotyping as a new tool for classification and monitoring of systemic autoimmune diseases. Clin Rev Allergy Immunol. 2017;5(10):017–8604.
35. Villani A-C, Satija R, Reynolds G, Sarkizova S, Shekhar K, Fletcher J, Griesbeck M, Butler A, Zheng S, Lazo S, Jardine L, Dixon D, Stephenson E, Nilsson E, Grundberg I, McDonald D, Filby A, Li W, De Jager PL, Rozenblatt-Rosen O, Lane AA, Haniffa M, Regev A, Hacohen N. Single-cell RNA-seq reveals new types of human blood dendritic cells, monocytes, and progenitors. Science. 2017;356(6335):eaah4573.
36. Klose CS, Artis D. Innate lymphoid cells as regulators of immunity, inflammation and tissue homeostasis. Nat Immunol. 2016;17(7):765–74.
37. Kim CH, Hashimoto-Hill S, Kim M. Migration and tissue tropism of innate lymphoid cells. Trends Immunol. 2016;37(1):68–79.
38. Dutzan N, Konkel JE, Greenwell-Wild T, Moutsopoulos NM. Characterization of the human immune cell network at the gingival barrier. Mucosal Immunol. 2016;9:1163–72.
39. Li H, Reeves RK. Functional perturbation of classical natural killer and innate lymphoid cells in the oral mucosa during SIV infection. Front Immunol. 2013;3:417.
40. Gladiator A, Wangler N, Trautwein-Weidner K, LeibundGut-Landmann S. Cutting edge: IL-17-secreting innate lymphoid cells are essential for host defense against fungal infection. J Immunol. 2013;190(2):521–5.
41. Gury-BenAri M, Thaiss CA, Serafini N, Winter DR, Giladi A, Lara-Astiaso D, Levy M, Salame TM, Weiner A, David E, Shapiro H, Dori-Bachash M, Pevsner-Fischer M, Lorenzo-Vivas E, Keren-Shaul H, Paul F, Harmelin A, Eberl G, Itzkovitz S, Tanay A, Di Santo JP, Elinav E, Amit I. The spectrum and regulatory landscape of intestinal innate lymphoid cells are shaped by the microbiome. Cell. 2016;166(5):1231–1246.e1213.

42. Eberl G. A new vision of immunity: homeostasis of the superorganism. Mucosal Immunol. 2010;3(5):450–60.
43. Kramer CD, Genco CA. Microbiota, immune subversion, and chronic inflammation. Front Immunol. 2017;8:255.
44. Levy M, Kolodziejczyk AA, Thaiss CA, Elinav E. Dysbiosis and the immune system. Nat Rev Immunol. 2017;17(4):219–32.
45. Medzhitov R. Recognition of microorganisms and activation of the immune response. Nature. 2007;449(7164):819–26.
46. Medzhitov R. Pattern recognition theory and the launch of modern innate immunity. J Immunol. 2013;191(9):4473.
47. Palm NW, Medzhitov R. Pattern recognition receptors and control of adaptive immunity. Immunol Rev. 2009;227(1):221–33.
48. Novak N, Haberstok J, Bieber T, Allam JP. The immune privilege of the oral mucosa. Trends Mol Med. 2008;14(5):191–8.
49. Eberl G. Immunity by equilibrium. Nat Rev Immunol England. 2016;16:524–32.
50. Baker DJ, Childs BG, Durik M, Wijers ME, Sieben CJ, Zhong J, Saltness RA, Jeganathan KB, Verzosa GC, Pezeshki A, Khazaie K, Miller JD, van Deursen JM. Naturally occurring p16Ink4a-positive cells shorten healthy lifespan. Nature. 2016;530(7589):184–9.
51. Childs BG, Baker DJ, Wijshake T, Conover CA, Campisi J, van Deursen JM. Senescent intimal foam cells are deleterious at all stages of atherosclerosis. Science. 2016;354(6311):472–7.
52. Demaria M, O'Leary MN, Chang J, Shao L, Liu S, Alimirah F, Koenig K, Le C, Mitin N, Deal AM, Alston S, Academia EC, Kilmarx S, Valdovinos A, Wang B, de Bruin A, Kennedy BK, Melov S, Zhou D, Sharpless NE, Muss H, Campisi J. Cellular senescence promotes adverse effects of chemotherapy and cancer relapse. Cancer Discov. 2017;7(2):165–76.
53. Kang TW, Yevsa T, Woller N, Hoenicke L, Wuestefeld T, Dauch D, Hohmeyer A, Gereke M, Rudalska R, Potapova A, Iken M, Vucur M, Weiss S, Heikenwalder M, Khan S, Gil J, Bruder D, Manns M, Schirmacher P, Tacke F, Ott M, Luedde T, Longerich T, Kubicka S, Zender L. Senescence surveillance of pre-malignant hepatocytes limits liver cancer development. Nature. 2011;479(7374):547–51.
54. Jurk D, Wilson C, Passos JF, Oakley F, Correia-Melo C, Greaves L, Saretzki G, Fox C, Lawless C, Anderson R, Hewitt G, Pender SL, Fullard N, Nelson G, Mann J, van de Sluis B, Mann DA, von Zglinicki T. Chronic inflammation induces telomere dysfunction and accelerates ageing in mice. Nat Commun. 2014;2:4172.
55. Pitiyage GN, Slijepcevic P, Gabrani A, Chianea YG, Lim KP, Prime SS, Tilakaratne WM, Fortune F, Parkinson EK. Senescent mesenchymal cells accumulate in human fibrosis by a telomere-independent mechanism and ameliorate fibrosis through matrix metalloproteinases. J Pathol. 2011;223(5):604–17.
56. Baar MP, Brandt RM, Putavet DA, Klein JD, Derks KW, Bourgeois BR, Stryeck S, Rijksen Y, van Willigenburg H, Feijtel DA, van der Pluijm I, Essers J, van Cappellen WA, van IJcken WF, Houtsmuller AB, Pothof J, de Bruin RW, Madl T, Hoeijmakers JH, Campisi J, de Keizer PL. Targeted apoptosis of senescent cells restores tissue homeostasis in response to chemotoxicity and aging. Cell. 2017;**169**(1):132–147.e116.
57. Chang J, Wang Y, Shao L, Laberge R-M, Demaria M, Campisi J, Janakiraman K, Sharpless NE, Ding S, Feng W, Luo Y, Wang X, Aykin-Burns N, Krager K, Ponnappan U, Hauer-Jensen M, Meng A, Zhou D. Clearance of senescent cells by ABT263 rejuvenates aged hematopoietic stem cells in mice. Nat Med. 2016;22(1):78–83.
58. Jeon OH, Kim C, Laberge RM, Demaria M, Rathod S, Vasserot AP, Chung JW, Kim DH, Poon Y, David N, Baker DJ, van Deursen JM, Campisi J, Elisseeff JH. Local clearance of senescent cells attenuates the development of post-traumatic osteoarthritis and creates a pro-regenerative environment. Nat Med. 2017;23(6):775–81.
59. Ebersole JL, Kirakodu S, Novak MJ, Exposto CR, Stromberg AJ, Shen S, Orraca L, Gonzalez-Martinez J, Gonzalez OA. Effects of aging in the expression of NOD-like receptors and inflammasome-related genes in oral mucosa. Mol Oral Microbiol. 2016;31(1):18–32.
60. Grice EA, Segre JA. Interaction of the microbiome with the innate immune response in chronic wounds. Adv Exp Med Biol. 2012a;946:55–68.
61. Grice EA, Segre JA. The human microbiome: our second genome. Annu Rev Genomics Hum Genet. 2012b;13:151–70.
62. Rosenfeld ME, Campbell LA. Pathogens and atherosclerosis: update on the potential contribution of multiple infectious organisms to the pathogenesis of atherosclerosis. Thromb Haemost. 2011;106(5):858–67.
63. Seoudi N, Bergmeier LA, Hagi-Pavli E, Bibby D, Curtis MA, Fortune F. The role of TLR2 and 4 in Behcet's disease pathogenesis. Innate Immunol. 2014;20(4):412–22.
64. Will CL, Lührmann R. Spliceosome structure and function. Cold Spring Harb Perspect Biol. 2011;3(7):a003707.
65. Zhang X, Yan C, Hang J, Finci LI, Lei J, Shi Y. An atomic structure of the human spliceosome. Cell. 2017;169(5):918–929.e914.
66. Silberring J, Ciborowski P. Biomarker discovery and clinical proteomics. Trends Anal Chem. 2010;29(2):128.
67. Wang Y, Springer S, Mulvey CL, Silliman N, Schaefer J, Sausen M, James N, Rettig EM, Guo T, Pickering CR, Bishop JA, Chung CH, Califano JA, Eisele DW, Fakhry C, Gourin CG, Ha PK, Kang H, Kiess A, Koch WM, Myers JN, Quon H, Richmon JD, Sidransky D, Tufano RP, Westra WH, Bettegowda C, Diaz LA, Papadopoulos N, Kinzler KW, Vogelstein B, Agrawal N. Detection of somatic mutations and

HPV in the saliva and plasma of patients with head and neck squamous cell carcinomas. Sci Transl Med. 2015;7(293):293ra104.
68. Martinez GL, Majster M, Bjurshammar N, Johannsen A, Figueredo CM, Bostrom EA. Salivary colony stimulating factor-1, interleukin-34, and matrix metalloproteinase-8 as markers of periodontal disease. J Periodontol. 2017;88:1–15.
69. Rangbulla V, Nirola A, Gupta M, Batra P. Salivary IgA, interleukin-1beta and MMP-8 as salivary biomarkers in chronic periodontitis patients. Chin J Dent Res. 2017;20(1):43–51.
70. Sahibzada HA, Khurshid Z, Khan RS, Naseem M, Siddique KM, Mali M, Zafar MS. Salivary IL-8, IL-6 and TNF-alpha as potential diagnostic biomarkers for oral cancer. Diagnostics (Basel). 2017;7(2):E21.
71. Wang A, Wang CP, Tu M, Wong DT. Oral biofluid biomarker research: current status and emerging frontiers. Diagnostics (Basel). 2016;6(4):45.
72. Rieckmann JC, Geiger R, Hornburg D, Wolf T, Kveler K, Jarrossay D, Sallusto F, Shen-Orr SS, Lanzavecchia A, Mann M, Meissner F. Social network architecture of human immune cells unveiled by quantitative proteomics. Nat Immunol. 2017;18(5):583–93.
73. Shapiro H, Thaiss CA, Levy M, Elinav E. The cross talk between microbiota and the immune system: metabolites take center stage. Curr Opin Immunol. 2014;30:54–62.
74. Takahashi N. Oral microbiome metabolism: from "who are they?" to "what are they doing?". J Dent Res. 2015;94(12):1628–37.
75. Washio J, Ogawa T, Suzuki K, Tsukiboshi Y, Watanabe M, Takahashi N. Amino acid composition and amino acid-metabolic network in supragingival plaque. Biomed Res. 2016;37(4):251–7.
76. Washio J, Takahashi N. Metabolomic studies of oral biofilm, oral cancer, and beyond. Int J Mol Sci. 2016;17(6):870.
77. Ślebioda Z, Szponar E, Dorocka-Bobkowska B. Vitamin D and its relevance in the etiopathogenesis of oral cavity diseases. Arch Immunol Ther Exp (Warsz). 2016;64:385–97.
78. Biosse Duplan M, Coyac BR, Bardet C, Zadikian C, Rothenbuhler A, Kamenicky P, Briot K, Linglart A, Chaussain C. Phosphate and vitamin D prevent periodontitis in X-linked hypophosphatemia. J Dent Res. 2016;96(4):388–95.
79. Buck MD, Sowell RT, Kaech SM, Pearce EL. Metabolic instruction of immunity. Cell. 2017;169(4):570–86.
80. Wei J, Raynor J, Nguyen TL, Chi H. Nutrient and metabolic sensing in T cell responses. Front Immunol. 2017;8:247.
81. Hassona Y, Cirillo N, Lim KP, Herman A, Mellone M, Thomas GJ, Pitiyage GN, Parkinson EK, Prime SS. Progression of genotype-specific oral cancer leads to senescence of cancer-associated fibroblasts and is mediated by oxidative stress and TGF-beta. Carcinogenesis. 2013;34(6):1286–95.
82. James EL, Michalek RD, Pitiyage GN, de Castro AM, Vignola KS, Jones J, Mohney RP, Karoly ED, Prime SS, Parkinson EK. Senescent human fibroblasts show increased glycolysis and redox homeostasis with extracellular metabolomes that overlap with those of irreparable DNA damage, aging, and disease. J Proteome Res. 2015;14(4):1854–71.
83. Freitag J, Berod L, Kamradt T, Sparwasser T. Immunometabolism and autoimmunity. Immunol Cell Biol. 2016;94(10):925–34.
84. Miyazawa M, Lopalco L, Mazzotta F, Lo Caputo S, Veas F, Clerici M, ESNS Group. The 'immunologic advantage' of HIV-exposed seronegative individuals. AIDS. 2009;23:161–75.

MIX
Papier aus verantwortungsvollen Quellen
Paper from responsible sources
FSC® C105338

If you have any concerns about our products,
you can contact us on
ProductSafety@springernature.com

In case Publisher is established outside the EU,
the EU authorized representative is:
Springer Nature Customer Service Center GmbH
Europaplatz 3, 69115 Heidelberg, Germany

Printed by Libri Plureos GmbH
in Hamburg, Germany